PUBLICATIONS OF THE ISRAEL ACADEMY

OF SCIENCES AND HUMANITIES

SECTION OF SCIENCES

———————

FAUNA PALAESTINA

MAMMALIA OF ISRAEL

FAUNA PALAESTINA

Y. Eshbol

Gazella gazella acaciae

FAUNA PALAESTINA

MAMMALIA
OF ISRAEL

by

HEINRICH MENDELSSOHN and YORAM YOM-TOV

Jerusalem 1999

The Israel Academy of Sciences and Humanities

Authors' Address:
Department of Zoology
George S. Wise Faculty of Life Sciences
Tel Aviv University, Ramat Aviv
69 978 Tel Aviv, Israel

ISBN 965-208-013-6
ISBN 965-208-145-0

Printed in Israel
at Keterpress Enterprises, Jerusalem

CONTENTS

PREFACE

The aim of this book is to present information gathered on the terrestrial mammals (including bats) of Israel, much of which was originally written in Hebrew and thus unavailable to zoologists not familiar with the language. We have also included some data gathered from various other areas in the Middle East, but have not presented all the information known about a species from elsewhere, as such information is available from other regional works. We have not included species which were once present in Israel but became extinct more than 70 years ago, such as the roe and the Mesopotamian fallow deer. Such species are only mentioned in the Introduction, in which we also briefly describe the marine mammals that are sighted along the coasts of Israel, either in the Mediterranean or the Red Sea. Taking a restricted view, there are presently 92 species of terrestrial mammals in Israel, in eight orders: they include six species of Insectivora, 32 Chiroptera (bats), one hare (Lagomorpha), 30 rodents, 18 carnivores, a wild ass (Perissodactyla) and four species belonging to the Artiodactyla. Most of these are native species, but one (the coypu) was introduced and another (the wild ass) has been re-introduced to the country recently.

The map of geographical areas in Israel and Sinai is given (see Figure 1, p. 33). The spelling of names of localities in Israel and Sinai is according to the maps published by the Survey of Israel.

ACKNOWLEDGEMENTS

Many people helped us to produce this work. We are indebted to Mr Arie Landsman and Ms Judy Shamoun who helped in measuring skulls and producing the tables; to Ms Neomi Paz for her careful editing of several drafts of the manuscript; To Dr Heather J. Bromley and Ms Ilana Ferber for editing the final version; to Ms Tzilla Shariv for her assistance with museum material; to Mr Amikam Shoob for photographing the skulls; Dr David Harrison helped in many ways; He and Dr P. J. J. Bates kindly allowed us to use the species and family keys in their book on *The Mammals of Arabia*. Dr Colin Groves commented and made several useful suggestions; Prof. Eitan Tchernov enabled us to work comfortably at the Zoological Museum of the Hebrew University, Jerusalem, and gave us advice. We are also grateful to the photographers who allowed us to use their photographs: Danny Afik, Azaria Alon, Leora and Ofer Bahat, Eyal Bar-Tov, Danny Bar-Shahal, Yossi Eshbol, Dror Havelena, Ilan Golani, Arie Landsman, Micha Livne, Amnon Loya, Lita Maman, Zeev Meshel, Uzi Paz, Roni Rado, the late Gail Rubin, Danny Simon, Amikam Shoob, Zvi Sever; their names are mentioned in the captions of the photographs. Prof. Francis Dov Por, secretary, and Prof. C. Clara Heyn, present chairperson of the *Flora et Fauna Palaestina Committee* of the Israel Academy of Sciences and Humanities, helped us in various ways. The Israel Academy of Sciences and Humanities financed the publication of this work. We are grateful to the following persons who read and commented on parts of the manuscript and allowed us to use their unpublished data: Prof. Zvi Abramsky (*Gerbillus andersoni, G. pyramidum*), Prof. Avinoam Danin (botany), Dr Lexs Farenhead (Cetacea), Prof. Akiva Flexer (geology), Prof. Jacob Friedman (botany), Prof. Zvi Garfunkel (geology), Dr Lea Gavish (*Sciurus anomalus*), Dr Eli Geffen (*Vulpes rueppellii*), Prof. Avraham Haim (*Acomys, Apodemus*), Mr Reuven Heffner (*Canis lupus*), Dr Micha Ilan (*Psammomys obesus*), Dr Boris Krasnov (*Gerbillus henleyi, Meriones crassus*), Ms Noga Kronfeld (*Acomys, Lepus*), Prof. Uzi Ritte (*Acomys, Apodemus*), Dr David Makin (Chiroptera), Dr David Saltz (*Equus hemionus*), Dr Zvi Sever (*Hystrix indica*), Dr Gregory Shenbrot (*Gerbillus henleyi, Meriones crassus*), Dr Benny Shalmon (Chiroptera), Prof. Amiram Shkolnik (Introduction, *Acomys*), Prof. Yoav Waisel (geomorphology, climate and botany), Mr Zohar Zook-Rimon (*Meriones tristrami, Nesokia indica*).

INTRODUCTION

GENERAL ASPECTS

THE STUDY AREA (Figure 1, p. 33)

The book covers the area of Israel (see Map of Geographical Areas, Fig. 1), i.e. the land area of about 28,000 km^2 between the Mediterranean Sea in the west and the Rift Valley in the east. This area forms part of the Levant which, relative to its size, is one of the richest and most diverse regions in the temperate part of the world regarding species. There are three main reasons for this diversity: the geological history of Israel; its geographical position as a meeting point between Asia, Africa and Europe; and the great geographical, climatic and edaphic diversity, which allow the presence of species with different ecological needs.

HISTORY OF MAMMALIAN RESEARCH IN ISRAEL

The history of biological research and the sources of information from Israel are dealt with in two of Bodenheimer's books (1935, 1950–1956). The first written records on the mammals of Israel are found in the Bible, where about 40 species of mammals, including several domesticated ones, are mentioned. Specific names are given mainly to large (ungulates and carnivores) or medium size mammals (hyrax, hare), while for smaller ones only a general name is given. For example, all small rodents are "mice" and all Chiroptera are "bats". Some of the species are mentioned in the context of description of clean (kosher) and un-clean animals (Leviticus 11), and not all can be identified today. While in this list only the species names are mentioned, in other places in the Bible the animals are mentioned together with a description of their behaviour or biology. These descriptions are often accurate and demonstrate a close knowledge of the animals, such as in the description of the wild ass (Job 39:5–8): "Who let the wild ass go free? Who untied his ropes? I gave him the steppes as his home, the salt flats as his habitat. He laughs at the commotion in the town, he does not hear the drivers shout. He ranges the hills for his pasture and searches for any green thing"; or the description of the coney (Hyrax): "The conies are but a feeble folk, yet make their home in the rocks" (Proverbs 30:26). Some of the animals are mentioned as symbols for various traits, such as the gazelles and leopards for swiftness and wolves for fierceness ("Asahel was as fleet-footed as a gazelle" II Samuel 2:18; "Their horses are swifter than leopards, fiercer than wolves" Habakkuk 1:8). However, some observations were erroneous, such as that in Leviticus 11:5: "The coney, though it chews the cud..". Obviously the writer mistook the threatening chewing movements as chewing the cud. Several of the species mentioned in the Bible, such as the lion, the bear, the wild ass, the roe and fallow deer and others, became extinct in Israel.

Another source of information on the animals of Israel during biblical times are the wall paintings in ancient Egypt, in which inhabitants of Canaan are depicted bringing presents to the Pharaohs including gazelles, Syrian bear and other mammals.

After the Bible, the main sources of information are: Josephus Flavius, Plinius, the New Testament, the Mishna and the Jerusalem Talmud, pilgrimage and travel reports of Europeans, and Arab geographers. Most of these are prescientific and deal mainly with large animals.

During the 19th century Israel was visited by many Europeans, some of whom were biologists who collected specimens and data. The most notable of these was Henry Baker Tristram, Canon of Durham, England, who visited Israel four times between 1858–1897 under the auspices of the Palestine Exploration Fund as part of the Survey of Israel. His works (Tristram, 1884, 1886, 1888) give detailed descriptions of the fauna and flora of Israel at the end of the 19th century, and were the first scientific reviews of the mammal fauna of this country.

Other European zoologists who contributed to the knowledge of mammals of Israel and the immediate neighbouring areas at from the end the last century until the end of the First World War, are H. C. Hart, J. C. Philips, R. I. Pocock, S. S. Flower, A. Nehring, and O. F. Thomas. During this period European museums initiated a major collecting effort for various groups of animals, including mammals. The main collectors of the time were Pater E. Schmitz and I. Aharoni, who collected chiefly for the Berlin Museum and the Baron Rothschild Collection in Britain. The Schmitz Collection, now in the Schmitz Girls College, Jerusalem, features some of the last specimens seen in the Middle East of several species, including the cheetah.

The opening of the Hebrew University of Jerusalem in 1925 and the establishment of its Department of Zoology headed by F. S. Bodenheimer started a new period for zoological research in Israel. This was followed by the establishment of the Biologic-Pedagogic Institute by Y. Margolin (1931) from which developed Tel Aviv University and its Department of Zoology. Other scientific institutions where mammological work is being carried out were later established in Oranim, Be'er Sheva' and Haifa. Many local and visiting scientists and students contributed to the knowledge on the distribution, systematics, ecology and behaviour of the mammals of Israel.

GEOLOGY

Israel is located on the northern margins of the Arabian-African continent, which once bordered the Tethys Sea, a position which greatly affected its geological development and present geomorphology. The geological history of Israel can be divided into three main stages, each with its characteristic tectonic regime (the following description is based largely on Garfunkel, 1988):

(a) The late Precambrian Pan African orogenic stage, when the crystalline basement and continental crust of the area were shaped.

(b) The early Cambrian to mid- Cenozoic platformal stage, when the area was part of a generally stable Arabo-African continent.

(c) The mid-Cenozoic to Recent rifting and continental breakup stage, during which the breakaway of Arabia from Africa took place, separating them by the Red Sea and Dead Sea Transform (Rift Valley). From the Early or Middle Miocene the northern part of the Arabian Plate moved about 105 km north in relation to the Israel-Sinai sub-plate, producing the Dead Sea Transform. The uplifting associated with the rifting produced the main physiographic features of Israel.

Stage 1. The crystalline basement is visible in Israel only in the Elat area, southern Negev, in a few uplifted outcrops, which together with similar outcrops in southern and eastern Sinai, southern Jordan and Arabia form the Arabo-Nubian Shield. These outcrops include metamorphic rocks (mainly schists and gneisses) intruded by various plutonic rocks (mainly quartz-dioritic to granitic). These rocks were shaped by tectonic and thermal events of the Pan-African orogeny, during the period from ca 800–600 million years ago. About half of the exposed basement rocks are relatively young granite plutons, some 600–580 million years old, which are very common all over the northern part of the Arabo-Nubian Massif. These deep–seated rocks are intruded by alkaline granites and still younger dike swarms. They are covered by unmetamorphosed, faulted and tilted volcanics and conglomeratic series.

Stage 2. The platformal stage began with a period of extensive erosion which produced a vast peneplain on the northern part of the Arabo-Nubian Massif. There were several sedimentation and erosion cycles between the Cambrian and the Eocene which produced a typical platform sedimentary cover which consists predominantly of sandstones and carbonates. During this period several important events are noteworthy: a) a rifting (faulting and magmatism) which included the formation of the Mediterranean continental margin in the Early Mesozoic, when extensive sedimentation and irregular subsidence took place; and b) uplifting and magmatism between the Late Jurassic and the Early Cretaceous; c) mild folding and faulting between the Late Cretaceous and the Neogene which produced a series of faults and broad asymmetric folds. The latter structures extend from the Euphrates River through northern Jordan to Israel, Sinai and Egypt, and comprise the Syrian Arc. The formation of the Syrian Arc in Israel created a series of mostly asymmetric anticlines whose axe near the Arabo-Nubian Massif (in the Negev) is parallel to that of the Massif (from south-west to north-east), and towards the north their direction changes to west-south-west to east-north-east.

During the platformal stage the area of Israel was mostly a part of an extensive, shallow marine basin where various sediments were accumulated, forming thinner layers near the Massif and thicker ones towards the north-west. In the extreme south of Israel, near and south of Elat, these layers have been eroded and the basement there is currently exposed to a height of up to 600–800 m above sea level, while near Yotvata, 40 km north of Elat, the basement plunges and disappears under marine and continental deposits which reach a thickness of a few km in the central Negev and about 4–6 kms further north. The sediments exposed around the shield are mainly coloured Nubian sandstone ranging in age from the Paleozoic to the Lower Cretaceous, whereas more to the north-west the marine deposits of the Mesozoic are mainly limestones, dolomites and marl. Deposition of limestone, chalk and marl is characteristic of the early Tertiary of Israel. The Mesozoic and early Tertiary carbonate sediments are the main rocks exposed in the present mountainous areas of Israel, with hard rocks typical to the mountains and softer ones on the slopes. Most of the present land of Israel rose from the sea during the Tertiary, and was uplifted during the third stage of the geological history of Israel, namely the rifting stage.

Stage 3: From the mid-Tertiary to Recent, the Near East, including Israel and the surrounding areas, were affected by rifting and faulting, accompanied by widespread basaltic volcanism, mainly on the eastern side of the Rift Valley. This tectonism produced the present configuration of Israel and its surrounding countries. In Israel it produced the three longitudinal, north-south strips consisting of a mountainous backbone, the deep Rift Valley in the east, and the low lying Mediterranean Coastal Plain in the west. The areas bordering the Red Sea were uplifted, and Late Cenozoic erosion there exposed the Precambrian basement of the Arabo-Nubian Shield. In contrast, the Coastal Plain and offshore regions of Israel have subsided several kms since the Neogene, accumulating sediments which wedge out inland and develop into littoral sandstones that extend inland as far as the Be'er Sheva' basin. Much of this sediment was transported by the Nile and redistributed along the continental margin by longshore currents in the Mediterranean Sea. In the north, the uplifted areas were crossed by the fault-controlled Yizre'el Valley, and by other faults north of it which are offshoots of the Dead Sea Transform. While these processes were in progress, plate convergence continued along the Alpine Orogenic zone further north, and Israel and the nearby countries became attached to the Eurasian landmass, but increasingly more separated from the African continent. The Dead Sea Transform accumulated a lateral motion which resulted in the formation of several basins which became filled with water to become lakes (the Dead Sea, the Sea of Galilee), and thick series of sediments filled the Rift Valley. The movement of the transform was accompanied by volcanic activity, mainly in the northern Jordan Valley and the Golan Heights.

During the Pleistocene, Israel experienced periods of wet pluvials and dry interpluvials which alternated many times throughout the last 1.8 million years and paralleled the glacial and interglacial periods further north. The pluvial periods were typified by a southward shift of the climatic belts accompanied by low sea levels, while the interpluvials showed opposite trends. The deposition of sand by marine currents along the Mediterranean shore continued and is the source of the present day coastal and western Negev dunes, and of sandstones that form low ridges along the Coastal Plain.

Faunal History

The above description is a brief summary of the geological history of Israel. It indicates that although land bridges between Africa and Eurasia may have existed occasionally before the Neogene, until this period Israel was mainly under African influence. Early Miocene exposures in the Negev revealed an essentially African fauna (mainly ungulates), but with some Eurasian carnivore and rodent representation. Biotic interchange between Africa and Eurasia continued during the rest of the Miocene with interruptions during periods of transgression of the Tethys Sea. At the end of the Miocene the faunas of Eurasia and Africa were very similar and consisted of an extremely diverse Savanna fauna which included some of the present Mediterranean elements.

"The Messinian Event" which occurred at the end of the Miocene, during which the Mediterranean went through a period of considerable desiccation for about a million years, marked the beginning of the formation of the Saharo-Arabian desert fauna. In contrast, the beginning of the Pliocene is marked by a cooler and wetter climate, but remains of mammals from this period found near Bethlehem include Savanna antelopes. Palaeontological excavations at 'Ubeidiya, 3 km south of Lake Kinneret (Sea of Galilee), have furnished extensive knowledge on the Early Pleistocene fauna (Tchernov, 1988). More than 50 mammalian species were identified and include animals which originated mainly from the Palaearctic such as the North African monkey (*Macaca sylvana*), from the east Mediterranean (*Arvicola, Mesocricetus, Spalax*), European (*Lutra, Vormela*), European and East Mediterranean (*Apodemus, Cricetulus, Felis, Lynx,* Cervidae), Asian (*Erinaceus*), Saharo-Arabian (*Gazella, Gerbillus, Oryx*), Ethiopian (Giraffidae, *Hippopotamus, Crocidura*) and Oriental (*Bos, Herpestes, Hystrix*) and even from the Nearctic (*Equus, Hypolagus, Canis*). Drier and warmer savanna conditions prevailed during the Mid-Pleistocene. The fauna discovered at Gesher Benot Ya'aqov included a proboscid (*Stegodon*), a rhino (*Dicerorhinus*) and an equid (*Equus*), while many Palaearctic elements disappeared. Later (150,000–250,000 years BP) remains from Oumm Qatafa and Give'at Shaul revealed another Palaearctic invasion, with *Cervus elaphus, Dama mesopotamica, Bos primigenius* and *Gazella gazella* together with several carnivores (*Hyaena hyaena, Ursus arctos*) and many rodents (Tchernov, 1988).

The fauna of the Upper Pleistocene has been determined from palaeontological excavations in caves in the Galilee and Mount Carmel. It included a mixture of African, European and Asian elements adapted to either eremic or Mediterranean conditions, but during this period until the Neolithic many of the larger species disappeared. These include three equids (*Equus caballus, E. hydruntinus* and *E. hemionus*), *Hippopotamus amphibius* which existed in the coastal rivers until historic times, the red deer *Cervus elaphus, Bos primigenius* and *Rhinoceros hemitoechus* which disappeared during the early Iron Age.

In summary: ever since the Miocene, the southern Levant, including Israel, has been a land bridge between Eurasia and Africa and constituted a playground for north-south shifting of eremian versus mesic faunal elements concomitant with the Quaternary climatic fluctuations. These events resulted in a complex of biota originating from the Palaearctic and the Palaeotropic.

GEOMORPHOLOGY AND SOILS

The area between the Mediterranean Sea and the Rift Valley is characterized by three longitudinal morphological units: the Coastal Plain, the central mountain range and the Rift Valley. The Coastal Plain ranges in elevation from sea level to 150 m above sea level (asl) and is covered mainly by various Pleistocene sandstones and sandy-loam flats. Its width ranges from several dozen kilometres in the south to a few hundred metres near Mount Carmel. Along the coast several north-south ridges of calcareous sandstone, known locally as kurkar, have developed. Between the ridges there are shallow valleys of reddish, sandy, clay-rich sediments and soil, known locally as hamra. The Mediterranean Sea deposits sand on the coast with greater amounts being deposited, and covering more area of land in the southern Coastal Plain than in the north. Hence, sand covers larger areas in the southern Coastal Plain than in its northern part. During periods of strong winds and heavy grazing the deposited sand becomes mobile and tends to form dunes. The kurkar ridges and the dunes are dissected by several small streams flowing from the east to the west. These streams drain the surplus runoff water from the central mountain ridge into the Mediterranean, and are almost dry during summer. The blockage of water flow by the kurkar ridges and the sand deposits along the coast prevents good drainage of the valleys between the ridges, thus enabling the creation of swamps and temporary winter rain pools.

Extending east from the Coastal Plain towards the mountain range, whose peaks range in elevation between 600–1200 m asl, lies a region of low relief, the Shefela. This is a plain of moderate slopes, created between the Middle to Late Miocene by sea abrasion and modified by younger erosion. East of the Shefela rises the central mountain ridge. The bedrocks of the mountain range consist mainly of upper Cretaceous marine carbonates and the soils are mostly terra rossa and brown rendzinas in the Mediterranean region and pale rendzinas and desert lithosols in the arid region. The western and eastern slopes of the range differ in steepness, being much steeper in the east of the range, where they border the Rift Valley. Along the coast of the Dead Sea the cliffs are as high as 400 m, and have taluses of fallen rocks and stones at the foot of the cliffs. The mountain range is dissected by several valleys (Bet HaKerem, Yizre'el and Bet She'an, Be'er Sheva') which separate the ridge into several distinctive mountains — the Upper and Lower Galilee, Gilboa, Samaria, Judea, Negev and Elat mountains. These valleys have heavy, alluvial grumosols in northern Israel and loessial serozems and brown soils in the arid areas in the south. In the arid zone the soils are mainly regs and calcarious lithosols. In northern and central Israel, parts of these valleys have been poorly drained, creating small swamps and temporary rain pools.

Within Israel, the Rift Valley comprises the Jordan Valley, Dead Sea and the 'Arava, which are part of the great Afro-Syrian Rift Valley system. This region is mostly below sea level with a lowest elevation of -400 m on the coast of the Dead Sea, rising from it towards both the north and south. Within Israel, the highest elevations of the Rift Valley are in the Hula Valley in the north and the 'Arava, about

90 km south of the Dead Sea, both about 200 m asl. There were originally three Recent lakes in the Israeli Rift Valley — the Ḥula, Kinneret (Sea of Galilee) and the Dead Sea. Ḥula Lake was drained during the 1950s and the level of the Dead Sea has dropped considerably due to the diversion of the Jordan and other rivers for irrigation during the last 30 years. Streams and rivers which flow to the Rift Valley from its western and eastern sides have become filled in with alluvial deposits and grumosols, and salt marshes have formed in areas of poor drainage near the Dead Sea and the 'Arava. The 'Arava also features patches of sand, which originated from erosion of Nubian sandstone of the Lower Cretaceous and earlier ages. Because sandy areas in the desert allow better penetration of water, they are relatively rich in vegetation and wildlife in comparison to the neighbouring regs and loess areas. In the Negev and Judean Deserts the wadis are filled with gravel and pebbles which enable flood water to be absorbed into the ground, sometimes forming shallow underground water aquifers which are exploited by plants and animals.

To the northeast of the Jordan River lie the Golan Heights. This is an elevated plateau rising from a height of 300 m asl east of Lake Kinneret to 1000 m asl near Mount Ḥermon, the northernmost area of Israel. During the Pliocene and Pleistocene the Golan was subjected to volcanic influence and presently most of it is covered with basalt. On its eastern side, near the border with Syria, the Golan is dotted with extinguished volcanoes, rising to heights of about 1200 m asl. The Golan is drained to the Jordan Valley and to the Sea of Galilee, and its short streams are often rocky and some run along steep ravines.

To the north of the Golan Heights is Mount Ḥermon, the highest mountain in our region (its peak is 2814 m asl). The Ḥermon ridge was formed by a large-scale folding and uplifting which exposed Jurassic rocks. Only about 90 km^2 of this mountain are controlled by Israel, nearly 10% of its total area. However, the highest point in this small area is 2200 m asl, which is much higher than the highest peak in Israel (Mount Meron, 1208 asl). The low temperatures, winter snow, strong winds and high rainfall form there a habitat which is unique in Israel and a home to various northern and alpine species.

CLIMATE
Precipitation
Israel has a Mediterranean climate characterized by mild temperatures in winter and a warm, dry summer (Jaffe, 1988). Most precipitation originates from low pressure zones to the north-west of the country (over the north-east Mediterranean Sea), and declines in amount from north to south, with temperatures increasing in the same direction; together these two phenomena create an arid zone in the south. Most precipitation is in the form of rain, with snow occuring only for a few days annually in the high mountains, particularly in the north of the country. The amount of precipitation is positively correlated with two main factors: the distance from the centre of the above-mentioned low pressure zones, and altitude. Hence, the largest amount

of rain falls in the mountains of the Upper Galilee, where mean annual rainfall is up to 1000 mm with a coefficient of variation of 25–30%, and the smallest amount falls in the extreme south (Elat) where mean annual rainfall is 25 mm with a coefficient of variation reaching 87%. The rainy season occurs between October and March and 66% of the annual precipitation falls between December and February. The number of annual rainy days also declines from north to south, being about 70 in the Upper Galilee and 10 in Elat. The Rift Valley lies in the rain shadow, and precipitation there is smaller in comparison with areas at the same latitude west of the central mountain range. For example, at latitude 31°45' (Judean Hills) mean annual rainfall is 360 mm west of the mountain range, 650 mm at its top (near Jerusalem), 190 mm 10 km to the east, and less than 100 mm near the coast of the Dead Sea.

Temperature

Temperatures in Israel are affected by several factors — mainly season, latitude, altitude and distance from the Mediterranean Sea. Temperatures are higher in summer than in winter while autumn and spring temperatures fall between summer and winter ones with the exception of occasional heat-waves called "Sharav" ("Khamsin" in Arabic) which originate from the south and east and affect the country between March and June and between September and November. The coldest month is January and the hottest is August: average daily temperatures for these months are 7°C and 24°C in the Upper Galilee (Ẓefat, 934 m asl), 13°C and 22°C in the Mediterranean Coastal Plain (Tel Aviv, 10 m asl), 11°C and 26°C in the Northern Negev (Be'er Sheva', 280 m asl) and 15°C and 32°C in the southern 'Arava (Elat, 12 m asl) in January and August, respectively. Temperatures are negatively related to latitude and altitude and are moderated by the influence of sea temperatures. Hence, the daily and annual ranges of temperature are only 10°C–11°C in the Mediterranean Coastal Plain, but 12°C and 18°C respectively, in the southern 'Arava Valley. Very low temperatures may occur, mainly in the high mountains and the 'Arava where night temperatures during winter may drop to several degrees below freezing, while extremely high temperatures of up to 54°C may occur along the Rift Valley.

Relative Humidity and Dew

Relative humidity is high along the Mediterranean Coastal Plain where daily averages range between 60%–70% throughout the year, and decreases to the east and south where daily averages during summer are as low as 25% (in Elat). Correspondingly, there are about 250 dewy nights along the Coastal Plain, but only very few such nights in the southern 'Arava Valley. In the central Negev there are up to 200 dewy nights annually, an outcome of the high elevation and the large difference between day and night temperature.

Evaporation

Potential evaporation in Israel increases with altitude and latitude. The average annual evaporation from free water surface is about 130 cm along the Mediterranean Coastal Plain and 270 cm in the southern 'Arava Valley.

Winds

The Mediterranean region is dominated by sea and land breezes, mainly during summer. Sea breeze starts in the morning, blows from the west at a velocity of 5–15 knots and calms at nightfall. Land breeze blows from the second half of the night in the opposite direction. This day and night regime is less pronounced in the inland hills, which experience frequent strong western winds throughout the day during all seasons. Along the slopes warm anabatic winds rise at daybreak and katabatic winds descend during the night. In the Northern Negev and the internal valleys the breeze regime dominates, but the direction of the wind changes to become parallel to the valley. Lake breezes occur in the Kinneret and the Dead Sea, influencing the coastal areas. Winter winds can reach up to 25 knots, and when blowing from the south they sometimes develop into dust and sand storms. Autumn and spring see the arrival of hot, dry winds called "khamsin", which come from the south or east.

Climatic Zones

In accordance with the Koeppen classification, three climatic regions are defined in Israel: the temperate Mediterranean (Cs), the semi-arid (BS) and the arid (BW). The temperate region covers about 40% of the country and is defined by its relatively higher annual rainfall (300–1000 mm), relatively low evaporation (less than 170 cm annually, up to 6 times that of precipitation), a mild and low range of temperatures (about 10°C–11°C), high relative humidity and as many as 250 dewy nights annually; the semi-arid region covers about 15% of the country; it receives between 100–300 mm precipitation annually, has a higher annual evaporation (170–200 cm, about 10 times that of precipitation), higher temperatures with larger daily and annual range (about 14°C), lower relative humidity and between 120–160 dewy nights; the arid region covers almost half of the country and is dry (average annual rainfall less than 100 mm), with high evaporation (up to 270 cm, more than 100 times that of precipitation), and is hot with particularly high temperatures during summer, and a large annual range of temperature (about 18°C), low humidity and almost no dew.

PLANT COMMUNITIES

The great geomorphological, pedological and climatological diversity of Israel is reflected in the diverse composition of its flora, which comprises more than 2,600 species of flowering plants, belonging to about 130 families. The principal plant communities in Israel are Mediterranean (mean annual rainfall above 350 mm), arid (mean annual rainfall is below 100 mm) or transition zone communities. These zones correspond to Eig's (1931–2) division of Palestine into the Mediterranean, Irano-Turanian and Saharo-Sindian (presently named Saharo-Arabian;

Danin and Plitmann, 1986) areas. In addition, a Sudanian element is recognised in the southern part of Israel, mainly along the 'Arava and Jordan valleys (Danin and Plitmann 1986). Psammophilous communities cover sandy areas in the Negev and along the Mediterranean coast. Only a few plant communities are characteristic of wetlands, springs or saline swamps. The following description is based largely on Waisel (1984) and Danin (1988).

The Mediterranean Zone
It is the relatively high rainfall in the Mediterranean mountain regions of Israel and the northern Coastal Plain that has presumably led to the development of a climax of sclerophyllous evergreen forest. Plant succession advances from herbaceous cover to a community (locally known as "batha") of dwarf shrubs (up to 0.5 m) and herbs, to a higher stage of shrubs (1–2 m) and small trees ("garrigue"), to scrub forest ("maquis") of medium-sized (up to 12 m) trees (Zohary, 1962). In most places the forest is not developed fully because of human interference, but mature stands of tall trees have been preserved in places that were considered sacred by the Arab population (chiefly near tombs of revered individuals). Some 750 km^2 of the Mediterranean zone of Israel were afforested by the British Mandatory and the State of Israel Forestry Departments and the Jewish National Fund, mostly by Aleppo pine (*Pinus halepensis*). These pine plantations are an ecological desert and besides *Rattus rattus* no mammal lives in them. Fruit plantations cover an additional area of approximately 1500 km^2, of which 600 km^2 are various citrus orchards and 600 km^2 are olive (*Olea europaea*) groves (Yom-Tov and Mendelssohn, 1988). Human interference in the past through burning (to improve grazing), cutting (for construction and fuel) and grazing has hindered the development of natural forests, and at present the Mediterranean zone of Israel is composed of a mosaic of various natural plant associations intermingled with man-made forests and pine plantations. About half of the tree species in the natural forest are evergreen, and have tough leaves (for example *Pistacia lentiscus, Quercus calliprinos, Laurus nobilis, Arbutus andrachne,* and *Ceratonia siliqua*). The rest are deciduous (for example *Pistacia palaestina, Quercus ithaburensis, Quercus boissieri* and *Crataegus azarolus*). The leaves of many of these trees are spiny, sclerophyllous and contain high amounts of secondary metabolites (tanins, phenols, etc.) which repel browsing by animals. Most species of dwarf shrubs such as *Sarcopoterium spinosum, Coridothymus capitatus* and *Phagnalon rupestre* replace their large "winter leaves" with smaller "summer leaves", while a minority, such as *Calicotome villosa, Lycium europaeum* and *Anagyris foetida,* are summer deciduous (Waisel, 1984).
The composition of the natural forest of Israel is influenced mainly by climate, lithology and soil type. Natural forests occupy about 350 km^2, mainly along the mountain ridges of the Galilee, Samaria, Judaea and on Mount Carmel. In these areas, the rockbeds of hard limestone and dolomite are covered by terra rossa soil. Such sites are characteristically covered by the *Quercus calliprinos–Pistacia palaestina* association. In the Upper Galilee, where the average annual precipitation

is the highest in Israel, *Quercus calliprinos* is accompanied by trees of *Q. boissieri* and *Arbutus andrachne* as well as by their companions such as *Rhamnus alaternus, R. punctatus, Eriolobus triobatus, Acer obtusifolium, Cratageus azarolus* and *Laurus nobilis.* Such forest types cast heavy shade, and the forest floor features only a few plants. Some of these constituents of the forest associations of the Galilee occur in the maquis of Mount Carmel. The maquis of the Judean Hills are drier and have arboreal companions of *Quercus calliprinos.* Open forests that are dominated by *Quercus boissieri* develop mainly on the relatively moist northern slopes of the Upper Galilee. They are accompanied by *Cercis siliquastrum, Pyrus syriaca, Prunus ursina* and *Crataegus azarolus..* Stands of *Quercus calliprinos* also develop on basalt or other volcanic substrata in the Golan, at elevations over 500 m above sea level. They are accompanied by *Q. boissieri, Crataegus monogyna, C. aronia* and *Prunus ursina.* Such woodland stands are more open and rich in ephemeral vegetation than in the Galilee. Another type of oak forest occurs on Mount Hermon, on the cold, wind-swept, slopes at elevations of 1300–1800 m. The typical arboreal plants of these communities are *Quercus boissieri, Q. libani, Juniperus drupacea, Acer microphyllum, Cotoneaster nummularia, Crataegus monogyna, Prunus ursina* and *Amygdalus korschinskii.*

Natural stands of *Pinus halepensis* exist in Israel only on some of the light rendzina soils of the Mediterranean zone. The pines are accompanied by another tree, *Arbutus andrachne* which is dominant where the soil has a high clay content.

Park forests (where tree canopies normally do not touch each other) exist in the Mediterranean zone of Israel in drier and warmer areas than those occupied by *Quercus calliprinos* woodland. Park forest dominated by *Quercus ithaburensis* occurs mostly in low regions (0–500 m above sea level) in the southern Golan Heights (Yahudia area), where it is accompanied by the trees *Pistacia atlantica* and *Ziziphus spina-christi*, by bushes of *Z. lotus* and by several herbaceous plants. This association appears on hard chalk with dark rendzina near Alonim–Shefaram and the Menashe Hills south of Mount Carmel, where it is accompanied by several arboreal species (*Styrax officinalis, Pistacia palaestina, Rhamnus lycioides* and *Quercus calliprinos*) as well as by the shrub *Majoran syriaca* . It has a rich herbaceous vegetation of Poaceae and Fabaceae. A park forest of *Q. ithaburensis* also existed formerly on hamra (sandy-loam) soil in the Sharon, between Mount Carmel and the Yarqon River, but today only a few trees remain there after this forest was destroyed by the Egyptian army of Muhamed Ali during the last century, and subsequently also by the Turkish army during the First World War (Waisel, 1984).

A park forest dominated by *Pistacia atlantica* and accompanied by *Q. ithaburensis,* exists on the hard limestone of the eastern Galilee between Har Kena'an and Metulla. This formation is less dense than the aforementioned park forests, and has specific arboreal companions, such as *Styrax officinalis, Pistacia palaestina, Rhamnus lycioides* and *Crataegus aronia.*

Park forest composed of *Ceratonia siliqua* and *Pistacia lentiscus* occupy large areas in the Mediterranean zone. They occur on limestone at low elevations (up to 300 m above sea level) of both western and eastern slopes of the central mountain ridge in the Galilee, Gilboa, Samaria, Carmel and Judea, as well as on calcareous sandstones (kurkar) in the Coastal Plain between Netanya and Mount Carmel. The composition of this community depends on climatic and edaphic factors. In the western Galilee and on Mount Carmel, the main components are *Pistacia palaestina* and wild olives (*Olea europaea*), while in southern Judaea a major component is *Rhamnus lycioides*. In the Sharon, the park forest is currently at the climax stage, in a succession sere with psammophylous and desert plants (*Artemisia monosperma, Retama raetam* and *Helianthemum stipulatum*) which developed into a Mediterranean shrub community of *Calicotome villosa, Rhamnus lycioides* and *Pistacia lentiscus* and eventually into the *Ceratonia–Pistacia* park forest.

Much of the Mediterranean zone of Israel is currently occupied by the early stages of the sere, namely by herbaceous associations, batha and garrigue. The dominant shrub in the batha is *Sarcopoterium spinosum*. On rendzina soils the typical shrubs are *Cistus incanus* and *C. salvifolius* as well as *Fumana* spp. and *Corydothymus capitatus*. The garrigue is typically composed of bushes of *Calicotome villosa, Pistacia lentiscus, Cistus incanus* and *C. salvifolius*. Stunted, multi-stem trees of *Quercus calliprinos* and *Pistacia palaestina* are common in areas where intensive grazing, felling and burning have taken place. In the Galilee these shrubs are accompanied by scattered bushes of *Anagyris foetida* .

The early stages of the Mediterranean succession seres are characterized by many geophytes, mostly in the batha communities. Only a few flowers, mainly orchids, appear in the dense forest. However, in open habitats, during spring, many of the geophytes bear beautiful flowers and form a rich display of colours. The dominant ones are tulips (*Tulipa agenesis*), *Allium* spp., *Asphodelus ramosus, Anemone coronaria, Asphodeline lutea, Narcissus tazetta*, various species of *Crocus, Iris*, and many orchids.

About 5,000 km^2 of Israel is cultivated, most of it in the Mediterranean zone. The most common crops are barley and wheat in winter and sunflower and cotton in summer. There are also large areas of citrus groves, orchards and vegetable fields.

The Transition Zone

The transition zone between the Mediterranean and desert zones stretches for more than 200 km from the north western Negev, through the Northern Negev and western Judean Desert and along the Jordan Valley up to Mount Hermon. The width of the transition zone varies from about 50 kms in the Northern Negev to less than 10 km on the eastern slopes of the Judean and Samarian Hills, widening again along the Upper Jordan Valley. In the north, the transition zone has a richer vegetation, including trees. This is due to a larger amount of rainfall and lower ambient temperatures. In the south the vegetation is poorer and lower, and is comprised of low shrubs and herbs.

In much of the transition zone between the Mediterranean and desert zones, where the climate is warmer and drier than that which supports park forest, one can find the savanna-like vegetation of spiny trees of East African origin. In several small areas in the Golan, from sea level to 200 m below sea level, on the eastern slopes of the lower Galilee south of the sea of Galilee and in isolated areas on the western slopes of the Judean Foothills in Emek HaElla, near Kefar Menachem, near Ramla, in Naḥal Tavor and in the Yizre'el Valley (near Nahalal), and near Ashdod in the southern Coastal Plain, there are stands of scattered *Ziziphus spina-christi* and *Acacia albida* (this second species does not grow in all the above mentioned areas). In the southeastern Galilee, an area similar in its geomorphology and climate to the lower Golan Heights, *Ziziphus spina-christi* is replaced by *Z. lotus* bushes. All these communities are accompanied by Mediterranean herbaceous vegetation, mainly belonging to the Poaceae, Fabaceae and Asteraceae.

The plant communities of most of the transition zone, where mean annual rainfall is 250–350 mm, are bathas of low shrubs with no trees. *Sarcopoterium spinosum* dominates the bathas of the north and central transition zone, as it does in the Mediterranean zone, where it is accompanied by *Phlomis brachyodon* and *Ballota undulata*. Other companions of this community depend on climate and substrate: *Thymelaea hirsuta*, *Astragalus bethlehemiticus* and *Euphorbia hierosolymitana* are found on hard limestone in the southern Judean Desert, and *Echinops polyceras*, *Alkanna strigosa*, *Ononis natrix* and *Arthemisia sieberi* on chalk.

Areas of the transition zone where mean annual rainfall ranges between 100–250 mm are referred to as "shrub-steppes". The most common plant dominating these areas on rocks is *Artemisia sieberi*, with several different companions in different sub-zones: *Thymelaea hirsuta* in the Negev close to the Mediterranean zone, *Noaea mucronata* in the Judean and Negev deserts, *Reaumuria negevensis* in the central Negev highlands and *Gymnocarpos decander* in the Northern Negev anticlines. Water conditions greatly affect the distribution of these communities: in areas where rain penetrates into rock fissures, they occur even where mean annual rainfall is only 50 mm, but the drier and warmer south facing slopes are often covered by the desert community of *Zygophyllum domosum* and *Gymnocarpos decander*. Where rock fissures occur closer to the Mediterranean zone they provide refuges for Mediterranean vegetation such as *Sarcopoterium spinosum*, the maquis vines *Ephedra campylopoda* and *Prasium majus* and the geophytes *Sternbergia clusiana* and *Narcissus tazetta*. Wide wadis of the central Negev highlands also provide favourable water conditions, and the climate is cooler than in other areas of the Negev. Hence, this area supports about 1,400 tall (4–5 m) *Pistacia atlantica* trees of Irano-Turanian origin, in addition to many smaller specimens of this species and the endemic *Amygdalus ramonensis*, *Rhus tripartita* and *Rhamnus dispermus*. In rainy years these areas become very spectacular, with a colourful explosion of red, yellow, violet, lilac and white flowers.

The uncultivated loess areas of the Northern Negev and southern Judean Deserts are dominated by the xerohalophytic shrubs *Hammada scoparia* in drier areas, and *Anabasis syriaca* in moister ones.

The vegetation at high elevations (above 1800 m) on Mount Ḥermon is characterised by a dense cover of low, spiny, rounded bushes ("cushion plants") with a strong Irano-Turanian element. Although the mean annual precipitation there is higher than in the transition zone further south, water availability is low due to snow cover for 4–5 months, high water infiltration into the soil and a very dry summer. Furthermore, strong winds have a desiccating effect on the plants. The vegetation on these slopes is composed of cushion-plants dominated by *Astragalus cruentiflorus* and *Onobrychis cornuta*, accompanied by *Acantholimon libanoticum*, *A. echinus* and *Astragalus echinus*. About 30 species of geophytes and several species of annuals grow in this community, gaining shelter inside the bushes.

The Desert Zone

The limiting factor for life in the desert is water, and its availability to plants is determined not only by precipitation, but also by micro-topography and soil type. The Israeli desert is defined as the area in which mean annual rainfall is less than 100 mm. It is part of the Saharo-Arabian desert belt, and species of this region dominate the Negev and Judean Deserts.

In desert areas which are close to the steppe zone, on hard limestone and dolomite where rain water penetrates into fissures, communities of *Zygophyllum dumosum* and *Gymnocarpus decander* prevail, while areas of soft rock, where there is strong runoff and little absorption of rain, are covered with the xerohalophyte shrubs *Suaeda asphaltica*, *Agathophora alopecuroides* and *Salsola tetrandra*. Since annual rainfall decreases towards the south and the east and solar radiation increases towards the south, the desert becomes increasingly drier in these directions. South and east of the country's water divide, most soils do not have sufficient water in the substratum to support perennial vegetation. The vegetation in these areas becomes more and more diffused towards east and south, and the slopes become bare. The poorest part in such areas is the top of the slope and the richest is the wadi bed. The composition and density of the vegetation increases in relation to the size of the catchment area, and there is a sequence of communities which become more diverse and denser along the wadi course. The top section of the wadis supports annual vegetation in rainy years; below this a community of small shrubs exists accompanied by some of the annuals of the higher section; further below a community of taller (1–2 m) shrubs exists, accompanied by species of the sections above; and the wide wadi bed itself supports *Acacia* trees (*A. gerrardii* at high elevations, *A. raddiana* in warmer sites and *A. tortilis* in the warmest sites of the 'Arava Valley). The composition of the shrub community in each of the sections varies greatly in relation to substrate: *Zygophyllum dumosum* dominates hard limestone slopes, *Hammada salicornica* dominates sandy wadis, *Artemisia sieberi* dominates silty and gravel wadi beds at high elevations and *Anabasis articulata* at lower elevations, *Reaumuria hirtella* and *Salsola tetrandra* are found in wadis on soft chalk, and *Retama raetam* and *Ochradenus baccatus* in wide wadis with sand and gravel.

The 'Arava Valley (together with the Dead Sea Area) is the warmest part of Israel and among its driest, with mean annual rainfall ranging from 25 mm in Elat to

100 mm near the Dead Sea. However, since this area lies below mountain ranges on its western (Negev) and eastern (Edom) sides, it has a relatively rich ground water catchment, mainly along the wadi beds. Some areas of the 'Arava (near Yotvata) resemble an East African savanna, with dense stands of tall, wide canopied *Acacia* trees of the three above-mentioned species. The trees are accompanied by a variety of bushes, depending on the substrate: *Haloxylon persicum* occurs on deep sand and wadis, *Anabasis articulata* on gravel plains (regs), *Hammada salicornica* and *Salsola cyclophylla* on sandy-gravel, and *Nitraria retusa* and *Alhagi graecorum* on salty soils. Some of these bushes are also the dominant plants in habitats which do not support *Acacia* stands. For example, *H. persicum* forms an open woodland of 2–4 m tall bushes on the deep sands of the southern 'Arava Valley.

Along the Rift Valley there are several springs which form oases. These oases support thermophilous Sudanian vegetation comprising several trees and bushes such as *Calotropis procera, Moringa peregrina, Salvadora persica, Cordia sinensistulipa, Balanites aegyptiaca* and the above-mentioned three species of *Acacia*.

Sand Vegetation

About 5% of Israel is covered by sands of three different origins: the Mediterranean coastal and the western Negev sand dunes, brought there by the Nile floods and the sea; the sandy areas of the 'Arava Valley, derived from the erosion of Nubian sandstone; and the internal dunes, which originate from the locally abundant Tertiary rocks. The shifting sand dunes along the coast are dominated by the perennial grass *Ammophila arenaria* and the western Negev dunes by *Stipagrostis scoparia*. When the sand stabilizes due to the addition of air-borne silt, the formation of a crust of cyanobacteria, soil algae, lichens and moss and a lack of grazing and trampling, these grasses are replaced by *Artemisia monosperma* as the second stage of the succession. This plant is often accompanied by other species of bushes and annuals, and in the depressions between the dunes by *Retama raetam*. The dominant species in the internal dunes of the Negev are *Anabasis articulata* and *Artemisia sieberi*. In the deep sands of the 'Arava *Haloxylon persicum* forms a shrubland of 2–4 m high bushes.

Wet Habitats

Israel is located on the margin of the Saharo-Arabian desert belt; hence its relative scarcity of wet habitats. The drainage of swamps and the pumping of water from rivers and springs has resulted in a further reduction of these habitats. However, riparian vegetation still exists along some streams in the Galilee, where *Platanus orientalis* grows in several places. *Salix acmophylla, Vitex agnus-castus* and several species of *Tamarix* which grow along streams are more common. *Populus euphratica* grows in warm places along the Jordan River and in desert oases. Several species form thickets along streams and near springs, the most common of which are *Rubus sanguineus, Arundo donax* and *Phragmites australis*. Several species of *Juncus* form carpets where there is a high water table, and *Typha domingensis* grows in swampy habitats.

BIOLOGICAL ASPECTS

ORIGIN OF MAMMALS AND THEIR DISTRIBUTION IN ISRAEL

The distribution of most mammals in Israel is, to a large extent, a reflection of the global range of each species. Animals with wide overall distribution (Cosmopolitan, Old World or Holarctic) are also widely distributed in Israel. For example, hares (Palaearctic and African) and wolves (Holarctic) occur, or occurred, throughout the country. On the other hand, animals with more restricted general distributions occur mainly in one zone in Israel. For example, 93% of Saharo-Sindian, Saharo-Arabian and Arabian eremic mammals which exist in Israel occur in the Negev and Judean Deserts, with some penetrating into the coastal sand dunes and the rest (7%) also occurring in the transition zone between the desert and the steppes. Seventy percent of temperate zone mammals, i.e. of European, Irano-Turanian or Mediterranean origin, occur in Israel in the relatively cool, northern part of the country. However, some Palaeotropic mammals (*Procavia capensis*, *Hystrix indica*, *Caracal cararcal*) occur throughout the country.

LATITUDINAL CHANGE IN THE NUMBER OF SPECIES

Several environmental factors, chiefly climatological but also edaphic, geographical and botanical are likely to affect species richness. In Israel, most climatological factors, e.g. temperature and precipitation, change in relation to latitude: from north to south temperature increases while precipitation declines. These two factors and possibly others affect many other environmental factors, making them all interdependent. In addition, correlation does not necessarily mean causation, and it is impossible to claim that any one specific factor is responsible for species richness. The most one could claim is that one factor is better correlated with species richness, and to make an educated guess as to which factor is the more important (Yom-Tov and Werner, 1996).

The richest areas in Israel (in terms of number of species) are transition zones between two or three of the Mediterranean, Irano-Turanian and Saharo-Arabian regions, namely the Northern Negev, the lower Jordan Valley and the mountain slopes which border it on the west, and Mount Hermon. These areas are also characterized by large numbers of soil types and plant communities. For mammals, the factors best correlated with species richness are the diversity of rainfall regimes, mean annual precipitation and the number of plant communities (in this order). These factors are highly correlated with latitude, hence there is a very high correlation between this factor and species richness. Within Israel, species richness declines from north to south, and its variation is predominantly explained by abiotic and vegetation factors (Yom-Tov and Werner, 1996).

VICARIANCE

Vicariance at subspecies, species and genus levels is frequent in Israel. It generally occurs along the north–south (and to a lesser extent west–east) gradient, along which one taxon geographically replaces its related counterpart. This kind of vicariance is evident in mammals as well as in other vertebrates. For example, the mountain gazelle, *Gazella gazella* , is common north and west of the 150 mm isohyet, while the desert gazelle, *G. dorcas*, occupies the deserts south and east of that isohyet. On the subspecies level vicariance occurs in several species: the leopard, *Panthera pardus tulliana* (now apparently extinct) was found in the north of Israel while *P. p. nimr* replaces it in the south; among rodents, in the mole rat *Spalax leucodon*, four chromosomal forms (2n = 60, 58, 54 and 52; Nevo, 1985) replace each other along a climatological gradient; in *Gerbillus pyramidum* (2n = 52 and 66; Zahavi and Wahrman, 1957) and *Acomys cahirinus* (2n = 36 and 38; Wahrman and Goitein, 1972) two vicariant chromosomal forms replace each other in Israel and Sinai.

While the above kind of vicariance is apparently due to climatological factors and generally pairs of taxa are involved, a second kind of vicariance occurs involving a pedological factor. In this kind of vicariance a trio of taxa is involved, typically with one form restricted to sandy soils and the other two following the usual north–south climatological gradient. Examples on the species level are the jirds (genus *Meriones*) and on the genus level the hedgehogs (Erinaceidae). *M. sacramenti* and *Hemiechinus auritus* are psammophilous, *M. tristrami* and *Erinaceus concolor* occur in the Mediterranean region of Israel and *M. crassus* and *Paraechinus aethiopicus* occur in the desert. Another trio comprises the foxes *Vulpes rueppellii* (on sandy soil in the Negev), *V. cana* (on steep rocky slopes in the desert) and *V. vulpes* all over the country (Yom-Tov, 1988).

THE EFFECT OF SUBSTRATE AND HABITAT ON DISTRIBUTION

The distribution of several species of mammals is strongly affected by substrate and habitat type. Israel has several psammophilous species, some of which are restricted to sands (*Meriones sacramenti, Gerbillus pyramidum, G. gerbillus*) while others may also occur on other light soils (*G. andersoni, Hemiechinus auritus, Vulpes rueppellii*), and still others prefer heavy soil (*Microtus socialis*) or well drained soils in the Mediterranean region (*Cricetulus migratorius, Meriones tristrami.*) Some mammals occur throughout the country in rocky habitats, on cliffs, taluses or among rocks (*Acomys cahirinus, Gerbillus dasyurus* and *Procavia capensis*), while others occur in such habitats only in the deserts (*Sekeetamys calurus, Acomys russatus, Vulpes cana* and *Capra ibex*). Forests do not cover a large area in Israel and only a few species are restricted to this habitat (for example *Dryomys nitedula* and *Apodemus sylvaticus*) and some which prefer forest may also occur in areas where the vegetational climax community is forest, but which at present have a much lower vegetational cover (for example *Apodemus mystacinus* and *Martes foina*). Being an arid country, it is not surprising that very few mammals e.g., *Arvicola terrestris* (now possibly extinct due to the drainage of the Hula swamp and lake) and *Lutra lutra*, are restricted to wet habitats.

The Effect of Ambient Temperature on Distribution

Ambient temperature in Israel declines from south to north. This factor has a considerable effect on animal distribution, as demonstrated by the distribution of rodents along the mountainous backbone of the country. Species whose distribution outside Israel is to the north and which are adapted to colder climes occur in the north of Israel, while species with a more southern distribution occur in its south. Thus, the Palaearctic snow vole *Microtus nivalis* and the Syrian squirrel *Sciurus anomalus* occur only as far south as Mount Ḥermon, the forest dormouse *Dryomys nitedula* as far south as the Upper Galilee, the common field mouse *Apodemus sylvaticus* as far south as Mount Carmel and the broad-toothed woodmouse *Apodemus mystacinus* as far south as the Judean Hills. In the Negev and Judean Deserts there are no rodents of mesic Palaearctic distribution, but only species whose world distribution lies within the Saharo-Arabian desert belt, such as Wagner's gerbil *Gerbillus dasyurus* (whose distribution extends to the Judean Hills and the Galilee) and the bushy-tailed gerbil *Sekeetamys calurus*. The black-tailed dormouse *Eliomys melanurus* occurs in the dry Central Negev mountains and the northern Golan Heights. Similar trends occur in other mammalian orders: among insectivorous bats, the Palaeotropic mouse-tailed bats (*Rhinopoma* spp.) occur mainly along the warm Rift Valley; none of the six Israeli species of the Palaearctic genus *Myotis* occur south of the Mediterranean zone of Israel; and Saharo-Arabian desert species such as Hemprich's long-eared bat *Otonycteris hemprichi,* the slit-faced bat *Nycteris thebaica* and Bodenheimer's pipistrelle *Pipistrellus bodenheimeri* do not occur in Israel north of the 100 mm isohyet. Among carnivores, the Palaearctic European badger *Meles meles* and the rock marten *Martes foina* occur only in the Mediterranean zone (but with a recent penetration into settled areas in the desert), while the eremic Blanford's fox *Vulpes cana* and the sand cat *Felis margarita* occur in Israel only in the deserts, each restricted to its own particular habitat.

Community Structure and Character Displacement

The presence of many closely related, sympatric species raises the question of how they share their habitat. This question has been addressed by several workers in Israel, who have approached the question from different angles. Horowitz (1989), Dayan and her co-workers (1989a, 1989b, 1990, 1991, 1992), Yom-Tov (1991), and Yom-Tov (1993) used a morphological approach to study the co-existence of two species of *Apodemus*, three guilds of carnivores, the psammophile gerbils and the insectivorous bats of the Dead Sea Area, respectively. On the other hand, a behavioural and ecological approach was used to study the co-existence of two species of *Acomys* in the oasis of 'En Gedi (Shkolnik, 1977) and of two species of *Gerbillus* in the sand dunes of the western Negev (Ziv *et al.*, 1993). The morphologists reported that co-existence was achieved by means of morphological character displacement, while the ethologists showed a temporal shift in activity, in which one species is active at a less favourable time than the other, while in sympatry. Shkolnik (1977) found that removal of the more dominant species (*Acomys cahirinus*) was followed by an extended activity of the originally diurnal *A. russatus*, including the cool hours of the night.

CLINAL VARIATION IN BODY SIZE AND PROPORTIONS – ZOOGEOGRAPHICAL RULES
The great climatological and morphological diversity of Israel is reflected in the great morphological dimorphism of several species of mammals. For example the hare and the wolf show a clinal decrease in body size from north to south in accordance with Bergmann's rule (Figure 2, p. 35) and in both species northern specimens are much bigger and heavier than the southern ones (Yom-Tov, 1966, Mendelssohn, 1982). A similar trend was also reported for the common spiny mouse *Acomys cahirinus* (Nevo, 1989) and less wide-spread species such as the mole rat *Spalax leucodon* (Nevo *et al.*, 1986). There is a clinal increase in the absolute and relative size of the limbs, ears (Figure 3, p. 35) and tail of hares from the north to the south of Israel, in accordance with Allen's rule.

Northern specimens of several species of mammals in Israel (for example the hare, the red fox and the wolf) tend to have darker fur than their conspecifics from the south, in accordance with Gloger's rule. Fur colour also changes in accordance with the colour of the soil: in the Sharon (the central Mediterranean Coastal Plain), hares which live on sand are lighter in colour than hares which live on brown (hamra) soil in the vicinity.

ACTIVITY PATTERNS OF MAMMALS IN ISRAEL
As might be expected for a country bordering a hot desert, most Israeli mammals are nocturnal. This is true for all species of the orders Chiroptera, Insectivora and for most Rodentia and the hare. However, some, normally nocturnal Mediterranean zone rodents are diurnal and may also be active during cool or cloudy days (for example the Levant and Snow voles, *Microtus socialis* and *M. nivalis*), even including two desert rodents (the golden spiny mouse *Acomys russatus* and the fat jird *Psammomys obesus*) In contrast, all ungulates (apart from the wild boar *Sus scrofa*, which is nocturnal and crepuscular) are mainly diurnal, but may become nocturnal if they are subjected to hunting or during the hot summer. Most carnivores are also nocturnal, but some (for example the red fox *Vulpes vulpes*, the golden jackal *Canis aureus*, the wolf *C. lupus*, the jungle cat *Felis chaus* and the caracal *F. caracal*), become crepuscular or even diurnal during the breeding season, when they are under pressure to provide food for their young. The Egyptian mongoose *Herpestes ichneumon* is the only carnivore which is mainly diurnal.

Israel has a relatively mild climate, but some species hibernate. This is true for the Israeli species of hedgehogs (*Paraechinus aethiopicus*, *Hemiechinus auritus* and *Erinaceus concolor*), for most insectivorous bats (Microchiroptera), the black-tailed dormouse *Eliomys melanurus* and the forest dormouse *Dryomys nitedula*.

REPRODUCTION OF MAMMALS IN ISRAEL
Most Israeli mammals give birth during spring (end of February to May), although

in species which are distributed throughout the country, desert populations start to breed earlier than the Mediterranean ones. For example, the red fox *Vulpes vulpes* gives birth starting in February in the desert and in April in the Mediterranean zone. Among the bats, desert species (*Pipistrellus bodenheimeri, Eptesicus bottae, Asellia tridens*) start breeding earlier (March–April) than the Mediterranean ones (for example *Myotis* spp. which give birth between April–June), and tropical species (*Rhinopoma hardwickei* and *R. microphyllum*) give birth even later (July–August). The tropical fruit bat (*Rousettus aegyptiacus*) has two birth peaks, one in spring and one in late summer (Makin, 1990).

In most Israeli mammals pregnancy starts immediately after mating, but insectivorous bats mate from the end of summer to the beginning of winter and give birth during spring or summer. The sperm is probably retained in the uterus until late winter, and gestation lasts about two months. Two mustelids, the European badger *Meles meles* and the marbled polecat *Vormela peregusna* have delayed implantation: they mate in April–May, but parturition takes place during the following January (*Vormela*) or February (*Meles*), about ten months later.

Many Israeli mammals have a strong breeding potential and, under conditions of plentiful food and water they extend the breeding season to summer and may even breed throughout the year. For example, fawns of the mountain gazelle, *Gazella gazella*, are normally born during April–May, but near agricultural settlements these gazelles breed throughout the year. This strong breeding potential is seen in captivity, where species such as the Levant vole *Microtus socialis guentheri*, most gerbils (*Gerbillus* spp.) and jirds (*Meriones* spp.), the bushy-tail jird *Sekeetamys calurus*, the black-tailed dormouse *Eliomys melanurus*, the common spiny mouse *Acomys cahirinus*, the hare *Lepus capensis*, the caracal *Felis caracal* and the honey badger *Mellivora capensis* breed throughout the year. In years with plenty of food these species extend their breeding season in the field for several months and also breed during summer.

Notes on Marine Mammals

Although this book deals chiefly with terrestrial mammals, we have also listed the marine mammals observed near and on the coasts of Israel, and have provided what little biological information there exists on these species in Israel.

The marine mammals observed near the coasts of Israel and its immediate neighbouring countries (Lebanon and Egypt) belong to three mammalian orders: Sirenia, Pinnipedia and Cetacea.

Sirenia (Plate V:1–4)

Several dugongs (*Dugong dugon*) were observed in the Gulf of Elat (Gulf of 'Aqaba) during the 1970s and early 1980s. Some of these were specimens which had become entangled in fish nets and suffocated, while others were live specimens which were observed swimming or grazing on sea grass near the western shore of the Gulf of Elat. In one case skeletal remains were found on the coast of Sinai, 40 km south of Elat. The largest group observed numbered nine specimens, whereas other observations were of smaller groups. The Zoological Museum of Tel Aviv University possesses parts, mainly skulls, of three males and two females. Their skull measurements are given below:

Skull measurements (mm) of *Dugong dugon hemprichi*

	Males (3)		Females (2)	
	Mean	Range	Mean	Range
GTL (F n = 1)	402	376–415	390	
CBL (F n = 1)	375		350	
ZB (M n = 2)	218	212–224	213	207–219
IC	72.4	71–75.5	70	66–74.5
UT	49.7	48–54	47.5	46–47
LT (M n = 2)	54.5	46–63	46.7	45.5–48
M (M n = 2)	281.5	261–302	268	256–280

PINNIPEDIA

Specimens of the Mediterranean monk seal (*Monachus monachus albiventer*) were occasionally observed near the coast of Israel, north of Haifa, before the Second World War. These specimens were apparently stragglers from a small population which bred at that time in coastal caves near Beirut, Lebanon. Monk seals continued to be observed near Beirut until 1976, but then disappeared, probably having been shot during the civil war.

CETACEA

Nine species of whales and dolphins have been observed in the Mediterranean or the Red Sea near the Israeli coasts, several of which were washed ashore. A single, apparently young specimen of sperm whale *Physeter catodon* (= *macrocephalus*) was washed onto the shore of Herẓliyya in 1934. This is the only known observation of this species from Israel.

A single fin whale *Balaenoptera physalus* was washed ashore near Ashqelon on the southern Mediterranean coast in 1956. This specimen was 16.5 m long, and its skeleton is kept in the Zoological Museum of Tel Aviv University. Another young specimen, 11.5 m in length, was found on the north-eastern coast of Sinai in 1980. Both specimens were very thin. In addition to these, several lower jaws of young specimens of this species were brought ashore entangled in fishermen's drag nets. These jaws were found during the 1930s and 1940s.

Cuvier's beaked whale *Ziphius cavirostris* is occasionally seen in the Mediterranean coastal waters of Israel, and several specimens were brought to the Zoological Museum of Tel Aviv University (ZMTAU). Three wounded specimens (apparently by marine mines) were washed onto the shore of Tel Aviv during the Second World War and several others during the 1960s. The mean greatest length of the skull found in the ZMTAU is 870 mm (n = 4; sd = 15, range 850–885) mm and the mean length of the mandible is 778 mm (n = 3; sd = 16, range 760–790 mm).

The most common cetacean near the coasts of Israel, both in the Mediterranean and the Red Seas, is the bottlenosed dolphin *Tursiops truncatus*. Several specimens of this species are washed onto the Mediterranean coast of Israel every year and more than 30 skulls have been brought to the ZMTAU. They are washed ashore almost all year round: 60% of them between July–October, and 28% during January and February. Groups of this species, sometimes numbering dozens of specimens, occasionally follow fishing boats, and at times specimens get entangled in fishing nets or injured by motor boats. They are also often observed in the Red Sea. The mean greatest length of the skull (GTL) of six females and the mandible (M) is 505 mm (sd = 22.5, range 478–535) and 430 mm (sd = 22, range 400–455), respectively, and for three males, mean GTL is 471 (sd = 16.6, range 453–485) and mean M is 396 mm (sd = 24.4 , range 370–418 mm). Specimens identified as *Tursiops gilli* and *T. aduncus* were observed in the Red Sea. However, the systematic status of these species is unclear, and they might be forms of *T. truncatus* (Leatherwood and Reeves, 1983).

The common dolphin *Delphinus delphis* is frequently observed along the Mediterranean coast, although less so than *Tursiops truncatus*. Only four specimens are preserved at the ZMTAU: the mean greatest length of the skull of these specimens is 425 (sd = 28.8, range 408–468 mm) and mean mandible length of three is 363 mm (sd = 23.4, range 346–390).

Four additional cetaceans were observed in the Red Sea near the shores of Sinai (L. Farenhead, personal communication). A single false killer whale, *Pseudorca crassidens* was caught from a small group in the southern Gulf of Elat and was kept for several months in a dolphinarium in Tel Aviv. Its GTL and M are 590 and 470 mm, respectively. On another occasion a flock of 15 specimens was observed in this area. Groups numbering 20–30 specimens of Risso's dolphin *Grampus griseus* have been observed near Tiran Island at the southern tip of Sinai. Two skulls from the Gulf of Elat are kept at the ZMTAU: their GTL is 495 and 500 mm, and M of one skull is 395 mm.

The spotted dolphin *Stenella attenuata* is very common in the Gulf of Elat and flocks of 20 to 100 specimens are frequently observed, while schools of several hundred have also been seen there. Only one female specimen of this species is kept at the ZMTAU: its GTL is 406 mm and M is 353 mm.

Single specimens of *Sousa plumbea* have been seen in the Gulf of Suez.

INTRODUCED SPECIES [see also p. 403]

Two species of mammals, the coypu (nutria) *Myocastor coypu* and the wild ass *Equus hemionus*, were successfully introduced into Israel. These two species are now part of the Israeli fauna and are dealt with in this book.

The Nature Reserve Authority also intends to introduce the Arabian oryx (*Oryx leucoryx*) to the ʻArava and the Negev, the fallow deer (*Dama dama mesopotamica*) to the Galilee and the roe deer (*Capreolus capreolus*) to the Carmel.

EXTINCT SPECIES (Plate IV:1–2)

Little is known about the dates of disappearance of various species in Israel. However, the red deer (*Cervus elaphus*) and the hippo (*Hippopotamus amphibius*) existed in Israel during historic times, apparently until about 3000 years ago, and the lion (*Panthera leo*) disappeared during the twelfth century (Bodenheimer, 1935). Some localities still bear its name. In the Galilee there is a "Tel el Assad" (hill of the lion in Arabic) and a village "Dir el Assad". More information is available from the nineteenth century onward. Several species of mammals disappeared from the area west of the Jordan River (presently Israel) between mid-1800 until the early 1900s, mainly due to habitat change and hunting, or a combination of both factors. They include the Arabian oryx (*Oryx leucoryx*), the roe deer (*Capreolus capreolus coxi*), the fallow deer (*Dama dama mesopotamica*), the cheetah (*Acinonyx jubatus*) and the Syrian bear (*Ursus arctos syriacus*). The Arabian oryx was found throughout the Arabian and Syrian Deserts and Bodenheimer (1953) reported that its hide was sold in Beʼer Shevaʻ market during the 1930s. There are no records of it from Israel after the early 1800s (Talbot, 1960), but a mounted skin of a specimen hunted in Jordan at the beginning of this century is in the Pater Schmitz collection, Jerusalem. Roe deer lived in the Coastal Plain, the Galilee, Mount Carmel and the Upper Jordan Valley (Tristram, 1888, Carruthers, 1909) and the last population in Israel was exterminated from Mount Carmel by Druze hunters who sold them to a German butcher in Haifa. The last specimen was apparently shot in 1912. Fallow deer inhabited the dense forest of the Galilee and were seen on Mount Tabor by Hasselquist (1757) and later by Tristram (1888). Antlers, apparently originating from Jordan, were sold in Jerusalem until 1925, and a broken antler of a young specimen was found by Eliezer Smoli on Givʻat Hamoreh in 1936. The cheetah was observed by Tristram (1866) near Mount Tabor and in the Galilee, but the latest observations from Israel are from December 1959. On 9 December a specimen was reported running parallel to the road in the ʻArava Valley near Naḥal Hyon, north of Yotvata. It ran for a few minutes in front of a lorry travelling at 80 km per hour, stopped near the road and was clearly seen by the drivers, who did not recognize the animal but gave a good description of it. When shown pictures of leopards and of cheetahs, they at once pointed at the cheetah. Twelve days later (22 December) at about 7 pm, a specimen ran in front of a jeep in the same area. The Syrian bear, a common animal in Israel during Biblical times, was very rare by the beginning of this century. Tristram (1888) reported seeing a specimen near Arbel, not far from the north-western shore of the Sea of Galilee, and the last specimens from the region were hunted on Mount Ḥermon during the First World War by German officers (Harrison and Bates, 1991) and by I. Aharoni. The water vole *Arvicola terrestris* has never been found alive in Israel, but remains of at least 30 skulls were found in barn owl (*Tyto alba*) pellets on the banks of the Ḥula Lake before it was drained during the 1950s (Dor, 1947). The water vole has not been found since the 1950s and was possibly eliminated due to the disturbance to its aquatic habitat when the Ḥula Lake and swamp were drained. However, no search for this species has taken place and suitable habitats still exist in the area.

THE CONSERVATION STATUS OF MAMMALS IN ISRAEL

Most Israeli mammals are protected by the Wild Animals Protection Law, enacted in 1954. The effective enforcement of the law by the Nature Reserve Authority has enabled several species of formerly over-hunted animals (*Gazella gazella*, *G. dorcas*, *Capra ibex*) to increase. Although most Israeli mammals are no longer threatened by hunting, there are still several species which are presently endangered, mainly by habitat destruction and by severe environmental pollution (Yom-Tov and Mendelssohn, 1988). Most species of insectivorous bats (Microchiroptera) in the Mediterranean region of the country are threatened by human activity. More than 30 species of Microchiroptera are presently known in Israel, and many caves used to be inhabited by thousands of roosting bats. The number of these bats decreased drastically during the 1950s and most species in the Mediterranean zone became rare or disappeared. This is true for all Mediterranean populations of *Rhinolophus* (*R. ferrumequinum*, *R. blasii*, *R. euryale*, *R. mehelyi*, *R. hipposideros*), *Asellia tridens*, all Israeli species of *Myotis* (*M. myotis*, *M. blythi*, *M. nattereri*, *M. capaccini*, *M. emarginatus*), *Miniopterus schreibersi* and *Eptesicus serotinus*. This decline may be attributed to three factors. The first is the fumigation of caves by the Department of Plant Protection of the Ministry of Agriculture. These fumigations were intended to control the number of fruit-eating bats (*Rousettus aegyptiacus*), which were considered as pests for some fruits and which often roosted together with insectivorous bats. The second factor is secondary poisoning. Noctuid moths consistute an important part of the diet of these bats, and the larvae of several species of moths are serious agricultural pests. Fields are sprayed with insecticides and pesticide residues accumulate in the fat of insectivorous bats and affect their survival, particularly in times of stress and towards the end of hibernation. The third factor is visitation of caves by hikers, who disturb roosting and hibernating bats, causing expenditure of fat reserves and desiccation. Two species of Mediterranean insectivorous bats have been less affected than others: *Pipistrellus kuhli* and *Tadarida teniotis*, which roost in hollow trees, wooden houses, old ruins and between walls. This indicates that fumigation and visitation of caves pose a more serious threat than secondary poisoning. Little is known about the bats of the desert areas of Israel, but their conservation status seems to be much better than that of the Mediterranean bats.

Habitat destruction is threatening several species of mammals. The rapid disappearance of the coastal dunes has decreased the habitat available to the endemic subspecies of jerboa (*Jaculus jaculus schlueteri*), two species of gerbils (*Gerbillus andersoni* and *G. pyramidum*) and the long-eared hedgehog (*Hemiechinus auritus*). The intensive agricultural use of the 'Arava sands has decreased the habitat of *Gerbillus nanus*, the sand cat (*Felis margarita*) and of the sand fox (*Vulpes rueppellii*). *Vulpes rueppellii* populations have also decreased due to competition with the larger red fox (*Vulpes vulpes*), whose populations in the 'Arava have increased considerably due to availability of food in garbage dumps and it has penetrated into sandy areas, in which the sand fox was formerly free from competition.

New, intensive agricultural practices, such as growing summer crops, the use of heavy tractors and deep ploughing have also affected the habitat (loess plains in the Northern Negev) of the greater Egyptian jerboa (*Jaculus orientalis*).

The intensive use of water in Israel has caused the drying up of most coastal streams and several streams in the catchment area of the Jordan River. This factor, together with serious water pollution, has been followed by a dramatic decrease of the otter (*Lutra lutra seistanica*) population.

The leopard population in the Negev and Judean Deserts (*Panthera pardus nimr*) is now threatened due to its small population size, estimated at less than 20 individuals, while its northern population (*Panthera pardus tulliana*) has been eliminated. The only population of ungulates threatened in Israel is that of *Gazella gazella acaciae* Mendelssohn *et al.*, 1997, the subspecies of the mountain gazelle in the 'Arava. In the early 1950s this population numbered several hundred (H. Mendelssohn, pers. obser.), but now there are only about 12 individuals, most of which are females (B. Shalmon, pers. comm.). Very little is being done to save this endemic subspecies, which seems to be doomed to extinction (Mendelssohn *et al.*, 1997).

RELATIONS BETWEEN WILD AND DOMESTICATED MAMMALS

Two species, the dog and the cat were domesticated in the Middle East and their ancestral stock, the wolf (*Canis lupus*) and the wild cat (*Felis lybica*) still live in our region. Both species interbreed in nature with the domesticated forms, and we estimate that at present there are hardly any pure wild cats in Israel. Feral domestic cats and hybrids, which live mainly on garbage and live in areas of high human population density, have a disastrous effect on wildlife, particularly reptiles, small mammals and ground nesting birds. Cases of mating of feral or domestic dogs with wolves are known from the Golan (during the early 1990s) and the Yizre'el Valley (in 1963/4) (Mendelssohn, 1982).

Recently there have been reports of hybrids of wild boar (*Sus scrofa*) and domestic pigs which were hunted in the western Negev, mainly in and near Naḥal Besor. Rangers of the Nature Reserves Authority are trying to exterminate these animals before they spread to other areas of the country.

Bedouin sometimes raise ibex kids in their goat herds. Ibex can breed with goats, and the offspring are fertile. An experimental herd of such hybrids is kept at Kibbutz Lahav in the Northern Negev.

SYSTEMATIC ASPECTS

SOURCES OF INFORMATION

Palaeontological studies carried out by D. M. A. Bate, G. Haas, M. Stekelis and later by O. Bar-Yosef and E. Tchernov (reviewed in Tchernov, 1988) shed light on the mammalian fauna of the Pleistocene.

Several regional works which include Israel or deal with neighbouring countries have been published during this century. Among them is B. Aharoni's work (1932) on the rodents of Palestine and Syria; works on the mammals of Egypt, including Sinai, by J. Anderson and W. E. de Winton (1902), K. Wassif (1954), K. Wassif and H. Hoogstral's monographs (1954), D. J. Osborn and I. Helmy (1980) and Qumsiyeh (1985); works on the mammals of Lebanon by R. E. Lewis and his co-workers (1962, 1967, 1968), S. I. Atallah's monograph on the mammals of the Eastern Mediterranean region and Qumsiyeh (1996) on the mammals of the Holy Land. The most important regional work are D. L. Harrison's three volumes on the mammals of Arabia (1964–1972, and updated by Harrison and Bates in 1991) which include Israel.

The palaeontological studies carried out in Israel were summarized by Tchernov (1988). External parasites of the mammals of Israel were studied by Theodor and Costa (1967).

At least some aspects of the biology of most species of Israeli mammals were studied during the second half of this century and a list of the more notable works is included, organised in systematic order.

Insectivora: *Erinaceus concolor* and *Hemiechinus auritus* were studied near Tel Aviv by Schoenfeld and Yom-Tov (1985), and in the laboratory by Dmie'l and Schwarz (1984).*Crocidura suaveolens* was studied in the laboratory by Hellwing (1970, 1971, 1973a, 1973b), in the Negev by Ivanitskaya *et al.* (1996a), and in the Coastal Plain by Rotary (1983) and Zafriri (1972).

Chiroptera: A survey of the Chiroptera was made between 1970–1975 by Makin (1977); the distribution and some aspects of the biology of the Microchiroptera in Israel was studied by Shalmon (1994 and personal communication) and co-workers; the bats of the Mediterranean zone of Israel were studied by M. Dor (personal communication); those of the Upper Galilee by Muskin (1993) and in the Dead Sea Area by Yom-Tov (1993) and Yom-Tov *et al.* (1992a). Population dynamics of *Rousettus aegyptiacus* was studied by Makin (1990); Behaviour and population dynamics of *Pipistrellus kuhli* by Barak and Yom-Tov (1989, 1991); *Pipistrellus bodenheimeri* was studied near the Dead Sea by Yom-Tov *et al.* (1992b). The diet of several species of Microchiroptera was studied by Whitaker *et al.* (1994).

Carnivora: The distribution of this group was studied by Ilani (1979) and community structure of several families by Dayan and her co-workers (1989a, 1989b, 1990, 1991, 1992). *Mellivora capensis* was bred in captivity by H. Mendelssohn and M. Stavy (personal observations); *Vormela peregusna* was studied by Ben David (1988); *Canis aureus* by Golani and Keller (1974), Golani and Mendelssohn

(1971), Macdonald (1979) and by Yom-Tov *et al.* (1995); *Canis lupus* by Mendelssohn (1982), in the Negev by Afik and Alkon (1983) and by R. Hefner (personal communication);*Vulpes vulpes* in the 'Arava Valley by Assa (1990); *Vulpes cana* by Mendelssohn *et al.*, (1987); in 'En Gedi and near Elat by Geffen and his co-workers (1990, 1992a, 1992b, 1992c, 1992d, 1994) and Geffen and MacDonald (1992); *Herpestes ichneumon* in the Coastal Plain by Ben-Yaakov and Yom-Tov (1983) and Hefetz *et al.* (1984); *Hyaena hyaena* in the Negev by Bouskila (1983, 1984), Macdonald (1978), Skinner and Ilani (1980) and Van Aarde *et al.* (1988); *Felis margarita* in the 'Arava Valley by Ilani (1986) and M. Abadi (1989); *Caracal caracal* in the 'Arava Valley by Weisbein (1989); *Panthera pardus* in the Judean Desert by Ilani (1986, reviewed by Ayal, 1990); its distribution in the Negev by Reichman and Shalmon (1992).

Hyracoidea: *Procavia capensis* was studied in the Galilee by Meltzer (1967).

Perissodactyla: *Equus hemionus*, an introduced population, was studied in the Negev by D. Saltz (personal communication) and by Saltz and Rubenstein (1995).

Artiodactyla: *Sus scrofa* was studied in the Galilee by Cnaani (1972) and on Mount Carmel by Rosenfeld (1996); *Gazella gazella* was studied by Mendelssohn (1974) and Yom-Tov and Ilani (1987); in the north of Israel and the Golan by Grau and Walther (1976), Baharav (1983), Ayal and Baharav (1985), Geva (1980) and by Shimshony *et al.* (1986); on the Carmel by Getraida and Pervolotzky (1990) and by Geffen (1995); in the Coastal Plain by Kooler *et al.* (1995); in the 'Arava by Shalmon (1988 and personal communication) and by Mendelssohn *et al.* (1997); *Gazella dorcas* by Mendelssohn (1974), Yom-Tov and Ilani (1987), Shkolnik (1988), Baharav (1980, 1983) and Ward and Saltz (1994); *Capra ibex* by Aronson (1982), Hakahm (1982), Baharav and Meiboom (1982), Greenberg-Cohen *et al.* (1994), Levi and Bernadsky (1991) and by Müller *et al.* (1995).

Lagomorpha: *Lepus capensis* was studied by Yom-Tov (1967) and Stavy *et al.* (1982).

Rodents: *Sciurus anomalus* was studied on Mount Hermon by Gavish (1995 and personal communication); the biology and reproduction of *Microtus socialis* was studied by Bodenheimer (1953) and Cohen-Shlagman *et al.* (1984a, 1984b); *Arvicola terrestris* skulls were found in the Ḥula Valley by Dor (1947); the cytology of the Gerbillinae was studied by Zahavi and Wahrman (1957), Wahrman *et al.* (1988), Wahrman and Gurevitz (1973), their diet by Bar *et al.* (1984), and their community structure by Yom-Tov (1991); *Gerbillus pyramidum* and *G. andersoni* were studied by Abramsky and his co-workers (1980, 1982, 1984, 1985, Ziv *et al.* 1993, Brand and Abramsky, 1987); *Gerbillus gerbillus* was studied in the 'Arava Valley by Brand and Abramsky (1987) and by Abramsky *et al.* (1985); *Gerbillus henleyi* was studied in the Negev by Shenbrot *et al.* (1994); *Gerbillus dasyurus* was studied in the Judean Hills by Ritte (1964) and in the Negev by Shenbrot *et al.* (1997); geographical variation of *Meriones tristrami* was studied by Chetboun and Tchernov (1983) and its biology by Bodenheimer (1949), Naftali and Wolf (1954), Abramsky and Sellah (1982) and Z. Zook-Rimon, (personal communication); *Meriones crassus* was studied by Krasnov *et al.* (1996); *Psammomys obesus* was studied in the Negev Highlands by Or (1974) and Ilan and Yom-Tov (1990), its physiology by Degen (1988) and Degen *et al.* (1991), and its reproduction in the laboratory by Shaham (1977); the systematics of *Apodemus* was studied by Fillippucci *et al.* (1989); the physiology of the two species existing in Israel was studied by Haim and Rubal (1992) and Haim *et al.* (1993); the biology of *A. mystacinus* was studied in the Galilee, the Judean hills and on Mount Carmel by Granot (1978), Ritte (1964), Yahav and Haim (1980) and Nizan (1997); the biology *of A. sylvaticus* was studied in the Galilee by Granot (1978); the feeding on pine cones by *Rattus rattus* was studied by Aisner and Terkel (1992); the reproduction of *Nesokia indica* was studied in the laboratory by Z. Zook-Ramon (personal communication); the relationships between *Acomys cahirinus* and *A. russatus* were studied by Shkolnik (1971), Mann (1986) and Shargal (1997), their cytology by Wahrman and Goitein (1972), their physiology by Shkolnik and Borut, 1966) and their morphological variation by Nevo (1989); *Acomys cahirinus* was studied in the Judean Hills by Ritte (1964); the physiology of *A. russatus* was studied by Shkolnik and Borut (1969), Haim and Borut (1975) and Haim (1991); *Spalax leucodon* was studied extensively by E. Nevo and his co-workers (1969, 1982) and by Y. Terkel and his co-workers (Rado *et al.*, 1992, Zuri and Terkel, 1996); *Dryomys nitedula* was studied by Nevo and Amir (1961); the reproduction of *Jaculus jaculus* and *J. orientalis* was studied in the laboratory by H. Mendelssohn (personal observations); the biology and behaviour of *Hystrix indica* were studied by Alkon and Saltz (1988a, 1988b, 1989) in the Negev Highlands and by Sever and Mendelssohn (1988, 1991) near Tel Aviv and in the laboratory.

STRUCTURE OF THE BOOK
In writing the book we have used primary sources of information (published original research) as well as regional or systematic studies on particular groups of mammals. Among the latter we have relied on Ellerman and Morrison-Scott (1951), Corbet (1978) and Harrison and Bates (1991). For comparisons with neighbouring countries we have used Harrison and Bates (1991) and Atallah (1977) on the mammals of the Middle East, Lay (1967) on the Mammals of Iran, Osborn and Helmy (1980) on the Mammals of Egypt and Qumsiyeh (1985) on the Bats of Egypt. Some groups of mammals, particularly bats, have hardly been studied in our region, and when available we have used information from other regions, particularly Europe. Especially helpful for bats were Ahlen (1990), Schober and Grimmberger (1987) and Stebbings (1977), who summarized data from many sources.
The discussion of each species starts with its English, scientific and Hebrew names. For systematic position and synonyms we have relied mainly on the recent book on *The Mammals of Arabia* (Harrison and Bates, 1991) and for species not mentioned there we have used Wilson and Reeder (1993). Since very detailed descriptions of most species are given by Harrison and Bates (1991), we only briefly describe the main morphological characteristics of a species, but note the variation and subspe-

cies found in Israel. Keys are provided for most groups, but not for large, easily identified species, such as those belonging to the Artiodactyla, and to Carnivora where there is only one Israeli species per family. The keys in this work are adopted, with minor modifications, from Harrison and Bates (1991). The description includes tables on body and skull characters of specimens deposited in Israeli museums, (the zoological museums of Tel Aviv University and The Hebrew University of Jerusalem, and the local collections at Bet Ussishkin, Dan, and at the Oranim Seminary, Qiryat Tiv'on).

Most measurements were made following Harrison and Bates (1991) and were carried out as follows:

Total length of body: from the tip of the snout to the root of the tail, above the anus.

Tail: from the root of the tail to its tip, but not including the terminal pencil.

Hind foot: from the proximity of the heel behind os calcis to the extremity of the pulp of the longest digit, not including the claw.

Ear: from the lower border of the external auditory meatus to the tip of the pinna, not including any tufts of hair.

Forearm (in bats): from the extremity of the elbow to the extremity of the carpus with the wing folded.

Greatest length of the skull (GTL): The greatest anterio-posterior diameter of the skull, taken from the most projecting point at each extermity.

Condylo-basal length (CBL): from the exoccipital condyle to the anterior extermity of the premaxillary.

Zygomatic breadth (ZB): the greatest width of the skull across the zygomatic arches.

Breadth of braincase (BB): the width of the braincase at the posterior roots of the zygomatic arches.

Interorbital constriction (IC): the narrowest width of the skull across the interorbital region.

Upper (maxillary) teeth row (UT): from the front of the upper canine to the back of the crown of the last upper molar.

Lower (mandibullary) teeth row (LT): from the front of the lower canine to the back of the crown of the last lower molar.

Mandibule (M): from the condyle of the skull to its most anterior projecting point, including the lower incisors, where necessary.

Tympanic bulla (TB): the greatest anterio-posterior diameter of the tympanic bulla.

We aimed at providing data for ten males and ten females of each species; for geographically very variable species, such as the hare, the wolf and the leopard, data are given for more than one population. For each character we give the mean, standard deviation (SD) of the mean and the range. However, when sample size is smaller than three specimens, SD is not given.

Species description is followed by its distribution. For information on distribution in the Palaearctic we used Corbet (1978), for Africa, Smithers (1983) and for the Middle East, Harrison and Bates (1991). Data on the distribution in Israel are from the zoological museums of the Hebrew University, Jerusalem, and Tel Aviv University, as well as from various other, smaller collections in Israel. For bats we have also used data collected by Makin (1977), Yom-Tov *et al.* (1992a, b) and Shalmon (1994, and personal communication), as well as unpublished data supplied by various researchers who caught and identified bats during their studies. Similarly, for species which are poorly represented in museums, we used information provided to us by reliable observers (e. g., E. Geffen for *Vulpes cana*, etc.). Distribution maps are provided for each species, and museum records are marked differently from other information. When available, we have also provided data on estimated population size in Israel.

Distribution is followed by data on fossil record. Such data are scarce for most mammals, and our main source was Tchernov (1988), who summarized these data recently. Fossil record is followed by a brief description of the karyotype, which is known for about a third of Israeli mammals. Karyotype is followed by habits, in which we give data on activity, habitats, preferred temperatures, hibernation (if occurring), social behaviour, enemies, senses, calls, etc. Next we deal with diet and feeding habits of the species. This is followed by reproduction. Here we describe the breeding season, mating behaviour, length of pregnancy and lactation, litter size, the rate of development of the young, age at sexual maturity and longevity. Where available, we have added data on reproduction in captive mammals, mainly from the Tel Aviv University Research Zoo (TAURZ). Reproduction is followed by description of parasites found, relying mainly on a review by Theodor and Costa (1967). Finally, we deal with relations with man, detailing damage or benefit of the species to humans, captive breeding and any mention of the species in ancient Jewish sources. Evidently, the information available on the various species differs in detail, and depends if the species is easily observed, makes frequent contact with man or if any research was carried out on it. Hence, much information is available, for example, on gazelles, but very little on most bats, and this is reflected in the species accounts.

Figure 1: Geographical Areas in Israel and Sinai

1. Upper Galilee; 2. Lower Galilee; 3. Carmel Ridge; 4. Northern Coastal Plain; 5. Valley of Yizre'el; 6. Samaria; 7. Jordan Valley and Southern Golan; 8. Central Coastal Plain; 9. Southern Coastal Plain; 10. Foothills of Judea; 11. Judean Hills; 12. Judean Desert; 13. Dead Sea Area; 14. 'Arava Valley; 15. Northern Negev; 16. Southern Negev; 17. Central Negev; 18. Golan Heights; 19. Mount Ḥermon; 20. Northern Sinai; 21. Central Sinai Foothills; 22. Sinai Mountains; 23. Southwestern Sinai

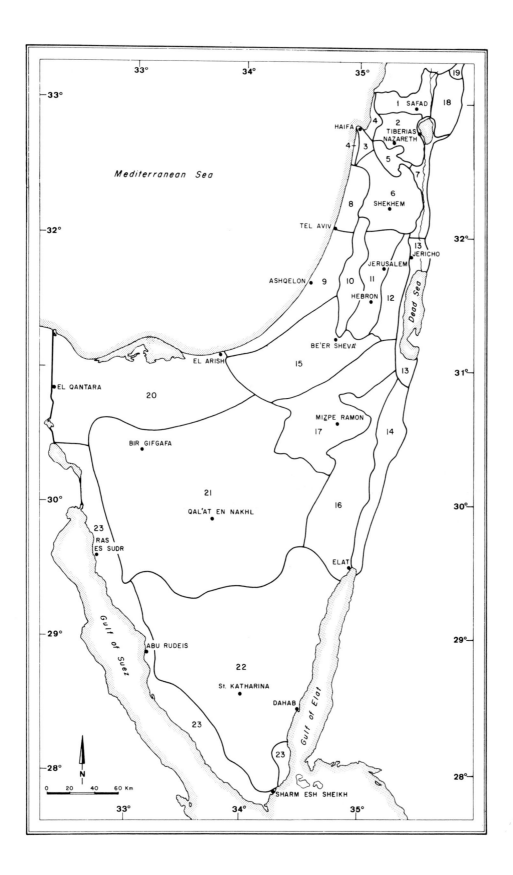

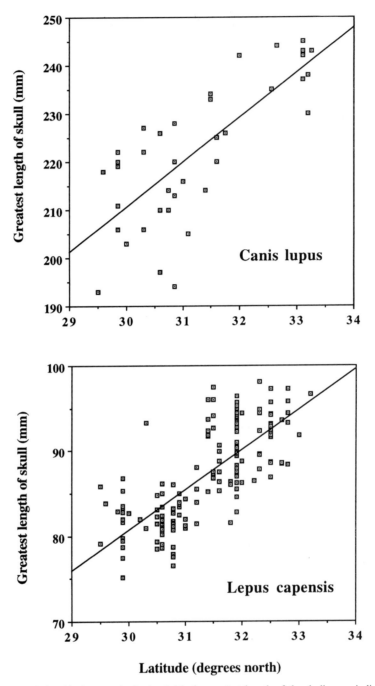

Figure 2: The relationship between body size (with the greatest length of the skull as an indicator) and latitude among the wolves (*Canis lupus*) and hares (*Lepus capensis*) in Israel.

For wolves: $Y = 9.33X - 69.40$, $r^2 = 0.580$, $p = 0.0001$, $n = 38$;

For hares: $Y = 47.34X - 61.56$, $r^2 = 0.584$, $p = 0.0001$, $n = 173$.

Lepus capensis

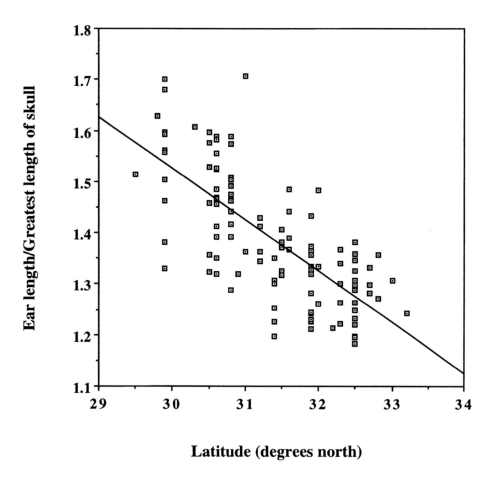

Figure 3: The relationship between latitude (X) and the relative size of the ear (Y) of hares in Israel.
$Y = 4.49X - 0.10$, $r^2 = 0.535$, $p = 0.0001$, $n = 114$.

SYSTEMATIC PART

INSECTIVORA אוכלי חרקים

This order is represented in Israel by two families: Erinaceidae and Soricidae.

ERINACEIDAE Hedgehogs קיפודיים

This family is represented in Israel by three genera, each with one species, which are largely parapatric: *Erinaceus concolor* inhabits the Mediterranean region, *Hemiechinus auritus* inhabits sand and other light soils and *Paraechinus aethiopicus* inhabits desert areas. The distribution areas of these species partly overlap in the Be'er Sheva' area.

Key to the Species of Erinaceidae in Israel
(After Harrison and Bates, 1991)

1.	Ears large, projecting above the spines	2
–	Ears small, not projecting above the spines.	**Erinaceus concolor**
2.	Prominent median bare patch dividing the spines on the forehead.	**Paraechinus aethiopicus**
–	No median patch as above	**Hemiechinus auritus**

ERINACEUS Linnaeus, 1758

Erinaceus Linnaeus, 1758. *Systema Naturae*, 10th ed. 1:52.

Erinaceus concolor Martin, 1838   East European Hedgehog
Figs 4–5; Plate VI:1–2   קיפוד מצוי

Erinaceus concolor Martin, 1838, *Proc. Zool. Soc. Lon.*, 1837: 103. Near Trebizond, Asia Minor.
Synonymy in Harrison and Bates (1991).

This species is the largest hedgehog living in Israel, although Israeli specimens are generally smaller than their conspecifics further north. The body is robust and the head pointed, with small ears and moveable snout.
Erinaceus concolor concolor is the subspecies occurring in Israel (Harrison and Bates, 1991).

Table 1
Body measurements (g and mm) of *Erinaceus concolor concolor*

	Body		Tail	Ear	Foot
	Mass	Length			
	Males (12)				
Mean	640	226	27	24	42
SD	190	13	5	3	4
Range	350–880	210–240	20–37	20–30	35–50
	Females (12)				
Mean	528	234	28	24	42
SD	186	26	7	4	3
Range	375–750	205–295	10–35	18–28	38–47

Males are significantly heavier than females. In the field body mass reaches a maximum in summer and minimum in winter, with an average difference of 12% (not significant). Maximal recorded body mass was 1 kg. There is a considerable variation in body mass and two pregnant females weighed 460 and 750 g respectively (Schoenfeld and Yom-Tov, 1985).

Table 2
Skull measurements (mm) of *Erinaceus concolor concolor* (Figure 4)

	Males (11)			Females (9)		
	Mean	Range	SD	Mean	Range	SD
CBL	55.1	51.3–57.9	2.4	54.9	53.3–57.6	1.4
ZB	34.3	31.5–37.6	2.0	34.1	31.4–35.7	1.5
MaxB	23.4	21.1–25.5	1.4	23.7	22.4–25.2	1.1
IC	13.8	13.1–14.6	0.4	13.9	13.0–14.8	0.6
UT	28.0	26.4–29.3	0.9	27.2	26.3–27.8	0.5
LT	26.1	24.4–27.1	1.0	25.1	23.0–26.4	0.9
M	43.6	40.5–45.2	1.7	43.4	41.2–45.6	1.3
MdB	28.1	26.0–30.1	1.1	28.5	26.8–29.6	0.9

Dental formula is $\dfrac{3.1.3.3}{2.1.2.3} = 36$.

The frontal upper incisors are larger than other incisors. Milk teeth are replaced at the age of one month.
Karyotype: Examined in Jordan, where 2N = 48, FN = 90 (Qumsiyeh, 1991).

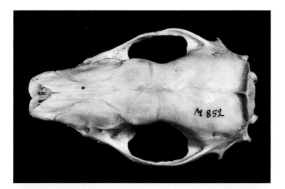

Figure 4: *Erinaceus concolor concolor.* Dorsal, ventral and lateral views of the cranium, and a view of the mandible.

1 cm

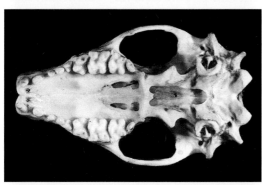

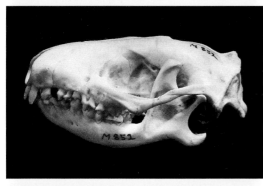

Figure 5: Distribution map of *Erinaceus concolor concolor.*

● – museum records.

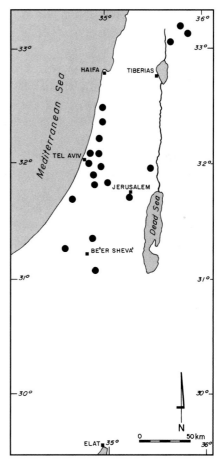

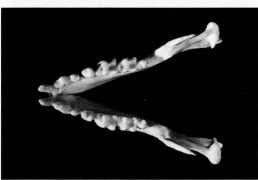

Distribution: This species occurs throughout most of Europe and Asia, with the exception of extreme desert, very cold or tropical areas (Corbet, 1978), and has been introduced to New Zealand.

Israel is at the southernmost limit of its distribution, where it lives in the Mediterranean region of the country in all habitats except sand dunes, although the conversion of sand dunes to agricultural areas has enabled it to invade such areas (Figure 5).

Fossil Record: The earliest remains of this species in Israel are from Umm Qatafa cave, and date to the Upper Paleolithic period (Tchernov, 1988).

Habits: The hedgehog was studied by Schoenfeld and Yom-Tov (1985) in a suburban area near Tel Aviv, and most data quoted here are from their study. The hedgehog is nocturnal, but nursing females have been seen to be active during the early hours of the morning and before dusk. During the day it hides under rocks or in dense vegetation, where it lines its burrow with dry leaves and straw (and with cloth found near human habitation). It tears the nesting material with its legs and teeth and tramples it until it forms a shallow scrape. Although it usually does not dig burrows, it sometimes uses ready-made burrows about 50 cm long. It makes use of several hiding places in its home range and uses them for several days, weeks or even months, before leaving for no apparent reason for another hiding place.

These hedgehogs are solitary, but do not defend a territory. An average home range is about 1.6 hectares (range 0.80–2.72, n = 4), with no difference between the sexes. Home ranges of several individuals may overlap, and in one case 10 individuals were found in a 11 m long and 2 m wide hedge. Individuals are frequently found foraging or resting at distances of a few metres from each other without any apparent interaction between them. A nightly walking distance of 300 m is usual, and it can walk at a maximum speed of 1 km/hr (measured for 5 min. Schoenfeld and Yom-Tov, 1985).

In Israel, although this species hibernates for short periods of several days, it was found to be active during winter nights even at temperatures of about 4°C. Under laboratory conditions the hedgehog starts hibernating at temperatures of 11°–12°C. Body temperatures during hibernation decrease to about 1.5°C above ambient temperature and stabilize at about 12°C, breathing rate decreases to 10 per minute and heart rate decreases from 190 to 20 beats/min. Metabolic rate during hibernation is about 32% of the basal metabolic rate. Spontaneous arousals occurred about three times per month, and an average hibernation bout lasted nine days (Dmie'l and Schwarz, 1984).

Hedgehogs are occasionally noisy animals: they growl when attacked, and males wail like a human baby when fighting other males.

The senses of smell and hearing are well developed, and sight is used for short distances.

Food: The hedgehog is omnivorous, and its diet includes snails, insects, reptiles (lizards and small snakes, including venomous ones), rodents and eggs, as well as plant material and fruits. The following items were found in stomachs and faeces of hedgehogs in Israel: snails (Helicidae), insects (Heteroptera, Coleoptera, Orthoptera, Hymenoptera) and Myriapoda, as well as lizard scales and remains of human food such as minced meat. They seek their food on the ground and by digging in vegetation and soil. While digging, hedgehogs sniff the substrate vigorously and make snorting noises. They find snails and slugs by following their slime trails.

Reproduction: During March–May a male and a female form a pair bond which can last a day or two, and they may stay together in the same hiding place during this period. Males follow females, making a characteristic wailing sound, and fighting each other. Courting bouts last 30–180 min (n = 10). Gestation lasts 5 weeks, and females give birth in one of their hiding places. In Israel, litter size is 3–5, and births normally occur between April and June. Very young hedgehogs are also found during September–October, and it is possible that females produce more than one litter a year. Newborns weigh about 10 g at birth, and are covered with soft, white spines. They open their eyes when 14–18 days old, and when 3 weeks old they follow their mother and begin to consume solid food. In one case the mother stayed with her young at the same nest for 18 days before moving them to a new resting place. At this age they are capable of rolling themselves into a ball. Independence is reached at 6 weeks, and adult size at 6 months. Hedgehogs mature sexually when one year old. In captivity they may live for 6 years, but in nature few live longer than 2–3 years.

Parasites: The spines which cover their backs prevent hedgehogs from grooming effectively, and they are often attacked by various external parasites. Infection rate was 45% for ticks (*Rhipicephalus sanguineus*) and 22% for fleas (*Archaepsylla erinacei*). It is not rare to find a specimen carrying 20 or more ticks, usually on the rear of the back or the back of the ears. Several specimens found in the Tel Aviv area were badly infected by skin fungus, apparently *Trichophyton mentagrophytes* (Schoenfeld and Yom-Tov, 1985). Theodor and Costa (1967) report that this species is also a host to the fleas *Pulex irritans, Ctenocephalides canis* and *Synosternus pallidus*, and the ticks *Hyalomma* sp., *Haemaphysalis erinacei* and *Ornithodoros tholozani*.

HEMIECHINUS Fitzinger, 1866

Hemiechinus Fitzinger, 1866. *Sitzungsberichte Math-Nat. K. Akad. Wiss.*, 1:565.

Hemiechinus auritus (Gmelin, 1770)   Long-eared Hedgehog
Figs 6–7; Plate VI:1, 3   קיפוד החולות

Erinaceus auritus Gmelin, 1770. *Novi Comment. Acad. Sci. Imp. Petropol.* 14:519. Astrakhan, southeastern Russia.
Synonymy in Harrison and Bates (1991).

This is the smallest of the three species of Israeli hedgehogs. The spines are short, yellow-grey, with longitudinal ridges and minute nodosities. The ears are very long with rounded tips and are the most characteristic feature of this species. The fur is white, but tipped with light brown, and covers the flanks and belly. The legs are relatively longer than in the other two species of Israeli hedgehogs.

Hemiechinus auritus calligoni Satunin, 1901 and *H.a. aegyptius* Fischer, 1829 are the two subspecies that intergrade in Israel (Harrison and Bates, 1991).

Distribution: The long-eared hedgehog occurs in north-west Africa (Libya and Egypt) and in Asia from Sinai and northern Arabia to India and Mongolia (Corbet, 1978).

In Israel it occurs mainly on sand dunes and other light soils along the Mediterranean Coastal Plain, Northern Negev and Judean Desert and the Golan (Figure 7). Recent stabilization of the coastal dunes due to irrigation and other human activities has resulted in the reduction of its distribution in Israel, and the expansion of *Erinaceus concolor*.

Fossil record: The earliest remains of this species found in Israel are from the Kebaran period (Tchernov, 1988).

Habits: The long-eared hedgehog was studied by Schoenfeld and Yom-Tov (1985) in a suburban area near Tel Aviv, and most data quoted here are from their study. It is nocturnal, but active mainly during the early hours of the night and early morning. During the day it hides in small caves and tunnels dug in the soil, chiefly below bushes, where the sand is more stable. Of the 37 resting places found in the study area, 26 (70%) were burrows, while the rest were depressions under stones. The tunnels are generally L-shaped structures, 10–50 cm long and 10–17 cm in diameter. In three cases the burrow was Y-shaped, with a female at one end and her young at the other. *Hemiechinus* can dig very quickly, and in one case a specimen was observed to dig 10 cm in soft soil in five minutes. Individuals can dig several burrows during one night, and change their resting place almost nightly. Burrows are used by more than one individual, but not concurrently. Burrows provide good insulation: while ambient temperatures ranged from 15°–36°C, the temperature in the burrow ranged between 20°–28°C, and differences of 10°C were recorded between ambient and burrow temperature during mid-day or late night. The presence of a hedgehog in the burrow did not greatly affect the burrow temperature.

Hemiechinus is more sensitive to cold than the European hedgehog, and probably all individuals hibernate for various periods during winter when ambient temperatures fall below 11°C. Very few individuals were caught during winter, and five individuals were found hibernating continuously for 5, 5, 8, 20, and 40 days, respectively. Under laboratory conditions they start hibernating at temperatures of 11°–12°C, body temperature decreases to about 1.5°C above ambient temperature and stabilizes at about 12°C. Metabolic rate during hibernation is about 17% of the basal metabolic rate. Spontaneous arousals occurred about 5 times per month, and an average hibernation bout lasted 5 days (Dmie'l and Schwarz, 1984).

These hedgehogs are also solitary, but do not guard a territory. Average home range is about 2.85 hectares for females (range 0.17–4.25, n = 6), and 4.97 hectares for males (range 1.64–9.97, n = 5). Home ranges of several individuals may overlap and individuals are frequently found foraging or resting at distances of a few metres from each other without any apparent interaction between them. A nightly walking distance of 1 km is usual, and it can walk at a maximum speed of 12 km/hr (measured for 5 min.). This species has a greater relative renal medullary thickness than the European hedgehog, and is capable of producing the highest urine concentration of the three Israeli hedgehogs (maximal concentration 4,010 mosmol/kg H2O (Yaakobi and Shkolnik, 1974).

When frightened *Hemiechinus* is less likely than other hedgehogs to roll up, and prefers to flee from danger.

These hedgehogs are occasionally noisy: they growl when attacked, and males wail like a human baby when fighting other males.

The senses of smell and hearing are well developed, and sight is used for short distances.

Food: This species is almost entirely insectivorous, but Atallah (1977) suggests that it also eats small vertebrates. The following items were found in stomachs and faeces of this species in Israel: snails (Helicidae), insects (Heteroptera, Coleoptera, Orthoptera, Hymenoptera) and Myriapoda (Schoenfeld and Yom-Tov, 1985). They find their food on the ground and by digging in vegetation and in the soil. While digging hedgehogs sniff the substrate vigorously, and make snorting noises. They find snails by following their slime trails.

Reproduction: Parturition takes place during spring and summer, and a female may deliver two litters per annum. Gestation lasts 36–37 days (Herter, 1965). Litter size is 1–5, newborns weigh about 5 g (Mendelssohn and Yom-Tov, 1987), and are 25–35 mm in length (Pitcher, 1976). The female is very sensitive to interference during lactation, and reacts by moving the young to a new burrow if disturbed. The eyes open about 21 days after birth (Harrison and Bates, 1991).

Parasites: The spines which cover their backs prevent hedgehogs from grooming effectively, and they are often attacked by various external parasites. However, this species is more flexible than other, larger species and is less affected. Infection rate was 25% by ticks (*Rhipicephalus sanguineus*) and 2.5% by fleas (*Archaepsylla erinacei*). It is not rare to find a specimen carrying 20 or more ticks, usually on the rear side of the back or the back of the ears. In addition, five worms (*Moniliformis monoliformis*, Acanthocephala) were found in the intestine of one specimen (Schoenfeld and Yom-Tov, 1985). This species is also a host to the flea *Synosternus pallidus* and the tick *Hyalomma* sp. (Theodor and Costa, 1967).

Table 3
Body measurements (g and mm) of *Hemiechinus auritus*

| | Body | | Tail | Ear | Foot |
	Mass	Length			
Males (9)					
Mean	182	161	19	35	31
SD	51	16	5	5	4
Range	103–247.5	140–185	14–25	30–43	25–36
Females (12)					
n = 11					
Mean	227	168	20	36	32
SD	66	17	5	4	4
Range	148–358	150–195	10–28	30–42	28–40

Males are significantly heavier than females. In the field body mass reaches a maximum in summer and minimum in winter, with an average difference of 15% (not significant). Maximal observed body mass was 390 g (Schoenfeld and Yom-Tov, 1985).

Table 4
Skull measurements (mm) of *Hemiechinus auritus* (Figure 6)

| | Males (3) | | | Females (10) | | |
	Mean	Range	SD	Mean	Range	SD
CBL	45.2	44.9–45.6		44.9	43.0–46.0	0.85
ZB (M n = 2)	26.05	26.0–26.1		25.9	23.7–27.7	1.0
BB	19.5	18.5–20.7		19.7	18.0–21.8	1.1
IC	11.1	10.8–11.4		11.1	10.3–11.8	0.4
UT (F n = 8)	22.4	21.9–22.8		22.4	21.2–23.9	0.8
LT (F n = 9)	20.15	19.8–20.5	(n = 2)	20.7	19.8–21.6	0.7
M (F n = 9)	34.3	34.2–34.5	0.15	35.3	33.9–36.3	0.8
MbB	21.8	21.4–22.0	0.32	21.95	21.2–22.4	0.35

Dental formula is $\dfrac{3.1.3.3}{2.1.2.3} = 36$.

The frontal upper incisors are larger than the other incisors.
Karyotype: 2N = 48, FN = 92 based on 14 specimens from Iraq (Bhatanger and El-Azawi, 1978).

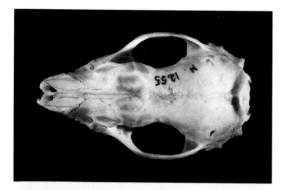

Figure 6: *Hemiechinus auritus*. Dorsal, ventral and lateral views of the cranium, and a view of the mandible.

1 cm

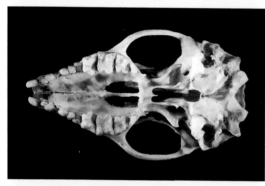

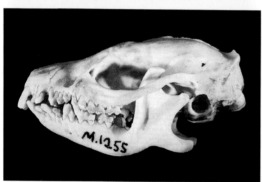

Figure 7: Distribution map of *Hemiechinus auritus*.

● – museum records.

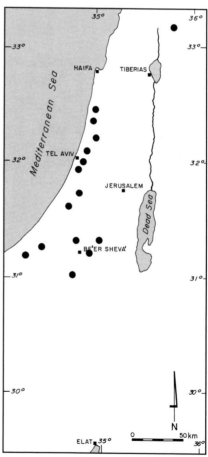

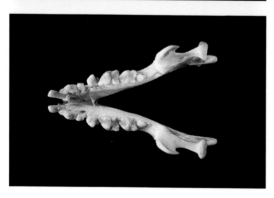

45

PARAECHINUS Trouessart, 1879

Paraechinus Trouessart,1879. *Reuve Mag. Zool. Paris*, ser. 3, 7:242.

Paraechinus aethiopicus (Ehrenberg, 1833)　　　　Ethiopian Hedgehog
Figs 8–9; Plate VI:1,4　　　　　　　　　　　　　　　קיפוד המדבר

Erinaceus aethiopicus Ehrenberg, 1833. In: Hemprich & Ehrenberg, *Symbolae Physicae Mammalium*, 2: sig. k. ecto (footnote). Dongola Desert, Sudan.
Synonymy in Harrison and Bates (1991).

Table 5
Body measurements (g and mm) of *Paraechinus aethiopicus pectoralis*

| | Body | | | | |
	Mass	Length	Tail	Ear	Foot
	Males (8)				
Mean	347	206	23	40	33
SD	66	25	4	3	4
Range	260–420	175–240	17–28	33–43	30–42
	Females (7)				
Mean	396	204	23	47	33
SD	129	14	5	5	2
Range	227–605	180–222	18–34	40–55	30–36

Table 6
Skull measurements (mm) of *Paraechinus aethiopicus pectoralis* (Figure 8)

| | Males (8) | | | Females (8) | | |
	Mean	Range	SD	Mean	Range	SD
CBL	50.9	50.2–51.5	0.4	51.0	44.2–54.2	3.1
ZB	29.3	28.1–30.9	0.9	30.1	31.6–38.6	1.0
MaxB	21.4	20.2–22.5	0.7	22.4	20.3–23.9	1.2
IC	11.7	11.2–12.4	0.4	12.0	11.2–12.7	0.6
UT	24.2	24.0–24.9	0.3	24.6	23.5–25.6	0.8
LT	21.9	21.5–22.3	0.3	22.4	21.1–23.4	0.7
M	37.6	36.5–38.2	0.7	38.7	37.5–40.2	1.2
MdB	27.3	26.0–28.3	0.8	27.7	26.3–28.7	0.8

Dental formula is $\frac{3.1.2-3.3}{2.1.2.3} = 34\text{--}36$.

The frontal upper incisors are larger than the other incisors.
Karyotype: $2N = 48$, $FN = 96$, based on six males collected from Saudi Arabia (Al-Saleh and Khan, 1985).

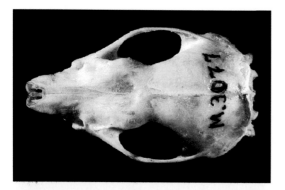

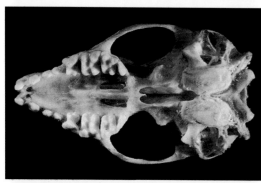

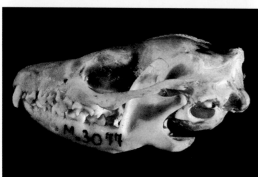

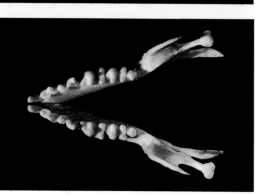

Figure 8: *Paraechinus aethiopicus pectoralis.* Dorsal, ventral and lateral views of the cranium, and a view of the mandible.

1 cm

Figure 9: Distribution map of *Paraechinus aethiopicus pectoralis* .

● – museum records.

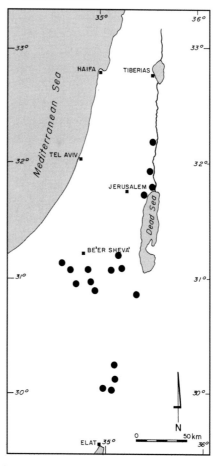

This hedgehog is characterized by a prominent bare patch in the middle of the fore-head which divides the frontal spines into two groups. There is a sharp contrast between the white ventral fur and the black dorsal side.

Paraechinus aethiopicus pectoralis Heuglin, 1861 is the subspecies occurring in Israel (Harrison and Bates, 1991).

Distribution: This hedgehog is distributed in North Africa from Morocco to Egypt and south to Somalia, and also occurs in desert areas of the Middle East (Corbet, 1978), mainly near and in oases.

In Israel it occurs in the Negev and Judean Deserts, and in the Jordan Valley as far north as Adam bridge (about 30 km north of the Dead Sea). It is the most desert adapted of Israeli hedgehogs, but its distribution overlaps that of *Hemiechinus* in the Northern Negev and *Erinaceus* in the Jordan Valley (Figure 9).

Habits: *Paraechinus* is nocturnal, but active also during early evening and morning hours. During the day it rests under rocks and in tunnels dug by itself or other animals.

In Saudi Arabia it hibernates during cool weather (December–February), often in earths of foxes or in disused tunnels dug by hares. However, even when hibernating for several weeks it usually emerges every few days to feed (Harrison and Bates, 1991). This species has a greater relative renal medullary thickness than the European hedgehog (Yaakobi and Shkolnik, 1974).

Its senses of smell and hearing are well developed, and sight is used for short distances.

Food: It is omnivorous, with a diet that includes snails, insects, reptiles (including small snakes), rodents and eggs, as well as plant material and fruits (Atallah, 1977).

Reproduction: The male actively courts the female. Gestation lasts about five weeks and the young are born between April and June, in a shallow burrow. Litter size ranges from 1–5, and a female may produce 2–3 litters annually in areas rich in food, such as oases and agricultural settlements. Newborns weigh 8–9 g, and reach 160 g at 10 weeks (Haltenorth and Diller, 1980).

Parasites: This species is infected by fleas (*Synosternus pallidus*) and ticks (*Haemaphysalis erinacei* and *Rhipicephalus sanguineus*) (Theodor and Costa, 1967).

SORICIDAE   Shrews   חדפיים

This family is represented in Israel by three species belonging to two genera, *Crocidura* (with two, but possibly three, species. See below), and *Suncus* (Figure 10).

Key to the Species of Soricidae in Israel
(After Harrison and Bates, 1991)

1.   Very small body weighing less than 3 g; skull normally with four small upper unicuspid
teeth.   **Suncus etruscus**

–   Greatest length of skull less than 20 mm. Hind foot less than 10 mm.

    Crocidura suaveolens

–   Greatest length of skull more than 19 mm. A dark line separates the white belly from
the dark, almost black back, and the tail is distinctly bicoloured. **Crocidura leucodon**

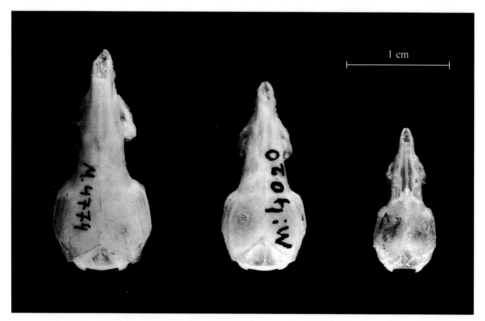

Figure 10: Cranium of *Crocidura leucodon judaica* (largest), *C. suaveolens monacha* (in the middle) and
Suncus etruscus etruscus (smallest).

CROCIDURA Wagler, 1832 חדף

Crocidura Wagler, 1832. *Isis*, p. 275.

This genus is represented in Israel by two species: *C. leucodon* and *C. suaveolens*. However, Ivanitskaya *et al.* (1996a) claim that a third species, *C. ramona*, is present in the Negev Desert. This species is similar to *C. suaveolens* in skull dimensions having a flatter skull and paler, grey-silver fur, but differs in chromosome number (2N = 28, FN = 42; Ivanitskaya *et al.*, 1996a).

Dental formula is $\dfrac{3.1.1.3}{1.1.1.3} = 28$.

Crocidura suaveolens (Pallas, 1811)   Lesser white-toothed shrew
Figs 10–12; Plate VII:1,4   חדף מצוי

Sorex suaveolens Pallas, 1811. *Zoographia Rosso-Asiat.*, 1:133. pl. 9, Fig. 2. Khersones, Crimea, southern Russia.
Synonymy in Harrison and Bates (1991).

This is the most common shrew in Israel. Males are heavier and larger than females, and the base of their tail is wider due to the development of glands there. The tail is longer than half the length of the body.

The colour of new fur is grey, but changes in time to brown and winter fur is denser and greyer than summer fur. The fur of young individuals is denser and shinier than that of adults. Specimens from the Galilee or areas with dark soil have darker fur than those from the south of Israel or from areas of lighter soils. Specimens from the Negev and Judean Desert generally have silver-grey fur, but other specimens from the same areas are much darker. For example, a specimen from Naḥal 'Arugot in 'En Gedi had dark fur, while two others, caught in the rocky desert above the wadi had silver-grey fur. The ventral side is lighter in colour than the dorsal side, but the change is gradual.

Crocidura suaveolens monacha Thomas, 1906 is the subspecies occurring in Israel (Harrison and Bates, 1991).

Distribution: This shrew is distributed over most of Europe (apart from the extreme north), the North African Mediterranean countries and the Middle East, and possibly also in Kashmir (Corbet, 1978).

In Israel it occurs in every habitat in the Mediterranean region, and specimens were also found in the Judean and Negev Deserts, all the way south to Elat (Figure 12). Its occurrence in the desert might be a result of range expansion due to agriculture. It prefers wet habitats, near streams and in dense vegetation where its population density is high.

Fossil record: The earliest remains of this species in Israel are from the Upper Paleo-lithic period (about 150,000 years BP, Tchernov, 1988).

Habits: The biology of this species was studied in the field by Rotary (1983) and Zafriri (1972). It is mainly nocturnal, but also active during the early morning and late evening, and throughout the day under cover of dense vegetation. In captivity it has bouts of activity during both day and night.

It is a solitary species. Males guard their territories which they defend against other males, but which may contain more than one territorial female which guards its own territory against other females. Males mark their territories with faeces and with a secretion from skin glands (hence the scientific name of the species, *suaveolens*, meaning in Latin "agreeable smelling"), located behind the ears and on the sides of the body.

During the day they hide under stones, and in summer also in crevices in heavy soil. Within their hiding places they build nests from dry grass and other vegetation. Shrews appear to be easy prey for owls, and shrew skulls are common in owl pellets. Shrews produce various calls. The young squeak and there is evidence that they also produce ultra-sounds. Their senses of smell and hearing are well developed.

Food: This shrew feeds mainly on insects and other invertebrates. It prefers insect larvae, earthworms, snails and isopods, but does not hesitate to attack and eat small vertebrates such as reptiles and mice, chiefly young ones. In cold weather it may feed most of the day, consuming as much food as its own mass. In summer it climbs shrubs at night, in order to reach aestivating snails (*Helicella, Theba*), gnawing at the shell and eating the snail on the spot.

Coprophagy (eating its own faeces) is common, and provides the animal with vita-mins produced by micro-organisms in the intestines.

Reproduction: The breeding biology of this species was studied in captivity by Hellwing (1970, 1971, 1973a, 1973b) and in the field by Zafriri (1972). It breeds dur-ing spring and early summer, but under favourable ambient weather conditions in captivity females gave birth during most of the year apart from the hot summer months, with a peak in March–May. During the mating season the testes become green. The penis has characteristic spines. Males court by sniffing the female, fol-lowing her, sniffing her genital area and scratching at it with their legs or teeth. This is followed by many successive copulations, which may number 300 per hour. Ovulation is induced by copulation, and at least 50 copulations are needed to induce it. After each series of copulations the male licks its penis. Sometimes the female mounts the male after copulations have taken place. Pregnancy lasts on average 29 days (range 24–32). Five days prior to delivery the female constructs an eight cm diameter, five cm high nest, made of leaves, straw and feathers, in which she will give birth and raise the young. Average litter size is four (range 1–7), and increases with the age of the female to reach a maximum at the ninth or tenth litter. Each newborn weighs about one g, and is seven mm long, but those born in litters larger than average are even smaller. The newborn is hairless, pinkish, and through its transparent skin one can see the internal organs (liver, intestines) and the milk in

Table 7
Body measurements (g and mm) of *Crocidura suaveolens monacha*

	Body		Tail	Ear	Foot
	Mass	Length			
Males (11)					
Mean	7	67	43	7	12
SD	1	6	3	1	1
Range	4.8–8.5	58–77	35.5–48	5.0–9.0	11–13.5
Females (12)					
Mean	6	62	41	7	11
SD	1	4	2.5	1	1
Range	3.35–7.5	55–70	38–45	6–10	10–12

Table 8
Skull measurements (mm) of *Crocidura suaveolens monacha* (Figure 11)

	Males (10)			Females (11)		
	Mean	Range	SD	Mean	Range	SD
GTL	19.1	18.7–19.9	0.4	18.6	18.0–19.2	0.4
Max B	5.9	5.6–6.2	0.2	5.7	5.3–6.1	0.2
IC	4.3	4.0–4.9	0.3	4.1	3.9–4.2	0.2
UT	8.3	7.9–8.7	0.2	8.1	7.9–8.5	0.2
LT	7.6	7.2–7.9	0.2	7.4	6.9–7.6	0.2
M	11.7	11.4–12.0	0.3	11.5	11.2–11.5	0.2
MdB	8.7	8.5–9.0	0.2	8.4	7.9–8.0	0.3

Karyotype: $2N = 40$, $Fn = 50$, based on specimens collected from the Eastern Mediterranean region, including six from the vicinity of Tel Aviv (Catzeflis *et al.*, 1985).

the stomach. The newborn doubles its birth weight after three days; when five days old its ears open, and at ten days the eyes open and the first teeth emerge. At this age the young get their first solid food regurgitated from their mother's mouth, and later the mother regurgitates the food in front of the young. When three weeks old the young reach 90% of adult weight and become independent. When the mother moves her young to another nest she either carries them in her mouth or they form a "row", in which one of the young grasps the rear of its mother's back with its teeth and the litter mates do the same to each other, forming a file led by the mother. Such rows are formed from the age of one week. There is a post-partum oestrus (females can become pregnant from a few hours to five days after giving birth). However, if they do not become pregnant within a few days after giving birth, the next oestrus occurs 18 days after delivery, and the following one after 50 days. In captivity, sex ratio at birth is 53 males to 47 females. Females have several litters a year.

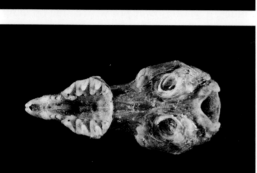

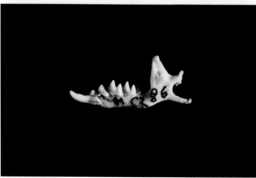

Figure 11: *Crocidura suaveolens monacha.* Dorsal, ventral and lateral views of the cranium, and a view of the mandible.

1 cm

Figure 12: Distribution map of *Crocidura suaveolens monacha.*

● – museum records.

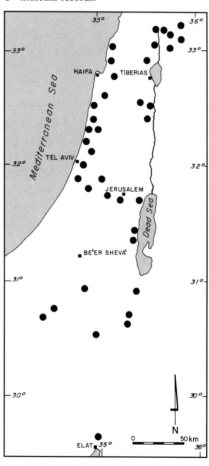

53

Under normal conditions the young reach sexual maturity at 6–8 weeks, but in very crowded, captive colonies this can be delayed to one year. Captive females may breed up to 19 times in their lifetime, but reach maximum fertility when 9–12 months old. In captivity they may live for 3.5 years, but in nature mean life expectancy is about 5 months, and only a few individuals live to the age of 1.5 years.

Parasites: This species is infected by fleas (*Leptopsylla algira* and *Nosopsyllus sincerus*) and ticks (*Ixodes redikorzevi* and *Rhipicephalus sanguineus*), as well as by *Trypanosoma* sp. (Theodor and Costa, 1967).

Relations with man: Shrews are easy to keep and breed in captivity. At Tel Aviv University Research Zoo they are kept on sand or wood shavings in 50x40x15 cm stainless steel boxes, and are provided with hides, a small flower pot with a side opening. Several families can be kept (and will breed) in one box. Food is provided twice daily, and the diet comprises fly larvae and a mixture of minced meat, hard boiled eggs, powdered milk and dry fish, fortified with vitamins. Water is provided ad libitum. When food is not sufficient they may become cannibalistic. Data from captive breeding presented here are from Hellwing (1970, 1971, 1973a,b).

Crocidura leucodon (Hermann, 1780)　　　　　Bicoloured white-toothed shrew

Figs 10, 13–14; Plate VII:2　　　　　　　　　　　חדף לבן שן

Sorex leucodon Hermann, 1780. In: Zimmermann, *Geographische Gesch. d. Menschen*, 2:382.
　Vicinity of Strasbourg, Bas Rhin, Eastern France.
Synonymy in Harrison and Bates (1991).

This is the largest shrew in Israel. Its head and body are relatively large and the tail is short, which distinguishes it from the lesser white-toothed shrew. A dark line separates the white belly from the dark, almost black back, and the tail is distinctly bicoloured, buffy whitish below and pale brown above.

Crocidura leucodon judaica Thomas, 1919 is the subspecies occurring in Israel (Harrison and Bates, 1991).

Distribution: This shrew occurs in mid-latitudes in Europe from France to the Caspian Sea and the Volga, and in Asia Minor and Lebanon (Corbet, 1978).

Very few specimens have been caught so far in Israel, and all were found in the Mediterranean region, on the Carmel, Judea, eastern Yizre'el Valley and the Sharon (Figure 14). It prefers densely vegetated areas.

Fossil record: The earliest remains of this species in Israel are from the Upper Palaeolithic period (about 150,000 years BP, Tchernov, 1988).

Habits: Very little is known about this species in Israel. It is a nocturnal and crepuscular species, with habits similar to the lesser white-toothed shrew.

Reproduction: In Europe it breeds in spring and gestation lasts 28–30 days. Litter size is 4–6. In captivity it breeds throughout most of the year and may live up to two years or more.

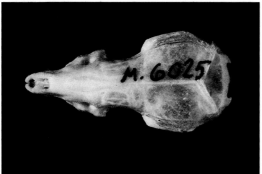

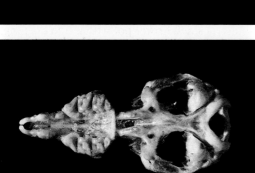

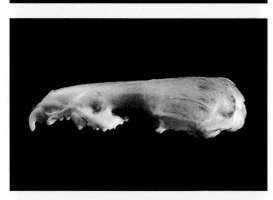

Figure 13: *Crocidura leucodon judaica.* Dorsal, ventral and lateral views of the cranium, and a view of the mandible.

1 cm

Figure 14: Distribution map of *Crocidura leucodon judaica.*

● – museum records.

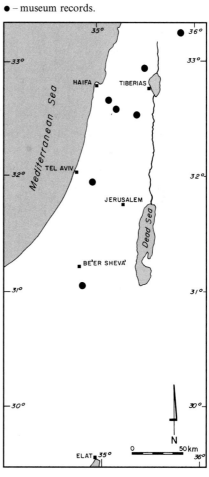

55

Table 9
Body measurements (g and mm) of *Crocidura leucodon judaica*

| | Body | | Tail | Ear | Foot |
	Mass	Length			
Males (5)					
	n = 3	**n = 3**	**n = 4**		
Mean	10	73	36	7	12
SD	4	2	2	0.5	0.7
Range	6.2–14.5	71–75	34–40	7–8	11.5–13
Females (1)					
			n = 2		
Mean	14	73	29.5	7.0	13
Range	14	73	29–30	7.0	13

Table 10
Skull measurements (mm) of *Crocidura leucodon judaica* (Figure 13)

| | Males (3) | |
	Mean	Range
GTL	20.0	19.2–20.5
Max B	5.9	5.6–6.3
IC	4.5	4.3–4.6
UT (n = 2)	9.45	9.4–9.5
LT (n = 2)	8.6	8.5–8.7
M	12.3	10.1–13.8

Karyotype: 2N = 28, based on specimens from Lesbos (Vogel and Sofianidon, 1996).

SUNCUS Ehrenberg, 1833

Suncus, 1833, Ehrenberg. In: Hemprich and Ehrenberg, *Symbolae Physicae Mammalium*, 2:folio k.

This genus is represented in Israel by one species, *Suncus etruscus.*

Suncus etruscus Savi, 1822   Savi's pygmy shrew
Figs 10, 15–16; Plate VII:3   חדף זעיר

Suncus etruscus Savi, 1822, *Nuovo Giron, de Letterati, Pisa*, 1:60. Pisa, Italy.

This shrew is the smallest mammal of Israel, as well as the smallest mammal world-wide. Its body is cylindrical, and the neck almost indistinguishable from the head and body. The ears are well developed and the eyes tiny. The tail is longer than half the body length with long hairs along its length, possibly having a sensory function while walking backwards in tunnels. The hair is about 2 mm long, the fur dense and silver-grey tinged with brown when new, turning brown when worn. The young are greyer than the adults.

Suncus etruscus etruscus is the subspecies occurring in Israel (Harrison and Bates, 1991).

Distribution: The pygmy shrew occurs in southern Europe, North Africa, and in large areas of Asia south of latitude 45°, as far south as Thailand, Malaya, Borneo and India, and in the Middle East as far south as Aden in the Arabian Peninsula. In Israel, it is not rare in the Mediterranean region, but only at low altitudes, often in more open areas than *Crocidura suaveolens*. Several specimens were found in the the Central Negev (Sede Boqer, Mitzpe Ramon and in western Makhtesh Ramon), and Tristram (1888) identified a specimen of *Sorex pygmaeus* (believed by Boden-heimer (1958) to be Savi's pygmy shrew) from Mar Saba in the Judean Desert (Figure 16).

Fossil record: The earliest remains of this species in Israel are from the Upper Palaeolithic period (about 150,000 years BP, Tchernov, 1988).

Habits: Very little is known about this species in Israel. It is nocturnal and crepuscular, and during the day it usually hides in burrows or above ground amongst dense vegetation. Because of its small size it is able to hide in small cracks in the ground, in insect holes, etc. In winter it can be found under stones which are exposed to the sun and provide warmth.

Its main predators are owls, but it is also taken by the great shrike (*Lanius excubitor*), which impales it on thorns of various trees and bushes. One specimen, active during the day on the ground, was attacked by a stone chat (*Saxicola torquata*). It was saved from the bird and housed in a matchbox. Because of its high metabolic rate, it had to be fed every two hours and either a mealworm or a small cockroach was sufficient to satiate it. When given several insects at once, it killed all, but ate

Table 11
Body measurements (g and mm) of *Suncus etruscus etruscus*

	Body		Tail	Ear	Foot
	Mass	Length			
		Males (4)			
n = 3					
Mean	1.3	53.5	26.0	5.1	7.3
SD		14.7	4.24	0.25	0.50
Range	1.2–1.5	43–75	23–32	5.0–5.5	7.0–8.0
		Females (7)			
n = 5			**n = 6**		
Mean	2.3	44.7	25.4	4.8	6.9
SD	0.55	2.4	1.6	1.5	0.92
Range	1.5–2.9	42–48	24–28	3.0–6.0	5.0–8.0

Table 12
Skull measurements (mm) of *Suncus etruscus etruscus* (Figure 15)

	(11)		
	Mean	Range	SD
GTL	13.3	12.8–14.2	0.4
MaxB	3.9	3.6–4.0	0.1
IC	2.9	2.7–3.0	0.1
UT	5.4	5.0–5.7	0.2
LT	5.0	4.5–5.3	0.2
M	7.7	7.3–8.5	0.3
MdB	5.9	5.7–6.0	0.15

No significant difference in size has been found between males and females. This sample includes three males, three females and five specimens of unknown sex.

Dental formula is $\frac{3.1.2.3}{1.1.1.3} = 30$.

Karyotype: 2N = 42, FN = 68, based on specimens from France (Meylan, 1968).

Figure 15: *Suncus etruscus etruscus.* Dorsal, ventral and lateral views of the cranium, and a view of the mandible.

1 cm

Figure 16: Distribution map of *Suncus etruscus etruscus.*

● – museum records.

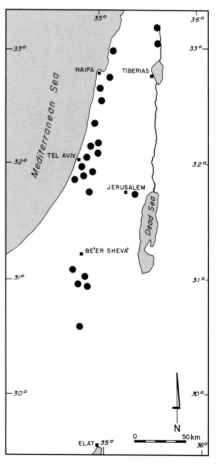

only one. The dead ones were not eaten later. When one meal was missed the shrew was found dead after four hours.

Food: The pygmy shrew is mainly insectivorous, but it also eats small snails and other small invertebrates. Due to its high body surface/mass ratio it has a very high metabolic rate, and its heart rate can reach 1300 beats/min.

Reproduction: In Israel it breeds during winter and spring. Gestation lasts 28 days. In Israel, only litters of three were found. The newborns each weigh 0.2 g at birth. The female delivers in a nest among tree roots, under stones and other cover. Only the female takes care of the young. In captivity longevity is up to 1.5 years.

CHIROPTERA Bats עטלפים

This order is represented in Israel by eight families: Pteropodidae, Rhinopomatidae, Emballonuridae, Nycteridae, Rhinolophidae, Hipposideridae, Molossidae and Vespertilionidae with at least 32 species in Israel.

Key to the Families of Chiroptera in Israel
(After Harrison and Bates, 1991)

1. Tail vestigial, not contained in an interfemoral membrane; terminal claw on index finger. **Pteropodidae: Rousettus aegyptiacus**
– Tail well developed and at least partly contained in an interfemoral membrane; no claw on index finger. 2
2. Tail emerges from the upper half of the interfemoral membrane near the mid-point. **Emballonuridae**
– Tail emerges from the edge, or is completely enclosed within the intefemoral membrane. 3
3. Tail emerges from the edge of interfemoral membrane with a considerable part of its length free. 4
– Tail enclosed in the interfemoral membrane to the tip or with only the tip protruding. 5
4. Tail thin, longer and mouse-like; distinct nasal inflations present on the rostrum of the skull. **Rhinopomatidae**
– Tail, relatively shortened and thickened; no nasal inflations present on the rostrum of the skull. **Molossidae: Tadarida teniotis**
5. No noseleaf present. 6
– Noseleaf present. 7
6. Muzzle simple, without a vertical median furrow bordered by cutaneous lappets. **Vespertilionidae**
– Muzzle with a vertical median furrow bordered by cutaneous lappets. **Nycteridae: Nycteris thebaica**
7. Noseleaf with a single vertical process (lancet) above a horseshoe. **Rhinolophidae**
– Noseleaf with either three vertical processes present above a horseshoe or with an erect transverse leaf lacking any vertical process. **Hipposideridae: Asellia tridens**

PTEROPODIDAE עטלפי פירות

This family is represented in Israel by one species, *Rousettus aegyptiacus.*

61

ROUSETTUS Gray, 1821

Rousettus Gray, 1821, *London Medical Repository*, 15:299.

Rousettus aegyptiacus (Geoffroy, 1810)　　　　　　　Egyptian Fruit Bat
Figs 17–18; Plate VIII:1　　　　　　　　　　　　　　　　עטלף פרי

Pteropus egyptiacus E. Geoffroy, 1810, *Annales Mus. Hist. Nat. Paris*, 15:96. (misprint) corrected to *aegyptiacus* in 1818, *Description de l'Egypte*, 2:134, pl. 3, Fig. 2. Great Pyramid, Giza, Egypt.
Synonymy in Harrison and Bates (1991).

This is the largest Israeli bat. Both first and second finger in the wing have a claw. The tail is short and not contained in the interfemoral membrane. The eyes are large. *Rousettus aegyptiacus aegyptiacus* is the subspecies occurring in Israel.
Distribution: The fruit bat is widely distributed in most of Africa, excluding the most extreme desert areas of the Sahara, and is found in the Middle East from Sinai to south-eastern Asia Minor and Cyprus and east to Sind (Corbet, 1978).
It is abundant in Israel, mainly in the Mediterranean region, but also occurs in the Negev Desert as far south as Elat (Figure 18), possibly mainly due to food availability in agricultural areas and from ornamental trees, mostly *Ficus* spp.
Fossil record: The earliest remains of this species in Israel is a molar dated to the Natufian period (10,000 years BCE) from the Lower Galilee (Tchernov, 1988).
Habits: The fruit bat was studied by Makin (1990), and most information from Israel is from his work. This species inhabits caves, preferably damp and dark ones, but also wells and ruins and even inhabited buildings, when they are not disturbed, e.g. they have been found in several temporarily unused underground parking areas in Tel Aviv. It is a highly gregarious animal, and each roosting site has between dozens and thousands of specimens hanging from the roof. They crowd tightly together and are noisy and restless. Before emerging from the roost they groom themselves with one foot and lick their fur (Smithers, 1983). They leave the caves after sundown and return before first light in the morning. When returning they are very noisy, as new arrivals seeking a place to hang disturb other individuals and there is much squabbling (Smithers, 1983). When disturbed during the day they tend to fly from place to place within the cave, but in small caves where there is no space they flee to the outside, even during the day. They do not hibernate in winter, but body mass increases towards autumn, reaching a maximum in October, and decreases during winter with a minimum in February. The mean difference between autumn and winter weight is about 10% (Makin, 1990).
Colonies may be mixed sex or uni-sex, or comprise mainly young specimens. Some specimens show site fidelity and return to the same cave every night, while others move between two or more caves which can be quite distant. For example, one specimen was caught within 3 months in two caves which were 40 km apart, and another

specimen was found 80 km from the first place in which it was caught. The largest recorded distance between roosting and feeding places during one day was 13 km. After emerging from their roost, fruit bats normally fly directly to their feeding ground and spend most of the night there until returning to the roost.

Their senses of hearing, sight and smell are well developed. This is the only genus of fruit bat which uses echolocation, mainly for orientation in caves. The calls, which sound like clicks, are produced by the tongue and emitted through the corners of the mouth with the jaws closed (Smithers, 1983). The calls are at frequencies below 20 kHz, and thus can be heard by humans. They use sight for navigation and smell for allocating food.

Average flight speed is 4.4 m/sec. Fruit bats are preyed on by falcons (Thomas, 1900) and owls (Qumsiyeh, 1985).

Food: As its name suggests, the fruit bat is mainly frugivorous. Before the development of orchards with many species of fruit and ornamental trees in Israel their main diet seems to have been the fruit of the sycamore fig (*Ficus sycomorus*) in summer and carobs (*Ceratonia siliqua*) in winter. They currently feed on a great variety of fruits, including fruits of various ornamental species of *Ficus* which are very common in Israel, as well as loquat, peach, fig, pear, banana, dates and other fruits. The yellow berries of the Persian lilac (*Melia azederach*), a common ornamental tree, the fruit of which ripen in winter, provides it with a staple winter diet. The fruit is manipulated by one of the bat's hind feet while it uses the other to cling to the branch while it eats the food, or the fruit is held against the chest by one of the hind feet (Smithers, 1983). Food availability is the cause of a major increase in populations since the establishment of the State of Israel. However, the present practice of picking fruit before ripening (for better shelf life) has decreased food availability to bats and the damage they do is minimal. Formerly, when fruit was picked when ripe, fruit bats caused considerable damage to peaches, pears, etc. It also feeds on leaves, buds and flowers. In 'En Gedi this species was seen feeding on flowers of the introduced baobab trees (*Adansonia*). Body mass may increase by 15% in one night after a good meal. The food is chewed, its juices swallowed, and the fibers spat out. The liquid faeces are squirted in flight and often soil houses, windows, cars, etc.

Reproduction: Females reach sexual maturity between 7–16 months old, and males when 14–20 months old. There is a post-partum oestrus. Most females usually give birth twice a year, in April and September. Births are more synchronized in April, when most females give birth during the first half of the month, than at the end of August to the beginning of September. 5%–10% of the females give birth between April and August. Gestation lasts about 120 days. The most common litter size is one, but twins have been recorded occasionally. The young are born with eyes closed and ears folded back. The eyes open and the ears become erect at about 10 days of age (Smithers, 1983). The mother carries the young attached to her chest for six weeks, until it weighs up to 60% of adult mass. At this stage its minimal forearm length is 68 mm and more, and it is left in the cave when the female forages. The

Table 13
Body measurements (g and mm) of *Rousettus aegyptiacus aegyptiacus*

	Body		Tail	Ear	Foot	Forearm
	Mass	Length				
	Males (12)					
Mean	130	146	10	24	19	92
SD	17	12	3	2	6	4
Range	87–150	130–165	7–17	21–27	16–22	88–99
	Females (12)					
Mean	100	121	10	23	18	93
SD	23	31	5	1	1	2
Range	88–150	118–160	6–16	20–24	16–25	88–96

Table 14
Skull measurements (mm) of *Rousettus aegyptiacus aegyptiacus* (Figure 17)

	Males (10)			Females (10)		
	Mean	Range	SD	Mean	Range	SD
GTL	43.8	41.1–45.4	1.3	42.5	40.9–43.9	1.0
CBL	42.0	39.3–43.5	1.3	40.8	38.8–42.2	1.1
ZB	26.4	24.1–27.6	1.3	25.6	24.5–26.9	0.8
BB	17.4	16.7–17.7	0.3	17.0	16.2–17.9	0.5
IC	8.1 (n = 9)	7.6–8.6	0.4	7.8	6.9–8.6	0.5
UT	16.8 (n = 8)	15.7–18.0	0.7	16.4	15.1–16.8	0.5
LT	18.2 (n = 8)	17.3–19.3	0.7	17.8	17.3–18.1	0.3
M	34.1	31.7–35.5	1.1	33.1	31.6–34.4	0.9

Dental formula is $\frac{2.1.3.2}{2.1.3.3} = 34$.

Karyotype: 2N = 36, Fn = 66, based on specimens from East Africa (Dulic and Mutere, 1973)

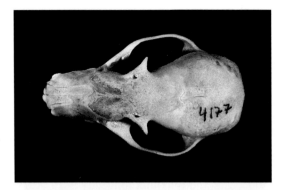

Figure 17: *Rousettus aegyptiacus aegyptiacus.* Dorsal, ventral and lateral views of the cranium, and a view of the mandible.

1 cm

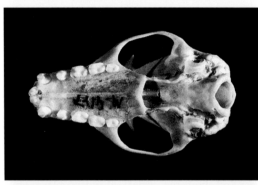

Figure 18: Distribution map of *Rousettus aegyptiacus aegyptiacus.*

● – museum records, ○ – verified field records.

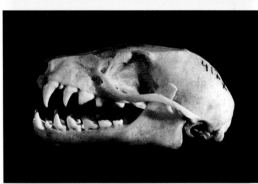

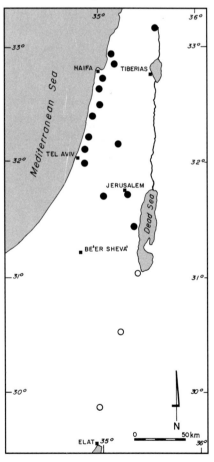

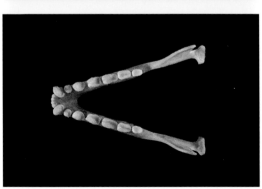

65

female recognizes the young by its smell (Kulzer, 1958). Lactation lasts at least nine weeks, but females with swollen teats were caught even several months after giving birth, indicating that lactation might last up to three months. The young sometimes lick the mouth of the mother after she has eaten fruit, and in this way supplement their diet with water and sugar, before later eating mouthfuls for themselves. By licking the mouth of the mother the young learn to identify suitable food. The young start flying when 9–10 weeks old and reach adult size when six months old. Females in good physical condition (generally adults, but not very old ones) give birth more frequently than other females.

About 50% of the young die within their first year, and average annual mortality later is 22%. Longevity in nature is 8–12 years and maximal longevity is 20 years; specimens with teeth eroded to the gums have been caught.

Parasites: The following ectoparasites were found on fruit bats in Israel: *Eucampsipoda hyrtlii* (Diptera Pupipara), *Ornithodoros salahi* and *O. tholozani* (Argasidae), *Ancystropus zeleborii, Meristaspis lateralis* and *Ichonoryssus flavus* (mites) (Theodor and Costa, 1967).

Relations with man: This species was (and sometimes still is) considered a pest due to its habit of eating fruit and defecating on buildings. Until the 1980s the Israel Ministry of Agriculture tried in vain to exterminate colonies by fumigation of caves with Gamaxen (Lindane) or Ethylene Dibromide. These campaigns hardly affected the fruit bat populations, because of their high breeding potential, the ample availability of food from ornamental and fruit trees, and the large variety of hiding places, but they did have a devastating effect on insectivorous bats in the Mediterranean region of Israel. As pest control officers knew only the term "bats", but did not differentiate between Macro- and Microchiroptera, they often fumigated caves in which only Microchiroptera roosted.

RHINOPOMATIDAE יזנוביים

This family is represented in Israel by two species of the genus *Rhinopoma*: *R. microphyllum* and *R. hardwickei*.

Key to the Species of Rhinopomatidae in Israel
(after Harrison and Bates, 1991)

1. Size larger, forearm 61–72 mm; greatest length of skull 18.7–21.6 mm; tail generally shorter than the forearm; tympanic bullae and nasal inflations relatively smaller.

   Rhinopoma microphyllum

- Size smaller, forearm 45–62 mm; greatest length of skull 15.0–19.6 mm; tail generally longer than the forearm; tympanic bullae and nasal inflations relatively larger.

   Rhinopoma hardwickei

RHINOPOMA Geoffroy, 1818   Mouse-tailed Bat

Rhinopoma E. Geoffroy, 1818. *Description de l'Egypte*, 2:113.

The tail is long and a large part of it protrudes beyond the uropatagial membrane. There is a connecting membrane between the ears and a small triangular ridge on the snout, hence its scientific name "rhino — nose and "pom" — a cover (Greek). The second digit has two phalanges.

Dental formula is $\dfrac{1.1.1.3}{2.1.2.3} = 28$.

Rhinopoma microphyllum (Brünnich, 1782)   Greater Mouse-tailed Bat
Figs 19–20; Plate VIII:2   יזנוב גדול

Vespertilio microphyllus Brunnich, 1782. *Dyrenes Historie Dyre-Sam. Uni. Nat.-Theater*, 1:50, pl. 6, figs. 1–4. Arabia and Egypt, type locality fixed as Giza by Koopman (1975: 366). Synonymy in Harrison and Bates (1991).

This is the larger of the two mouse-tailed bats in Israel. Females are smaller than males. The tail length is equal to or shorter than the forearm, which is 59–72 mm long.
Rhinopoma microphyllum microphyllum is the subspecies occurring in Israel (Harrison and Bates, 1991).
Distribution: It is recorded from isolated areas around the Sahara (Mauritania, Morocco, Nigeria, Sudan and Egypt) and in the Middle East from Israel east to Afghanistan and south to Arabia. It occurs also in India and Sumatra (Corbet, 1978).

67

Table 15
Body measurements (g and mm) of *Rhinopoma microphyllum microphyllum*

| | Body | | Tail | Ear | Foot | Forearm |
	Mass	Length				
			Males (10)			
	n = 2	n = 11				
Mean	20.5	68.7	57.9	20.3	13.8	65.9
SD	7.7	5.6	1.5	1.7	3.7	
Range	17–24	57.5–75.0	53.0–70.0	17.8–22.0	10.0–17.0	59.0–71
			Females (5)			
	n = 2			n = 4	n = 4	
Mean	16.5	65.6	56.8	18.6	15.3	65.2
SD		4.6	3.1	2.7	2.6	2.7
Range	16.0–17.0	59.0–70.0	52.0–60.0	14.5–20.0	13.0–19.0	64.0–70.0

Forearm as long as 72.4 mm was measured in males by B. Shalmon (pers. comm.).

Table 16
Skull measurements (mm) of *Rhinopoma microphyllum microphyllum* (Figure 19)

| | Males (2) | | | Females (4) | | |
	Mean	Range	SD	Mean	Range	SD
GTL	21.1	21.0–21.2		20.4	20.0–20.7	0.3
CBL	19.0	19.0		18.8	18.0–19.4	0.7
ZB	12.5	12.2–12.7		12.1	11.6–12.6	0.5
BB	8.9	8.9		8.7	8.1–9.4	0.6
IC	2.6	2.5–2.7		2.8	2.3–3.6	0.6
UT	7.6	7.5–7.7		7.2	7.0–7.4	(n = 3)
LT	8.5	8.4–8.5		8.0	7.9–8.2	0.2
M	15.0	14.9–15.0		14.8	14.2–15.4	0.6
TB	5.1	5.0–5.2		4.6	4.6–4.6	0

Karyotype: 2N = 42, FN = 66, based on two males from Jordan (Qumsiyeh and Baker, 1985).

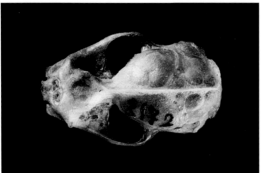

Figure 19: *Rhinopoma microphyllum micro-phyllum.* Dorsal, ventral and lateral views of the cranium, and a view of the mandible.

1 cm

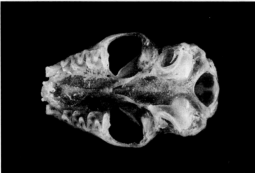

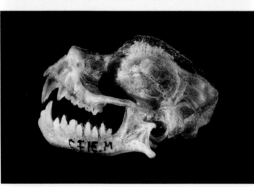

Figure 20: Distribution map of *Rhinopoma microphyllum microphyllum.*

● – museum records, ○ – verified field records.

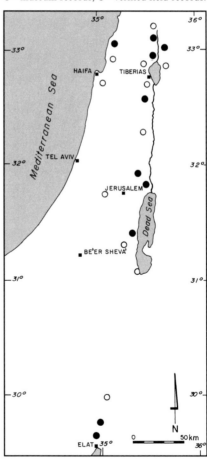

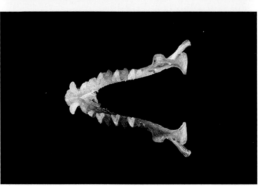

In Israel it occurs mainly in the Rift Valley, where specimens were found in Yotvata (50 km north of the Red Sea) and along the shores of the Dead Sea to southern Golan Heights and the Ḥula Valley, but also in the Galilee and 'Akko (Acre), the Judean Hills west of Jerusalem, in the Carmel and near 'Arad (Figure 20). Until the 1950s it was a very common species, but is rare today because of cave fumigation which began in 1958.

Habits: It roosts in caves and crevices, frequently together with *R. hardwickei*. The long tail is used as a tactile organ, informing the bat about the available space when it moves backwards into narrow crevices. In some caves it is not found all year round, giving the impression that it is migratory, possibly for short distances (Makin, 1977). Calls are produced at frequencies ranging from 22–37 kHz, with maximum energy at 27 kHz (Schmidt and Joergmann, 1986). Dor (1947) found three skulls in pellets of the barn owl *Tyto alba*. In autumn large amounts of fat are deposited in the rear part of the body, including the uropatagium. It has a strong, characteristic undulating flight pattern.

Food: It feeds on insects which it catches in flight (Wilson, 1973), sometimes above water.

Reproduction: Communal copulations of this species and of *Taphozous nudiventris* were observed in November in a large cave in Wadi Qilt, near Jericho (H. Mendelssohn, pers. obser.). The behaviour was similar in both species. The females hung from the roof of the cave and the males, with much vocalization from both sides, flew to the females, clinging to them and copulating. Each individual appeared to copulate several times. Makin (1977) found a female with a newborn in July, 1976. In India males and females occur together in early April, and parturition probably occurs in June (Brosset, 1962).

Parasites: The following ectoparasites were found on this bat: *Nycteribosca alluaudi* and *Ascodipteron namrui* (Diptera Pupipara), and *Ornithodoros tholozani* (Argasidae) (Theodor and Costa, 1967).

Rhinopoma hardwickei Gray, 1831   Lesser Mouse-tailed Bat
Figs 21–22; Plate VIII:3–4   יזנוב קטן

Rhinnopoma hardwickei Gray, 1831, *Zoological Miscellany*, 1:37. India.
Synonymy in Harrison and Bates (1991).

This is the smaller of the two mouse-tailed bats in Israel. Females are smaller than males. The tail is longer than the forearm whose length is 47–62 mm.
Rhinopoma hardwickei arabium Thomas, 1913 is the subspecies occurring in Israel (Harrison and Bates, 1991).
Distribution: It is recorded from the periphery of the Sahara (Mauritania, Morocco, Niger, Sudan, NE Kenya and Egypt) and in the Middle East from Israel east to Afghanistan and south to Arabia. It also occurs in India and Thailand (Corbet, 1978).
In Israel it is found mainly in the Rift Valley, from near Elat and the Dead Sea to southern Golan Heights and the Ḥula Valley, but also in the Galilee and the Judean Hills (M. Dor, pers. comm.; Makin, 1977. Figure 22). Until the 1950s it was a very common species, but is scarcer today, although still more numerous than *R. microphyllum*..
Habits: It roosts in caves and crevasses, frequently together with *R. microphyllum*., in groups which range in size from less than ten to several hundred. Both species prefer large caves with many crevices, to which they retire when threatened. About a thousand were observed emerging from a cave in the Gilboa' in 1990 (Y. Yom-Tov, pers. obser.). In a sample from Timna' (southern 'Arava Valley) body mass peaked in August (mean = 16.3 g, range 11.5–21.5, n = 16) and was at its lowest in March (mean = 10.1 g, range 7.5–11, n = 18), indicating that this species accumulates large fat reserves for its hibernation (B. Shalmon, pers. comm.). Ambient temperature in one hibernation cave ranged from 24°–28°C (B. Shalmon, pers. comm.). In some caves it is not found all year round, giving the impression that it is migratory, possibly for short distances (Makin, 1977). Calls are produced at frequencies ranging from 18–80 kHz, with maximum energy at 32 kHz (Schmidt and Joergmann, 1986). It has a strong, characteristic undulating flight pattern.
Food: It feeds on insects which it catches in flight (Wilson, 1973), sometimes above water.
Reproduction: Males with large testes were observed between March–May. Pregnant females and non-volant young were found in a cave in 'En Gedi in May 1988, and females carrying young were caught near Jericho on 3–4 July and during August 1988 (Yom-Tov *et al.*, 1992a, B. Shalmon, pers. comm.).
Parasites: The following ectoparasites were found on this bat: *Nycteribosca alluaudi* and *Ascodipteron rhinopomatos* (Diptera Pupipara) (Theodor and Costa, 1967).

Table 17
Body measurements (g and mm) of *Rhinopoma hardwickei arabium*

	Mass	Body length	Tail	Ear	Foot	Forearm
			Males (10)			
Mean	10	65	61	17	10	56
SD	2	5	6	4	4	2
Range	7–16	59–75	52–70	15–21	7–15	52–58
			Females (10)			
Mean	8	58	62	17	11	55
SD	3	5	3	3	3	2
Range	6–12	52–65	59–70	10–20	7–15	51–58

Forearm as long as 61.8 mm and weight of 21.5 g were recorded in males (B. Shalmon, pers. comm.).

Table 18
Skull measurements (mm) of *Rhinopoma hardwickei arabium* (Figure 21)

	Males (10)			Females (10)		
	Mean	Range	SD	Mean	Range	SD
GTL	17.9	17.1–18.6	0.45	17.6	17.0–18.2	0.4
CBL	15.9	15.2–16.4	0.39	15.8	15.1–16.1	0.3
ZB	10.4	9.7–10.8	0.38	10.2	9.8–10.4	0.2
BB	7.6	7.0–8.1	0.36	7.8	7.3–8.3	0.3
IC	2.8	2.6–3.1	0.16	2.8	2.6–3.1	0.2
UT	6.1	5.7–6.4	0.23	6.0	5.7–6.3	0.2
LT	6.4	6.1–6.8	0.20	6.4	6.0–6.7	0.2
M	12.3	11.9–12.9	0.31	12.2	11.8–12.8	0.3
TB (M n=9)	4.4	4.2–4.7	0.16	4.1 (n=3)	4.0–4.2	0.1

Karyotype: 2N = 36, FN = 68, based on two specimens from Samaria (Qumsiyeh and Baker, 1985).

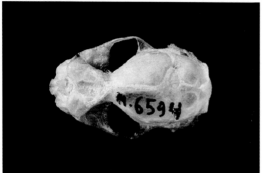

Figure 21: *Rhinopoma hardwickei arabium.* Dorsal, ventral and lateral views of the cranium, and a view of the mandible.

1 cm

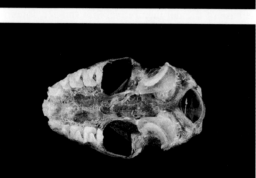

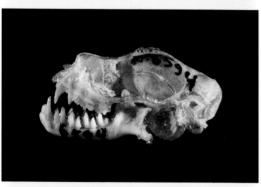

Figure 22: Distribution map of *Rhinopoma hardwickei arabium.*

● – museum records, ○ – verified field records.

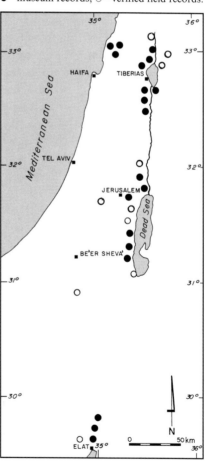

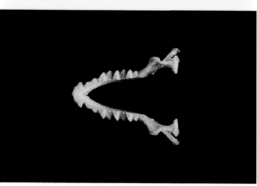

73

EMBALLONURIDAE אשמניים

This family is represented in Israel by two species of the genus *Taphozous* : *T. nudiventris* and *T. perforatus*.

Key to the Species of Emballonuridae in Israel
(After Harrison and Bates, 1991)

1. Fur on the belly does not extend back to the base of the tail membrane. Size larger, forearm 64–83 mm; greatest length of skull 24.8–31.8 mm.   **Taphozous nudiventris**

–   Fur on the belly extends back to the base of the tail membrane. Size smaller, forearm 58–66 mm; greatest length of skull 19.7–21.6 mm.   **Taphozous perforatus**

TAPHOZOUS Geoffroy, 1818   Sheath-tailed Bat

Taphozous Geoffroy, 1818. *Description de l'Egypte*, 2:113.

The tail emerges from the tail membrane behind its upper mid-point and resembles a spear thrown into the ground at an angle, hence the scientific (Greek) name of the family (emballo — to throw, oura — tail). The muzzle is without a noseleaf. The tragus is short and its distal part is club-shaped.

Dental formula is $\frac{1.1.2.3}{2.1.2.3} = 30$.

Taphozous nudiventris Cretzschmar, 1830 vel 1831   Naked Bellied Tomb Bat
Figs 23–24   אשמן גדול

Taphozous nudiventris Cretzschmar, 1830, In: Rüppell, *Atlas Reise Nordl. Afr., Saugeth.* p. 70.
Giza, Egypt.
Synonymy in Harrison and Bates (1991).

This is the largest among the Microchiroptera in Israel and the larger of the two species of its genus in Israel. Forearm length ranges from 64–83 mm. Because of its elongated head it appears somewhat similar to a small *Rousettus*. Its chin, throat and sides of face are hairless and the belly and lower back are quite naked, hence the scientific name of this species (nudiventris).
Taphozous nudiventris nudiventris is the subspecies present in Israel (Harrison and Bates, 1991).
Distribution: It occurs in the Sahel zone south of the Sahara from Senegal to Somalia and Sudan, in Arabia and the Middle East from Israel east to Afghanistan (Corbet, 1978).

In Israel it occurs along the Rift Valley, where it was caught near the Sea of Galilee and also observed near the Dead Sea (Yom-Tov *et al.*, 1992a). It has also been collected in Yagur, 'Atlit, Yizre'el and Zevulun Valleys (Makin, 1977, M. Dor, pers. comm. Figure 24). Prior to the already mentioned fumigation of caves it was not rare in caves in the Coastal Plain.

Habits: This bat roosts in colonies in crevices and fissures in cliffs, and sometimes also in buildings and tombs (thaphos — a grave in Greek). It has a strong, swift and usually high flight pattern (Harrison, 1977). In Iraq it hibernates after accumulating a considerable amount of fat (Al-Robaae, 1968), but in warmer climates (e. g., India) it does not do so (Kingdon, 1974). In Israel, in autumn, it develops fat reserves on the rear part of its body, abdomen, thighs and uropatagium. The fat is obvious and clearly seen through the thin abdominal hair (nudiventris = bare ventral part), which tends to be shed at this time of the year. Remains of this species were found in barn owl (*Tyto alba*) pellets in Israel (Dor, 1947).

Food: It feeds on high flying, relatively large insects.

Reproduction: Its reproduction was studied by Al-Robaae (1968) in Iraq. Insemination occurs during autumn, but the egg is fertilized only several days before emerging from hibernation. Gestation lasts nine weeks, and a single young is born in May, and carried by the mother for several weeks. It is capable of flying when about seven weeks old. Communal copulations of this species and *Rhinopoma microphyllum* were observed in November in a large cave in Wadi Qilt, near Jericho (H. Mendelssohn). The behaviour was similar in both species. The females hung from the roof of the cave and the males, with much vocalization from both sides, flew to the females, clinging to them and copulating. Each individual appeared to copulate several times.

Parasites: The following ectoparasites were found on this bat: *Eoctenes intermedius* (Polyctenidae) and *Argas confusus* (Argasidae) (Theodor and Costa, 1967).

Table 19
Body measurements (g and mm) of *Taphozous nudiventris nudiventris*

	Body		Tail	Ear	Foot	Forearm
	Mass	Length				
			Males (6)			
	n = 4	**n = 5**	**n = 4**	**n = 5**		**n = 5**
Mean	29	89	27	21	17	71
SD	18	9	7.2	2	1	4.0
Range	18–56	83–105	16–31	19–24	15–19	65–76
			Females (3)			
Mean		78	31	19	19	73.5
Range		74–85	27–36	18–20	18–21	70–76.5

Table 20
Skull measurements (mm) of *Taphozous nudiventris nudiventris* (Figure 23)

	Males (3)		Unknown sex (2)	
	Mean	Range	Mean	Range
GTL	28.1	26.9–28.8	28.6	28.1–29.2
CBL	24.5	24.1–25.1	26.8	26.3–27.2
ZB	15.5	(n = 1)	16.6	16.5–16.7
BB	12.6	12.0–13.0	13.0	12.9–13.0
IC	5.5	5.4–5.6	5.2	4.9–5.4
UT	11.1	11.0–11.3	11.3	11.0–11.5
LT	13.0	12.7–13.2	12.9	12.8–12.9
M	20.8	20.4–21.1	20.8	20.7–20.8

Karyotype: 2N = 42, FN = 64, based on two specimens from Egypt (Yaseen *et al.*, 1994).

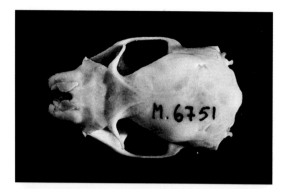

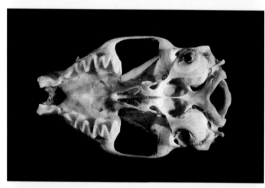

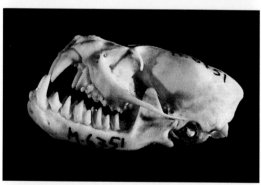

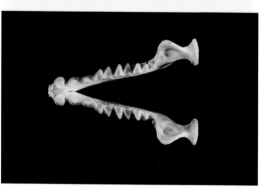

Figure 23: *Taphozous nudiventris nudiventris.* Dorsal, ventral and lateral views of the cranium, and a view of the mandible.

1 cm

Figure 24: Distribution map of *Taphozous nudiventris nudiventris.*

● – museum records, ○ – verified field records.

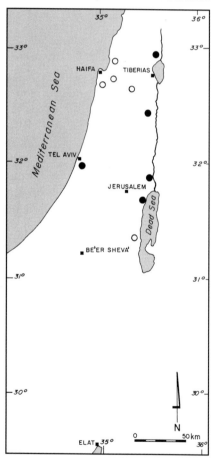

Taphozous perforatus Geoffroy, 1818   Egyptian Tomb Bat
Figs 25–26   אשמן קטן

Taphozous perforatus Geoffroy, 1818. *Description de l'Egypte* 2:126. Egypt.
Synonymy in Harrison and Bates (1991).

Smaller than *T. nudiventris* , with a forearm length ranging from 58–66 mm. Unlike
T. nudiventris, the pelage covers the underparts and extends to the tail and the prox-
imal part of the interfemoral membrane.
Taphozous perforatus haedinus Thomas, 1915 is the subspecies occurring in Israel
(Harrison and Bates, 1991).

Table 21
Body measurements (g and mm) of *Taphozous perforatus haedinus*

	Body Mass	Body Length	Tail	Ear	Foot	Forearm
	n = 1	n = 3	n = 4	n = 4	n = 5	n = 5
Mean	33	82	25	15	13	64.5
SD	2.5	3.7	1.0	1.4	0.7	
Range		80–85	21–30	14–16	12–16	64–65

The sample (n = 5) includes a male, three females, and a specimen of unknown sex.

Table 22
Skull measurements (mm) of *Taphozous perforatus haedinus* (Figure 25)

	Males (1) Mean	Females (2) Mean	Females (2) Range
GTL	19.7	21.35	21.3–21.4
CBL	18.3	19.05	19.0–19.1
ZB	—	12.4	12.4
BB	9.7	10.3	10.2–10.4
IC	5.8	4.75	4.7–4.8
UT	8.1	8.6	8.6
LT	8.4	9.3	9.3
M	14.3	15.9	15.6–16.2

Karyotype: 2N = 42, FN = 64, based on two specimens from Egypt (Yaseen *et al.*,
1994).

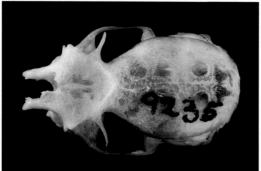

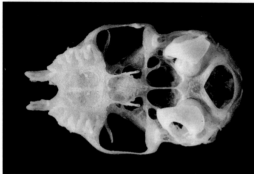

Figure 25: *Taphozous perforatus haedinus.* Dorsal, ventral and lateral views of the cranium, and a view of the mandible.

1 cm

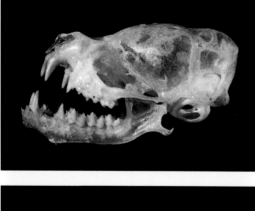

Figure 26: Distribution map of *Taphozous perforatus haedinus.*

● – museum records.

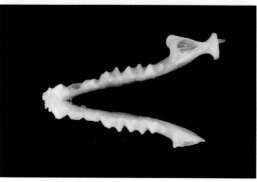

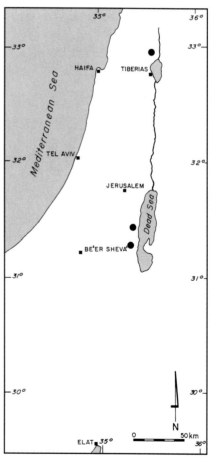

Distribution: It occurs in Africa south of the Sahara (excluding central Africa), Egypt, Arabia and the Middle East from Israel east to India (Corbet, 1978).

In Israel it apparently occurs along the Rift Valley, where it was caught near the Sea of Galilee (Naḥal 'Ammud) and the Dead Sea (Yom-Tov and Shalmon, 1989. Figure 26).

Habits: It roosts in colonies and singly in crevices and fissures in cliffs, and sometimes also in buildings and tombs (thaphos — a grave in Greek) (Harrison and Bates, 1991). It has a strong, swift and usually high flight pattern. A female collected at 'En Boqeq on 12 December 1988 had a considerable fat layer (Yom-Tov and Shalmon, 1989). In Iraq the number of animals inhabiting the summer quarters begins to diminish from the first week of November (Al-Robaae, 1968), and in southern Iraq they hibernate (Harrison, 1964). At 'En Boqeq individuals produced calls at a range of 27–34 kHz, with maximum energy at 30–31 kHz (Shalmon, 1994).

Food: It feeds on high flying, relatively large insects. Madkour (1977) found large quantities of moth scales in the stomachs of these bats, and suggests that they feed on the imagine stage of the cotton leaf worm *Spodoptera littoralis*.

Reproduction: Virtually nothing is known from Israel, but in Zimbabwe two females carried a single foetus each (Smithers, 1983).

NYCTERIDAE לילניים

This family is represented in Israel by one species, *Nycteris thebaica.*

NYCTERIS Cuvier & Geoffroy, 1795   Slit-faced Bat

Nycteris Cuvier & E. Geoffroy, 1795. *Magasin Encyclopédique*, 2:186.

Nycteris thebaica Geoffroy, 1818   Egyptian Slit-faced Bat
Figs 27–28   לילן

Nycteris thebaicus Geoffroy, 1818. *Description de l'Egypte*, 2:119, pl. 1. No. 2. Egypt.
Synonymy in Harrison and Bates (1991).

This family is characterized by the unique muzzle which has a deep median furrow with cutaneous projections on its sides. The tragus is well developed. The last tail vertebra is bifurcated.

Nycteris thebaica thebaica is the subspecies occurring in Israel (Harrison and Bates, 1991).

Distribution: This bat occurs in most of the savanna zone south of the Sahara south to South Africa, in North Africa from Morocco to Egypt, and in Arabia (Corbet, 1978).

In Israel it occurs in the Rift Valley from Elat to the Bet She'an and Yizre'el Valleys (Figure 28), but it is doubtful if it is a resident in the Bet She'an Valley where it was found only during winter (Makin, 1977).

Habits: It inhabits savanna, semi-desert and desert habitats, but is also found in oases in desert regions (Harrison, 1964). It roosts in caves, wells, hollow trees and houses (Smithers, 1983).

Its flight is slow and low (about 1 m) above ground. However, it has great maneuverability, enabling it to forage between bushes.

This is a "whispering" bat, whose calls are weak and barely recorded at distances greater than a few metres. Calls are at frequencies between 60–105 kHz, with most energy at 75 kHz (Shalmon, 1994).

Food: It feeds on insects, mainly Orthoptera and Lepidoptera (Yerbury and Thomas, 1895; Fenton, 1975) and other invertebrates which it catches either on the ground (scorpions, beetles and Solifugae) (Rosevear, 1965; Smithers, 1983) or while flying (termites). Larger prey is taken to a resting place and eaten there.

Reproduction: Copulation takes place between March–April and parturition takes place in maternal colonies during June. In Egypt, Gaisler *et al.* (1972) collected 16 females on 30 April 1969, each gravid with a large embryo, and Qumsiyeh (1985) found nursing young from April to July, at which time maternal colonies have few males.

Table 23
Body measurements (g and mm) of *Nycteris thebaica thebaica*

| | Body | | Tail | Ear | Foot | Forearm |
	Mass	Length				
	Males (11)					
	n = 9	n = 9			n = 10	n = 10
Mean	9	58.5	53	31	9	44.5
SD	0.75	5.0	2.0	1.9	0.8	1.6
Range	8–10.5	50–65	49–56	29–36	8–11	43–48
	Females (5)					
	n = 2					
Mean	10	59.5	53	32	9	46
SD	0.3	0.7	1.9	1.9	0.55	0.55
Range	9–10	59–60	51–56	29–34	9–10	45–46

Forearm as long as 48.2 mm and weight of 12.2 g were recorded in females by B. Shalmon (pers. comm.). Of the 10 specimens caught in the Dead Sea Area between April and December, females were significantly heavier than males (Yom-Tov *et al.*, 1992a).

Table 24
Skull measurements (mm) of *Nycteris thebaica thebaica* (Figure 27)

| | Males (5) | | | Females (4) | | |
	Mean	Range	SD	Mean	Range	SD
GTL	19.5	19.5–19.6	0.0	19.5	19.2–19.7	0.2
CBL	17.2	16.9–17.4	0.2	17.3	17.1–17.4	0.2
ZB	11.1	11.0–11.2	0.1	11.2	11.1–11.3	0.1
BB	8.7	8.3–9.3	0.45	8.7	8.3–9.3	0.45
IC	4.8	4.4–5.0	0.25	4.7	4.6–4.7	0.1
UT	6.8	6.8–6.9	0.05	6.9	6.8–7.0	0.1
LT	7.2	7.0–7.4	0.2	7.3	7.2–7.4	0.1
M	12.8	12.6–13.0	0.15	12.9	12.8–13.1	0.1

Dental formula is $\dfrac{2.1.1.3}{3.1.2.3} = 32$.

Karyotype: 2N = 42, FN = 78, based on two specimens from Kenya (Peterson and Nagorsen, 1975).

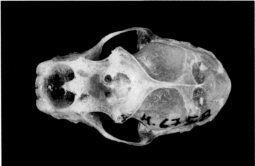

Figure 27: *Nycteris thebaica thebaica.* Dorsal, ventral and lateral views of the cranium, and a view of the mandible.

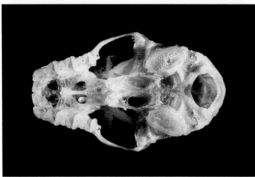

1 cm

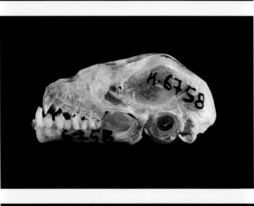

Figure 28: Distribution map of *Nycteris thebaica thebaica.*

● – museum records, ○ – verified field records.

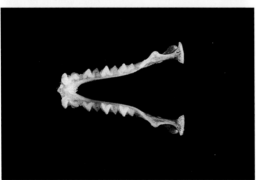

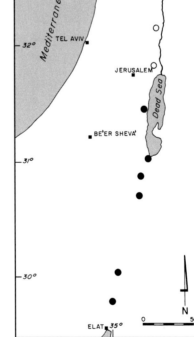

RHINOLOPHIDAE פרספיים

This family is represented in Israel by six species of a single genus *Rhinolophus*:

RHINOLOPHUS Lacépède, 1799   Horseshoe Bat

Rhinolophus Lacépède, 1799. *Tableaux des div. sousdiv. ordres et genres des Mammifères*: 15.

This genus is represented in Israel by six species: *R. ferrumequinum, R. clivosus, R. hipposideros, R. euryale, R. mehelyi* and *R. blasii*. Most species inhabit the Mediterranean region and only *R. clivosus* inhabits desert areas, while *R. hipposideros* occurs throughout the country.

This genus is characterized by the muzzle being covered by a noseleaf consisting of an erect posterior lancet, a lower horizontal horseshoe surrounding the nostrils and a perpendicular median sella. This is reflected in the scientific name "rhino" — nose, "lophos" — crest (Greek). The uropatagium is well developed and the whole tail is included in it. There is no tragus, but the antitragal lobe is developed.

Dental formula is $\dfrac{1.1.2.3}{2.1.3.3} = 32$.

Key to the Species of Rhinolophidae in Israel
(After Harrsion and Bates, 1991)

1.	Superior connecting process of the sella blunt in side view.	2
–	Superior connecting process of the sella acutely pointed in side view.	4
2.	Large species, forearm 53–61 mm; greatest length of skull 23.1–24.8 mm.	
	Rhinolophus ferrumequinum	
–	Medium or small species, forearm less than 54 mm; greatest length of skull less than 22.0 mm.	3
3.	Medium sized species, forearm 45–53 mm; greatest length of skull 19.3–21.3 mm.	
	Rhinolophus clivosus	
–	Small sized species, forearm 34–39 mm; greatest length of skull 14.2–15.8 mm.	
	Rhinolophus hipposideros	
4.	No marked contrast between crown areas of the first and third lower premolars.	
	Rhinolophus blasii	
–	Marked contrast between crown areas of the first and third lower premolars.	5
5.	Lancet essentially triangular without marked concavity at its base. Greatest length of skull 17.8–19.8 mm. **Rhinolophus euryale**	
–	Lancet abruptly tapered with marked concavity at its base. Greatest length of skull 19.2–19.8 mm. **Rhinolophus mehelyi**	

Rhinolophus ferrumequinum (Schreber, 1774)   Greater Horseshoe Bat
Figs 29–30; Plate IX:1–2   פרסף גדול

Vespertilio ferrumequinum Schreber, 1774. *Saugethiere*, 1:174, pl. 62 upper figs. France.
Synonymy in Harrison and Bates (1991).

The largest of the horseshoe bats ("ferrum-equinum", meaning a horseshoe in Latin) in Israel. The forearm length ranges from 53–61 mm. In side view, the upper part of the sella is elevated into a blunt point. The colour of the back is uniformly light greyish or light brown and the underparts are lighter in colour.

Rhinolophus ferrumequinum ferrumequinum is the subspecies occurring in Israel (Harrison and Bates, 1991).

Distribution: This bat occurs in the southern Palaearctic region from Britain to Japan, including Africa north of the Sahara (Corbet, 1978).

In Israel it is found in the Mediterranean region, where it was very common until the 1950s (Figure 30). It is still the most common of all horseshoe bats in Israel, but its populations have been greatly reduced by the fumigation of caves and it is now endangered. Only a few dozen specimens were found during a two year study (1990–1992) in the Upper Galilee (Muskin, 1993), and a similar number in the Carmel in a survey during 1992 and in the Golan during 1995 (B. Shalmon pers. comm.).

Habits: It roosts singly or in colonies numbering dozens to hundreds, in caves, ruins and tunnels, which it often shares with other horseshoe bat species. It hangs suspended by its feet and when asleep the wings are folded so closely over the head and body that it resembles a hanging pear (Harrison and Bates, 1991). In Europe it hibernates for about 200 days in winter quarters to which it migrates from summer quarters (Hill and Smith, 1985). A similar phenomenon takes place in Israel and it returns to its summer quarters in April. Winter quarters were not found in Israel (Makin, 1977), but a few specimens emerged from a cave in Mount Meron during winter (January 1991, December 1993. Y. Muskin and B. Shalmon, pers. comm.). A specimen caught by M. Dor on 18 February 1946 had food in its stomach (Makin, 1977). In Israel, constant frequency signals of long duration at maximum frequency of 83–84 kHz were recorded in Israel (FM at range 40–103 kHz. Muskin, 1993), with a short drop in frequency at the end.

Food: It feeds on insects and other invertebrates, with beetles which are found on vegetation forming a large proportion of its diet. It catches its food mainly while flying but also from the ground. Large prey is often taken to a perch, but most food is eaten on the wing (Stebbings, 1977). It generally forages quite low, about 2–5 metres above the ground.

Reproduction: In the Golan females give birth in May–June (B. Shalmon, pers. comm.). In Lebanon they probably give birth in May–July (Atallah, 1977), and in Israel maternal colonies with young were observed in Herzliyya in June and August (M. Dor, pers. comm.).

Table 25
Body measurements (g and mm) of *Rhinolophus ferrumequinum ferrumequinum*

| | Body | | Tail | Ear | Foot | Forearm |
	Mass	Length				
	Males (10)					
	n = 6				n = 9	n = 7
Mean	15	63	35	22	11	57
SD	3.3	3.1	3.7	1.5	1.5	2.0
Range	11.0–19.0	55.0–65.0	30.0–40.0	20–25	8.0–12.0	53.5–59.0
	Females (11)					
	n = 5	n = 9	n = 10	n = 10		
Mean	15.6	63.6	37.5	21.4	11.7	55.2
SD	3.2	6.7	5.3	1.9	1.9	3.0
Range	11.0–19.0	50.0–72.0	30.0–47.0	16.7–24.2	8.0–15.0	52.0–61.0

Forearm as long as 60.3 mm (males) and weight of 20.5 g (males) and 24.5 g (females) were recorded (B. Shalmon, pers. comm.).

Table 26
Skull measurements (mm) of *Rhinolophus ferrumequinum ferrumequinum* (Figure 29)

| | Males (3) | | Females (4) | | |
	Mean	Range	Mean	Range	SD
GTL (F n = 3)	24.2	23.9–24.8	24.1	24.0–24.2	0.1
CBL	21.4	20.8–22.1	21.2	20.2–22.0	0.8
ZB	12.3	12.0–12.7	12.5	12.3–12.6	0.1
BB	10.0	9.5–10.5	9.9	9.5–10.5	0.5
IC	3.0	2.9–3.1	2.8	2.5–3.1	0.2
UT	8.6	8.7–9.1	8.6	8.2–8.9	0.3
LT	9.5	9.0–10.0	9.3	9.0–9.5	0.2
M	16.2	15.2–17.3	16.4	16.2–16.8	0.3

Karyotype: 2N = 58, FN = 60, based on specimens from Jordan (Qumsiyeh *et al.*, 1986).

Parasites: The following ectoparasites were found on this bat: *Nycteribosca akollari, Stylidia biarticulata, S. biloba, S. integra* and *Penicillidia dufourii* (Diptera Pupipara), and *Rhinolophopsylla unipectinata* (Siphonaptera) (Theodor and Costa, 1967).

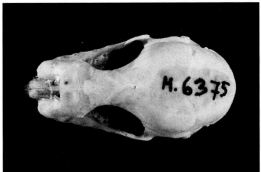

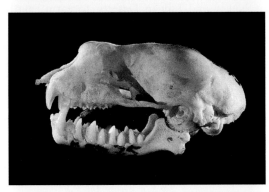

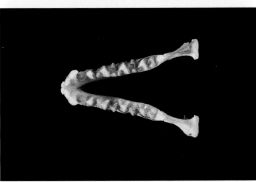

Figure 29: *Rhinolophus ferrumequinum ferrumequinum*. Dorsal, ventral and lateral views of the cranium, and a view of the mandible.

1 cm

Figure 30: Distribution map of *Rhinolophus ferrumequinum ferrumequinum*.

● – museum records, ○ – verified field records.

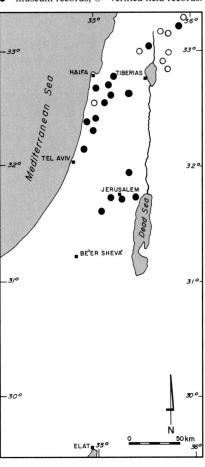

87

Rhinolophus clivosus Cretzschmar, 1828   Cretzschmar's Horseshoe Bat

Figs 31–32; Plate IX:3   פרסף הנגב

Rhinolophus clivosus Cretzschmar, 1828, In: Rüppell, *Atlas Reise Nordlich Africa, Saugeth.* 47.

   Mohila (= Muwailia), Red Sea Coast, approx. 27°49'N 35°30'E, Arabia.

Synonymy in Harrison and Bates (1991).

This is a medium-sized horseshoe bat whose forearm length ranges from 45–53 mm. In side view, the upper part of the sella is bluntly pointed, similar to that of *R. fer-rumequinum*. The colour of the back is light greyish-brown and the underparts are lighter in colour.

Rhinolophus clivosus clivosus is the subspecies occurring in Israel (Harrison and Bates, 1991).

Table 27
Body measurements (g and mm) of *Rhinolophus clivosus clivosus*.

	Body		Tail	Ear	Foot	Forearm
	Mass	Length				
Sex unknown (3)						
Mean	8	54	29	19	8	49
Range	8	51–58	23–35	18–21	7–9	47–50

In a sample of 13 specimens from the Negev and Judean Deserts, forearm length ranged from 46.9–51.0 mm and body mass from 9–13.5 g (n = 10) (B. Shalmon, pers. comm.).

Table 28
Skull measurements (mm) of *Rhinolophus clivosus clivosus* (Figure 31)

	Males (3)		Females (3)	
	Mean	Range	Mean	Range
GTL	19.9	19.6–20.1	20.0	19.4–20.7
CBL	17.4	17.1–17.9	17.7	17.0–17.3
ZB	10.1	9.8–10.3	10.3	10.3–10.4
BB	8.7	8.3–9.1	9.0	9.0–9.1
IC	2.7	2.5–2.9	2.5	2.5–2.5
UT	7.1	6.8–7.5	6.9	6.6–7.2
LT	7.6	7.5–7.8	7.6	7.4–7.8
M	13.3	13.2–13.4	13.7	13.4–14.0

Karyotype: 2N = 58, FN = 62 (Dulic and Mutere, 1974).

Distribution: It has a wide distribution in Africa, where it occurs north of the Sahara and the savanna region south of it all the way to southern Africa (Smithers, 1983). In Asia it occurs in Arabia and east to Turkestan and Afghanistan (Corbet, 1978). In Israel it occurs in the Negev and Judean Deserts (Figure 32).

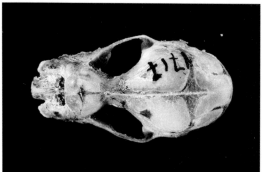

Figure 31: *Rhinolophus clivosus clivosus.* Dorsal, ventral and lateral views of the cranium, and a view of the mandible.

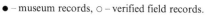

1 cm

Figure 32: Distribution map of *Rhinolophus clivosus clivosus.*

● – museum records, ○ – verified field records.

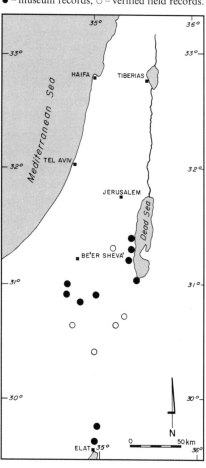

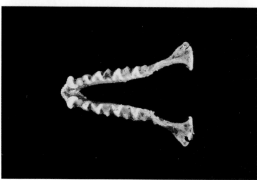

Habits: This species roosts singly and in small groups in caves and cellars although in southern Africa colonies may number several thousands (Herselman, 1980). Calls are produced at frequencies ranging from 44–104 kHz, with maximum energy at 90 kHz (Shalmon, 1994).

Food: It is insectivorous. Five pellets examined from 'En Gedi contained remains of Coleoptera, (probably Chrysomelidae), Hymenoptera, Lepidoptera, Scarabaeidae and Lygaeidae (Whitaker *et al.*, 1994).

Reproduction: A pregnant female with a single foetus was caught in the western Negev in March 1975, and in July volant young and one young still clinging to its mother were found in the same locality (Makin, 1977). In the Dead Sea Area, eight of nine females caught during May were pregnant and significantly heavier than males caught at the same time, and two young were netted in July (Yom-Tov *et al.*, 1992a). Births appear to take place in May.

Rhinolophus hipposideros (Bechstein, 1800)   Lesser Horseshoe Bat
Figs 33–34   פרסף גמדי

Vespertilio hipposideros Bechstein, 1800. In: Pennat, *Uebersicht Vierf. Thiere.*, 2:629. France. Synonymy in Harrison and Bates (1991).

This is the smallest horseshoe bat ("hippos" meaning a horse in Greek, "sider" meaning made of iron in Latin) in Israel and the second smallest bat in the country. The length of the forearm ranges from 34–39 mm. The upper part of the sella is barely distinguishable in side view and the lower part projects downwards and forwards, unlike any other horseshoe bat in Israel.

Rhinolophus hipposideros minimus Heuglin, 1861 is the subspecies present in Israel (Harrison and Bates, 1991).

Distribution: It occurs from Ireland, Iberia and Morocco through southern Europe and North Africa to Arabia, Sinai, Lebanon, Turkestan and Kashmir (Corbet, 1978).

In Israel it is found in the Mediterranean region, where it was once very common in the Upper Galilee (Muskin, 1993), in the Judean hills near Jerusalem, Herẓliyya, Ramat Yoḥanan (M. Dor, pers. comm.) and also in the Northern Negev and 'Arava Valley ('En Yahav), and the Dead Sea Area (Yom-Tov *et al.*, 1992a. Figure 34). It probably occurs throughout the country, but is now rare and B. Shalmon (personal communication) estimated their present (1996) number as 350 individuals.

Habits: It roosts singly in caves, tunnels, buildings and ruins. In Israel it has been recorded at 104 kHz in the Galilee (Muskin, 1993) and Judean Hills, 108 kHz in the Carmel and 110 kHz in the Golan and the Negev (Shalmon, 1994).

Food: It feeds on small insects (moths were recorded), but also on cave spiders which it catches by listening for its prey while hanging from a branch or rock, and then attacking swiftly, like a flycatcher.

90

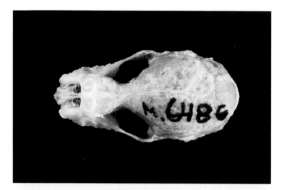

Figure 33: *Rhinolophus hipposideros minimus*. Dorsal, ventral and lateral views of the cranium, and a view of the mandible.

1 cm

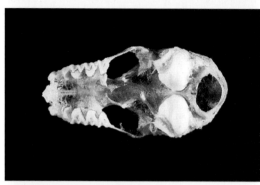

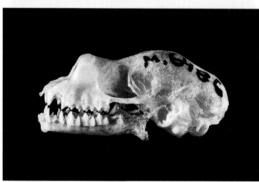

Figure 34: Distribution map of *Rhinolophus hipposideros minimus*.

● – museum records, ○ – verified field records.

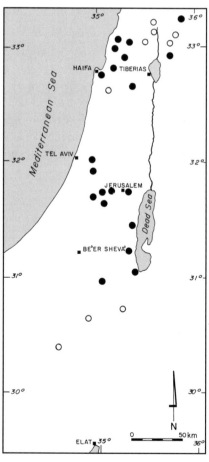

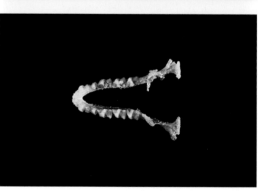

91

Table 29
Body measurements (g and mm) of *Rhinolophus hipposideros minimus*

| | Body | | Tail | Ear | Foot | Forearm |
	Mass	Length				
			Males (10)			
	n = 2	n = 8				n = 12
Mean	4	39	24	15	7	36.5
SD	0.7	2.9	1.8	1.4	0.8	1.1
Range	3.1–4.1	34.0–42.0	21.0–26.7	13.0–17.5	5.0–7.7	35.0–38.1
			Females (12)			
	n = 2	n = 9		n = 9		n = 11
Mean	2.6	40	25	14	7	36.5
SD	0.6	5.4	1.7	1.2	0.6	1.0
Range	2.1–3.0	33.0–48.0	23.0–28.2	12.0–15.7	6.0–7.9	35.0–38.5

Weight of 4.5 g was recorded in both males and females (B. Shalmon, pers. comm.).

Table 30
Skull measurements (mm) of *Rhinolophus hipposideros minimus* (Figure 33)

| | Males (5) | | | Females (7) | | |
	Mean	Range	SD	Mean	Range	SD
GTL (M n = 4)	15.0	14.7–15.4	0.3	15.1	14.2–15.7	0.6
CBL (M n = 4)	13.3	13.1–13.4	0.1	13.2	12.4–13.8	0.5
ZB	7.5	7.3–7.7	0.1	7.3	6.8–7.6	0.3
BB	6.8	6.6–7.2	0.2	6.6	6.3–7.2	0.3
IC	1.9	1.8–2.0	0.1	1.6	1.4–1.9	0.2
UT	5.2	5.1–5.2	0.1	5.0	4.6–5.3	0.3
LT (F n = 6)	5.4	5.3–5.5	0.1	5.2	4.9–5.5	0.2
M (F n = 6)	9.6	9.5–9.8	0.1	9.6	9.1–9.9	0.3

Karyotype: $2N = 58$, $FN = 60$, based on specimens from Jordan (Qumsiyeh *et al.*, 1986).

Reproduction: In Israel, a volant young was caught in August 1989 in Ne'ot HaKikar (Yom-Tov *et al.*, 1992a). Sexual maturity is reached at one year of age. Maximal longevity recorded in nature was 21 years (Schober and Grimmberger, 1987).

Parasites: The following ectoparasites were found on this bat: *Nycteribosca kollari, Nycteribia vexata, Stylidia biarticulata S. biloba* and *Penicillidia dufourii* (Diptera Pupipara) (Theodor and Costa, 1967).

Rhinolophus euryale Blasius, 1853 Mediterranean Horseshoe Bat
Figs 35–36 פרסף בהיר

Rhinolophus euryale Blasius, 1853. *Archiv Naturgesch.*, 19 (1): 49. Milan, Italy.
Synonymy in Harrison and Bates (1991).

A medium-sized horseshoe bat very similar to *R. clivosus*. Its forearm length ranges
from 41–50 mm. Unlike *R. clivosus*, in side view, the upper part of the sella is horn-
shaped and pointed. The sides of the lancet are triangular in shape and not concave
as in *R. mehelyi*.

Rhinolophus euryale judaica Anderson and Matschie, 1904 is provisionally described
as the subspecies present in Israel (Harrison and Bates, 1991).

Distribution: It occurs in the Mediterranean region of Europe and North Africa and
east to Turkestan and Iran (Corbet, 1978). There is some confusion between medium
size horseshoe bats in the Middle East, particularly between *R. euryale* and *R. mehelyi*.

Harrison and Bates (1991), who recently examined the available material, concluded
that in Israel this species was collected from 'Adullam Cave, Herẓliyya, Rosh
HaNiqra and the Galilee (Makin, 1977, M. Dor, pers, comm.; Figure 36). It
seems to have once occurred all over the Mediterranean zone of Israel, but today
it is known only from a very few caves in the Upper Galilee (Muskin, 1993). B. Shalmon (personal communication) estimated their number in Israel as 30 individuals.

Habits: It roosts singly (Atallah, 1977) or in closely packed groups of up to several
hundred in damp caves which may be shared with other bat species (Lay, 1967). In
Iran it hibernates during winter (DeBlase, 1980). It forages amongst trees and
bushes. Its flight is slow and fluttering. It is very agile and able to hover. The maximum energy of its calls is at 105 kHz (range 50–112. Shalmon, 1994).

Food: It feeds on moths and other insects.

Reproduction: The only report is of two pregnant females which were found on 3
April 1950 in Herẓliyya (M. Dor, pers. comm.).

Parasites: The following ectoparasites were found on this bat: *Nycteribosca kollari*
and *Stylidia biarticulata* (Diptera Pupipara), and *Eyndhovenia euryalis* (mites)
(Theodor and Costa, 1967).

Table 31
Body measurements (g and mm) of *Rhinolophus euryale judaica*

	Body		Tail	Ear	Foot	Forearm
	Mass	Length				
			Males (7)			
	n = 5					**n = 6**
Mean	10	51	23	18	9	47
SD	2.2	5.9	1.4	0.9	1.0	3.7
Range	7.8–13.0	45–62	21–25	17–20	8–11	41–50
			Females (2)			
Mean	9.5	44.0	23.0	17.0	8.5	45.5
Range	9–10		22–24		8–9	45–46

Table 32
Skull measurements (mm) of *Rhinolophus euryale judaica* (Figure 35)

	Males (4)			Females (1)
	Mean	Range	SD	Mean
GTL	19.2	18.8–19.8	0.6	19.3
CBL (n = 3)	16.6	16.2–16.9	0.4	17.1
ZB	9.6	8.9–10.2	0.7	9.7
BB	9.0	8.6–9.4	0.4	8.9
IC	2.6	2.3–3.0	0.3	2.5
UT	6.6	6.0–6.9	0.4	6.4
LT	7.1	6.7–7.3	0.3	6.7
M	12.4	12.6–12.9	0.6	12.6

Karyotype: $2N = 58$, $FN = 60$, based on specimens from Europe (Zima and Kral, 1984).

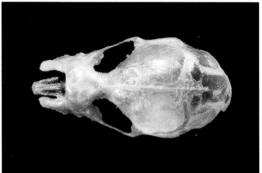

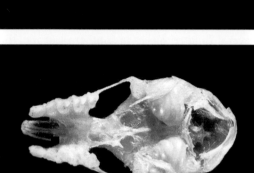

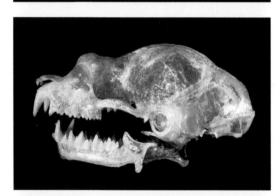

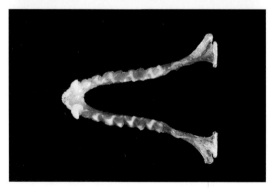

Figure 35: *Rhinolophus euryale judaica.* Dorsal, ventral and lateral views of the cranium, and a view of the mandible.

1 cm

Figure 36: Distribution map of *Rhinolophus euryale judaica.*

● – museum records, ○ – verified field records.

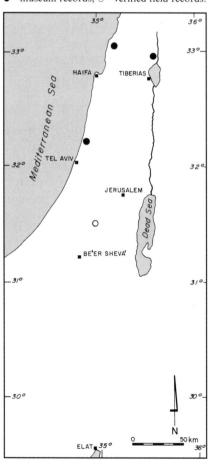

95

Rhinolophus mehelyi Matschie, 1901   Mehely's Horseshoe Bat
Fig 37   פרסף חיוור

Rhinolophus mehelyi Matschie, 1901. *Sitzungsberichte Ges. naturf. Freunde Berl.*: 225.
Bucarest, Romania.

A medium-sized horseshoe bat whose forearm length ranges between 48–51 mm.
The lancet is pointed with a marked concavity at the base. The upper part of the
sella is horn-shaped and pointed.
There are no specimens of this species from Israel in Israeli museums.
Distribution: This bat is known from several isolated places in southern Europe, the
Mediterranean, North Africa from Morocco to Egypt (Corbet, 1978), and further
east in the Middle East, Asia Minor through to Afghanistan (Harrison and Bates,
1991).
In Israel it was reported from several localities (Figure 37) in the Mediterranean
region (DeBlase, 1972). Today it is very rare everywhere, and possibly extinct in

Israel. However, there is some confusion
between medium size horseshoe bats in
the Middle East, particularly between *R.
euryale* and *R. mehelyi*, and the precise distribution of these two species still needs to
be determined.

Habits: It roosts in caves, sometimes in
very large colonies (30,000 were recorded
in one cave in Iran; Niazi, 1976), and
often with other bat species. It emerges
at dusk and forages between trees and
bushes. It is a low flying species. A specimen kept as a pet in an apartment flew
no higher than two metres. This observation may, however, in fact relate to *R. euryale*.

Food: It preys on moths and other insects.
Reproduction: Nothing is known from
Israel.

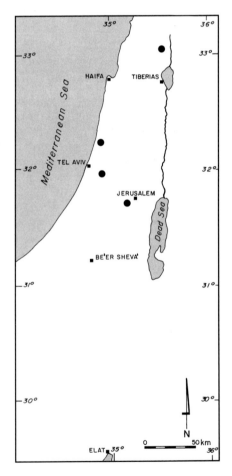

Figure 37: Distribution map of *Rhinolophus mehelyi
mehelyi*.

●–museum records.

Rhinolophus blasii Peters, 1866   Blasius' Horseshoe Bat
Figs 38–39   פרסף מצוי

Rhinolophus blasii Peters, 1866. *Monatsberichte K. preuss. Akad. Wiss. Berlin*, p. 17. Italy.

A medium-sized horseshoe bat whose forearm length ranges from 43–48 mm. It is very similar to *R. euryale*, but with distinctive differences in the form of the noseleaf. In side view, the upper part of the sella is more prominent than in *R. euryale* and bluntly pointed. The ears are generally smaller than in *R. euryale*.

Rhinolophus blasii blasii is provisionally described as the subspecies occurring in Israel (Harrison and Bates, 1991).

Distribution: This species is known from Italy through south-east Europe and south-west Asia to east Afghanistan. In Africa it occurs in Morocco and in eastern Africa from Eritrea to southern Africa (Corbet, 1978).

It was once very common in the Mediterranean zone of Israel (Figure 39), and M. Dor (pers. comm.) reports hundreds in caves in Tiv'on, Herẓliyya and Jerusalem. It was also reported from the Galilee and other localities in the Mediterranean region (M. Dor, pers. comm.). Today it is on the verge of extinction, and only 40–80 were reported by Muskin (1993) from one cave in the Upper Galilee and none from the Carmel (B. Shalmon, pers. comm.). B. Shalmon estimated their number in Israel as 30 individuals.

Habits: It roosts singly or in small groups in caves, hanging from the roofs in small clusters. In Iran it hibernates in large groups of several hundreds (Lay, 1967). In the Galilee it produces constant frequency signals at 94 kHz (Muskin, 1993).

Reproduction: Nothing is known from Israel.

Parasites: The following ectoparasites were found on this bat: *Nycteribosca kollari* and *Stylidia biarticulata* (Diptera Pupipara) (Theodor and Costa, 1967).

Table 33
Body measurements (g and mm) of *Rhinolophus blasii blasii*

| | Body | | Tail | Ear | Foot | Forearm |
	Mass	Length				
			Males (5)			
	n = 1			**n = 4**	**n = 4**	**n = 4**
Mean	6.7	54	25	15	9.5	45
SD		4.2	4.7	0.9	0.5	1.1
Range		48.0–58.0	20.7–30.0	14.6–16.6	9.0–10.0	43.7–46.0
			Females (5)			
Mean		58.4	25.1	16.0	9.7	46.4
SD		2.2	1.5	1.8	0.22	0.45
Range		56.0–62.0	23.0–27.1	13.6–18.6	9.3–9.8	45.9–47.1

Table 34
Skull measurements (mm) of *Rhinolophus blasii blasii*. (Figure 38)

| | Males (3) | | | Females (1) |
	Mean	Range	SD	Mean
GTL	18.4	17.8–19.0		17.0
CBL	15.7	15.1–16.6		16.1
ZB	9.0	8.7–9.1		8.9
BB	8.5	8.2–8.8		8.6
IC	2.5	2.4–5.6		2.3
UT	6.3	6.1–6.7		6.2
LT	6.7	6.4–6.9		6.8
M	11.7	11.5–12.0		11.6

Karyotype: 2N = 58, FN = 60, based on specimens from Jordan (Qumsiyeh *et al.*, 1986).

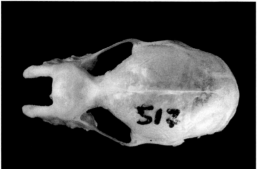

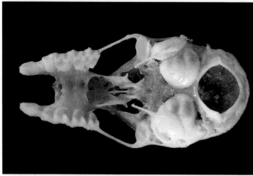

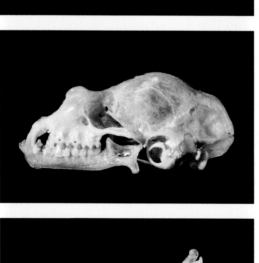

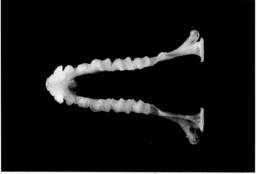

Figure 38: *Rhinolophus blasii blasii.* Dorsal, ventral and lateral views of the cranium, and a view of the mandible.

1 cm

Figure 39: Distribution map of *Rhinolophus blasii blasii.*

● – museum records, ○ – verified field records.

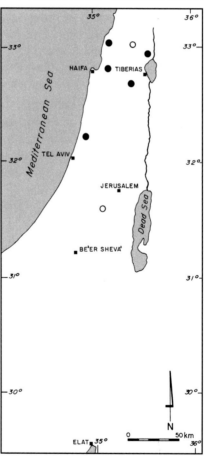

HIPPOSIDERIDAE פרספוניים

This family is represented in Israel by a single species, *Asellia tridens*. The scientific name of the family means a horseshoe ("hippos" — a horse in Greek, "sider" — made of iron in Latin).

ASELLIA Gray, 1838   Leaf-nosed Bat

Hipposideros Gray, 1831, *Zoological Misc.*, 1:37.

Asellia tridens (Geoffroy, 1813)   Trident Leaf-nosed Bat
Figs 40–41; Plate IX:4; Plate X:1   פרספון

Rhinolophus tridens E. Geoffroy, 1813. *Annales Mus. Hist. Nat. Paris*, 20:265. Egypt.
Synonymy in Harrison and Bates (1991).

The noseleaf consists of a lower horizontal horseshoe and an erect transverse leaf with three vertical projections, the central one pointed and the side ones blunt, hence its scientific name (tridens — three teeth). There is no tragus but a large anti-tragus is present.

Asellia tridens tridens is the subspecies present in Israel (Harrison and Bates, 1991).
Distribution: It occurs in North West and East Africa, and east through Arabia to Pakistan (Harrison and Bates, 1991).
In Israel it is found mainly along the Rift Valley from the Mount Ḥermon to Elat, but there are also reports from Jerusalem, Be'er Toviyya, and Jaffa (Yafo) (M. Dor, pers. comm.; Makin, 1977. Figure 41).
Habits: It roosts in colonies numbering tens to several hundred individuals in caves, ruins, cellars and old houses. In Ne'ot HaKikar, a colony of about 40 individuals inhabited a flooded underground shelter. In Timna', a colony was found to roost in an old mine where ambient temperatures range between 25–27°C (B. Shalmon, pers. comm.). Body mass in autumn (September–November) is much heavier (mean 12.2 g, SD = 2.3, range 14–16, n = 21) than in spring and summer (March–August; mean 9.8 g, SD = 1.2, range 8–11, n = 35), indicating that this species accumulated large fat reserves before hibernation. Large seasonal changes in colony size were reported for several colonies in Israel: D. Makin (pers. comm.) found a large population roosting in a cave in Samaria and another near the Dead Sea; in a deserted building near the Yarmouq River, a colony of several thousand declined from May towards summer, and in a building in Jaffa there was a colony of several hundred every October. Similar phenomena were reported elsewhere in the Middle East (summarized by Harrison and Bates, 1991). Hence, it is possible that in Israel they have short migrations from summer to winter quarters, as they do in Iraq where they disperse to winter quarters from mid-September to mid-November, returning to their summer quarters in April (Al-Robaae, 1966).

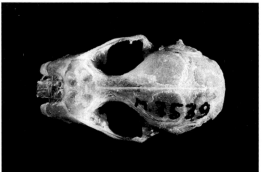

Figure 40: *Asellia tridens tridens.* Dorsal, ventral and lateral views of the cranium, and a view of the mandible.

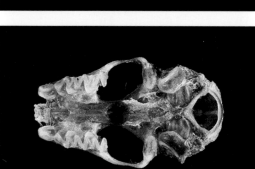

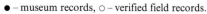

1 cm

Figure 41: Distribution map of *Asellia tridens tridens.*

● – museum records, ○ – verified field records.

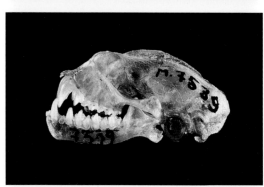

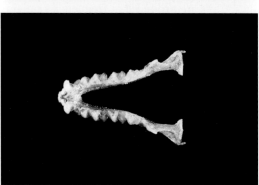

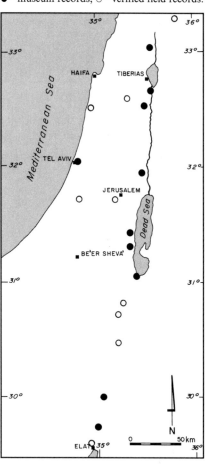

101

Table 35
Body measurements (g and mm) of *Asellia tridens tridens*

	Body		Tail	Ear	Foot	Forearm
	Mass	Length				
	Males (11)					
Mean	8	54	20	18	8	50
SD	2.0	5.0	3.5	2	1.5	1.0
Range	4.5–10	46–60	14–25	14–20	5–9	48–52
	Females (11)					
						n = 7
Mean	9	55	22	18	7	50
SD	2.0	4.0	3.0	2.0	2.0	1
Range	6.3–11.1	50–62	18–27	15–20	5–9	49–52

Forearm as long as 54.0 mm and weight of 16.5 g were recorded in males (B. Shalmon, pers. comm.).

Table 36
Skull measurements (mm) of *Asellia tridens tridens* (Figure 40)

	Males (9)			Females (9)		
	Mean	Range	SD	Mean	Range	SD
GTL	18.6	17.6–19.2	0.60	18.6	18.0–19.4	0.4
CBL	17.0	16.2–17.6	0.56	17.0	16.4–17.6	0.4
(M n = 8), (F n = 8)						
ZB (M n = 8)	10.2	9.4–10.7	0.49	10.3	9.2–11.0	0.6
BB (M n = 8)	7.4	6.8–7.9	0.45	7.8	7.0–8.9	0.6
IC	2.4	2.0–2.6	0.20	2.4	2.2–2.6	0.1
UT (M n = 8)	6.9	6.4–7.1	0.23	6.7	6.2–7.1	0.3
LT (F n = 7)	7.5	6.7–8.0	0.35	7.2	7.0–7.7	0.3
M	12.8	11.5–13.7	0.60	12.8	12.3–13.4	0.4

Dental formula is $\frac{1.1.1.3}{2.1.2.3}$ = 28.

Its flight is low and swift, like a large butterfly, and it is capable of rapid turns and twists. It forages among trees and bushes. Calls are produced at frequencies ranging from 109–130 kHz, with maximum energy at 121 kHz (Pye, 1972, Shalmon, 1994), the highest frequency recorded for an Israeli bat. In Libya, Booth (1961) reported that the sooty falcon (*Falco concolor*) hunts this bat.

Food: It feeds on insects which are taken on the wing. Whitaker *et al.* (1994) examined 12 pellets of this species from Nawit Pools near the Dead Sea and found that remains of Lepidoptera comprised the majority of their contents, but there were also remains of Diptera, mainly chironomids. Another sample of 147 pellets collected from an abandoned mine near Elat contained (in decreasing order) Coleoptera, Hymenoptera, Orthoptera, Lepidoptera, Diptera and Odonata. In Ḥazeva they were found feeding on large beetles and moths (B. Shalmon, pers. comm.).

Reproduction: Pregnancy lasts 9–10 weeks. In Iraq, parturition takes place from early June, and lactation lasts 40 days (Al-Robaae, 1966). M. Dor (pers. comm.) found a pregnant female in Jaffa in May 1951. In the Negev parturition probably takes place in May-June, and of seven females netted in July, six were lactating (Yom-Tov *et al.*, 1992a). A very young specimen was observed by Makin (1987) in early June in the Negev, and two volant, young females were caught in Ne'ot HaKikar on 2 August 1989 (Yom-Tov *et al.*, 1992a).

Parasites: The following ectoparasites were found on this bat: *Nycteribia schmidlii, Raymondia setosa* and *Penicillidia dufourii* (Diptera Pupipara) (Theodor and Costa, 1967).

MOLOSSIDAE אשפיים

This family is represented in Israel by a single species, *Tadarida teniotis*.

TADARIDA Rafinesque, 1814   Free-tailed Bat

Tadarida Rafinesque, 1814, *Precis Som.*, p. 55.

Tadarida teniotis (Rafinesque, 1814)   European Free-tailed Bat
Figs 42–43; Plate XI:2   אשף

Cephalotes teniotis Rafinesque, 1814. *Precis Som*, p. 12. Sicily.
Synonymy in Harrison and Bates (1991).

One of the largest insectivorous bats of Israel. The wings are long and narrow and forearm length ranges from 57–70 mm. Most of the tail projects out of the uropatagium which is supported by strong calcars. The ears have round tips and are not connected by a membrane across the head. There is a small, almost square tragus and a large antitragus. The face is dog-like, the upper lip pendulous and has a series of vertical furrows. The pelage is soft and ashy grey on the back, somewhat lighter on the underparts.

Tadarida teniotis rueppelli Temminck, 1826 is the subspecies present in Israel (Harrison and Bates, 1991).

Distribution: This bat occurs in the Mediterranean region of Europe and North Africa, and east through Asia Minor and the Middle East to Japan (Corbet, 1978). In Israel it is distributed throughout the country and was reported from the Rift Valley from the Ḥula to Elat, Jerusalem, Tel Aviv, Karmi'el, Herẕliyya, Lod, Sede Boqer, the Galilee the Golan and other localities (Figure 43).

Habits: This is a resident, occupying a wide range of habitats, from desert to Mediterranean and human settlements. Together with Kuhl's pipistrelle it is one of the two most common insectivorous bats in Israel. It roosts in groups ranging from dozens to hundreds in caves and buildings where it seeks narrow crevices and fissures. In

Table 37
Body measurements (g and mm) of *Tadarida teniotis rueppelli*

| | Body | | Tail | Ear | Foot | Forearm |
	Mass	Length				
			Males (8)			
	n = 3	**n = 7**		**n = 6**		**n = 7**
Mean	24	84	49	26	11	61
SD		4.0	2.1	2.7	1.4	0.5
Range	20–30	78–90	44–51	22–30	9–12	61–62
			Females (11)			
	n = 10					**n = 8**
Mean	19	82	50	26	12	61
SD	3.5	4.7	3.0	2.7	3.6	2.3
Range	14–26	76–90	45–56	21.5–31	9–20	57–64

Weight of 37.5 g was recorded in males (B. Shalmon, pers. comm.).

Table 38
Skull measurements (mm) of *Tadarida teniotis rueppelli* (Figure 42)

| | Males (4) | | | Females (6) | | |
	Mean	Range	SD	Mean	Range	SD
GTL	24.4	23.8–24.8	0.4	24.1	23.8–24.5	0.3
CBL	23.1	22.2–24.0	0.9	23.3	22.8–23.8	0.4
ZB (n = 7)	14.0	13.3–14.6	0.5	13.8	13.6–14.1	0.2
BB	11.6	10.8–12.4	0.75	11.6	10.7–12.2	0.8
IC	4.8	4.7–5.0	0.1	4.9	4.7–5.0	0.1
UT (n = 7)	8.8	8.7–9.0	0.1	8.9	8.5–9.0	0.2
LT	9.5	9.2–9.9	0.3	9.5	9.4–9.8	0.2
M	17.5	17.0–17.8	0.4	17.2	17.0–17.4	0.2

Dental formula is $\frac{1.1.2.3}{3.1.2.3} = 32$.

Karyotype: 2N = 48, FN = 76, based on specimens from Yugoslavia (Dulic and Mrakovcic, 1980).

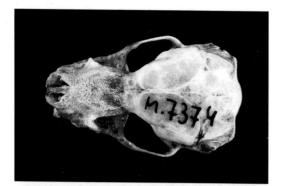

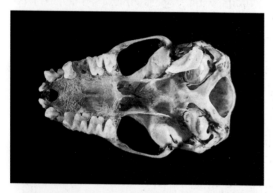

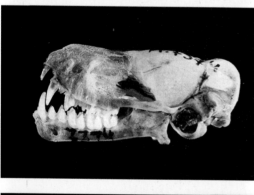

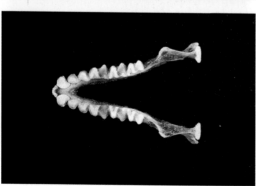

Figure 42: *Tadarida teniotis rueppelli.* Dorsal, ventral and lateral views of the cranium, and a view of the mandible.

├─── 1 cm ───┤

Figure 43: Distribution map of *Tadarida teniotis rueppelli.*

● – museum records, ○ – verified field records.

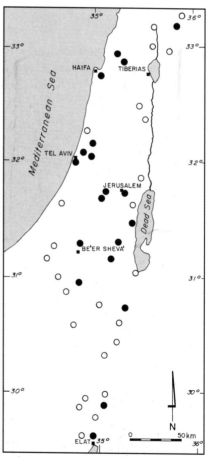

105

a cave near 'Adullam there were seasonal fluctuations in colony size: the bats arrived in March and left in May, returning again in August (Makin, 1977). It has scent glands which produce a strong smell, making their roosts distinctive (Makin, 1977). It flies quite high (Gaisler and Kowalski, 1986) and fast, but does not have great manoeuverability. Near the Dead Sea it was observed to hunt at least 50 m above ground near hotels and street lights, and has been reported to hunt at several hundred m above ground. Its calls are loud and produced at frequencies ranging from 10–60 kHz, with maximum energy at 14–18 kHz (Schober and Grimmberger, 1987, Shalmon, 1994). This is one of the few insectivorous bats in Israel whose calls can be heard by humans.

Food: It feeds on insects which it takes on the wing. Nine samples (450 pellets) from Karmi'el (Lower Galilee) contained mainly Lepidoptera, mostly moths, but also Coleoptera, especially ground beetles (Carabidae and Scarabaeidae), and Orthoptera (Gryllidae) (Whitaker *et al.*, 1994).

Reproduction: In Israel parturition takes place in June (records are from 1994 for a colony of hundreds of females in Mashabbe Sade and from a single female in Karmi'el, B. Shalmon, pers. comm.). In Lebanon parturition probably takes place in mid-June, and lactation of the single young lasts until early October (Harrison and Lewis, 1961). The young is carried on its mother's back, not clinging to the belly, unlike most other bats.

Parasites: The following ectoparasites were found on this bat: *Nycteribia pedicularia* (Diptera Pupipara) (Theodor and Costa, 1967).

VESPERTILIONIDAE נשפונײם

This family is represented in Israel by eight genera (*Nyctalus, Myotis, Eptesicus, Pipistrellus, Otonycteris, Plecotus, Barbastella* and *Miniopterus*) with at least 18 species. The muzzle is without noseleaf, the ears have a tragus and the tail is entirely enclosed in the uropatagium.

Key to the Genera of Vespertilionidae in Israel
(After Harrison and Bates, 1991)

1.	Two incisors in each upper jaw.	2
–	One incisor in each upper jaw.	8
2.	Four cheekteeth behind the canine in each upper jaw. (Penis not bent into an inverted L).	**Eptesicus**
–	More than four cheekteeth behind the canine in each upper jaw.	3
3.	Six cheekteeth behind the canine in each upper jaw.	**Myotis**
–	Five cheekteeth behind the canine in each upper jaw.	4
4.	Three lower premolars in each ramus.	5
–	Two lower premolars in each ramus.	6
5.	Ears very large, 31–38 mm in height, joined across the forehead.	**Plecotus**
–	Ears small, 9–12 mm, and separated on the forehead.	**Miniopterus**
6.	Ears face forwards, they are joined across the forehead with their anterior margins close together above the muzzle.	**Barbastella**
–	Ears widely separated and do not face forwards.	7
7.	Tragus expanded distally into a club-shaped extremity. Fifth finger short, barely exceeding the third or fourth metacarpal in length.	**Nyctalus**
–	Tragus not noticeably expanded distally. Fifth finger longer than the combined metacarpal and first phalanx in digits three and four. (In *P. savii*, only four post-canine teeth may be present in each upper jaw; in this species the penis is bent into an inverted L, see 2 above).	**Pipistrellus**
8.	Size large, forearm exceeds 55 mm, ears proportionately large, exceeding 30 mm.	**Otonycteris**

NYCTALUS Bowdich, 1825   Noctule Bat

Nyctalus Bowdich, 1825. *Excursions in Madeira and Porto santo*: 36.

Nyctalus noctula (Schreber, 1774)   Common Noctule Bat
Figs 44*, 44   רמשן לילי

Vespertilio noctula Schreber, 1774. *Die Säugethiere in Abbildungen nach der Natur.*, 1:166, pl. 52. France.
Synonymy in Harrison and Bates (1991).

* Photographs added in proof [see Fig. 44*].

This is a large sized bat characterized by its long, narrow wings. The distal end of the last vertebra of the tail protrudes about 2–3 mm from the uropatagium. The muzzle is broad and has a marked blackish swelling on the upper lip. The ears are short and the tragus mushroom-shaped. The fur is dense and is bronze-brown in colour, lighter on the underparts.

Nyctalus noctula lebanoticus Harrison, 1962 is apparently the subspecies occurring in our region (Harrison and Bates, 1991).

Table 39
Body measurements (mm) of *Nyctalus noctula lebanoticus*

Body Length	Tail	Ear	Foot	Forearm
Sex unknown (1)				
69	52	18	12	53

Only one skull (TAU M.9432) of this species available in Israeli museums. Its skull measurements are (Figure 44*): GTL = 19.0 mm, ZB = 13.2 mm, BB = 10.8 mm, IC = 5.2 mm, UT = 7.3 mm, LT = 7.8 mm, M = 14.5 mm.

In a sample of females caught on the Golan on 16 June 1996 mean forearm length was 54.1 mm (SD = 1.1). Mean body weight of 7 pregnant females was 35.6 g (SD = 2.2) and that of other 10 females 27.1 g (SD = 1.9).

Dental formula is $\frac{2.1.2.3}{3.1.2.3} = 34$.

Karyotype: 2N = 42, FN = 54, based on specimens from Europe (Dulic *et al.*, 1967).

Distribution: This species is widely distributed throughout most of Europe, Morocco, the Middle East (where it is very rare, but recorded from Lebanon and Oman; Harrison and Bates, 1991) and east to Turkestan, China and Japan (Corbet, 1978).

In Israel it has been found only a few times (Figure 44): by Festa (1894) from Jabal Qarantal near Jericho (the specimen was lost), and in the Ḥula Nature Reserve (Harrison and Makin, 1988), a breeding colony of about 500 females was observed in the Golan (B. Shalmon, pers. comm.) and was heard several times flying above a sewage pond in the Upper Galilee (Muskin, 1993).

Habits: In the Galilee its highest intensity of calls was at 20 kHz (Muskin, 1993).

Food: It feeds on large insects (moths, cockchafers) caught and eaten in flight.

Reproduction: A lactating female was caught in the Golan on 22 June 1995, and a breeding colony of females, some pregnant and others just after birth, was found in the Golan on 16 June 1996 (B. Shalmon and Y. Yom-Tov, pers. observ.). Maximum recorded age is 12 years (Schober and Grimmberger, 1987).

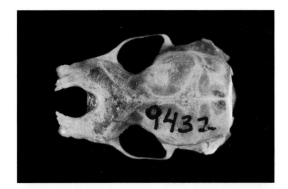

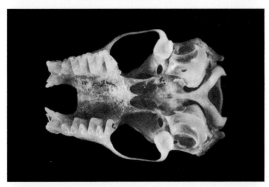

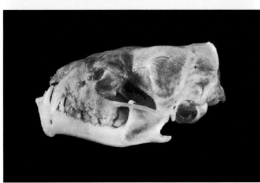

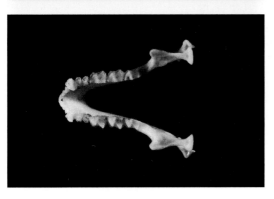

Figure 44*: *Nyctalus noctula lebanoticus.* Dorsal, ventral and lateral views of the cranium, and a view of the mandible.

1 cm

Figure 44: Distribution map of *Nyctalus noctula lebanoticus.*

● – museum records, ○ – verified field records.

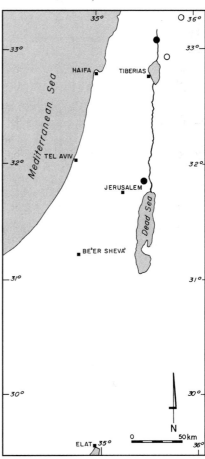

109

MYOTIS Kaup, 1829   Mouse-eared Bat

Myotis Kaup, 1829. *Skizzirte Entw.-Gesch. u. Naturl. syst. Europ. Thierwelt*, 1:106.

This genus is represented in Israel by six species: *M. myotis, M. blythii, M. emarginatus, M. capaccinii, M. mystacinus* and *M. nattereri*, all of which inhabit only the Mediterranean region of Israel. *M. mystacinus* was recently discovered on Mount Ḥermon (B. Shalmon, pers. comm.). Only one specimen is present in an Israeli museum, hence it is not included in the key below. The ears are long (Mys — a mouse, otis — an ear in Greek), and have a well developed long and sharply pointed tragus. The tail is about equal in length to the hind feet.

Dental formula is $\dfrac{2.1.3.3}{3.1.3.3} = 38.$

Key to the Species of Myotis in Israel
(After Harrison and Bates, 1991)

1. Large species, forearm length exceeds 55 mm; greatest length of skull exceeds 22 mm.
   2

- Small species, forearm length less than 45 mm; greatest length of skull less than 17.5 mm.
   3

2. Larger, greatest length of skull exceeds 24.8 mm. **Myotis myotis**
- Smaller, greatest length of skull less than 23.0 mm. **Myotis blythii**

3. Feet very large, three-quarters of tibial length. Outer edge of each tibia hairy.
   Myotis capaccinii

- Feet small, half or less of tibial length. Outer edge of each tibia not hairy. 4

4. Interfemoral membrane with a dense fringe of hairs between the tip of each calcar and the tail. **Myotis nattereri**

- Interfemoral membrane without a dense fringe of hairs between the tip of each calcar and the tail. Ears with a marked indentation on their outer borders.
   Myotis emarginatus

Myotis myotis (Borkhausen, 1797)   Greater Mouse-eared Bat
Figs 45–46   נשפון גדול

Vespertilio myotis Borkhausen, 1797. *Deutsche Fauna*, 1:80. Thuringia, Germany.
Synonymy in Harrison and Bates (1991).

This is a large bat, with a wingspan of 35 cm. It is difficult to distinguish between this species and *M. blythii*. It is more heavily built than *M. blythii*, larger (forearm length more than 60 mm), has longer ears and a relatively shorter tail.

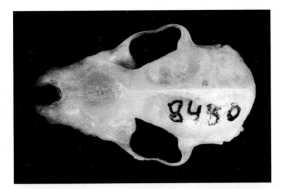

Figure 45: *Myotis myotis macrocephalicus.* Dorsal, ventral and lateral views of the cranium, and a view of the mandible.

1 cm

Figure 46: Distribution map of *Myotis myotis macrocephalicus* .

● – museum records, ○ – verified field records.

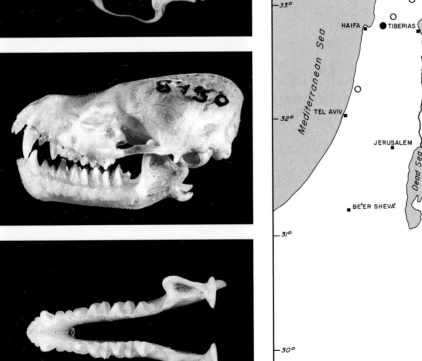

Table 40
Body measurements (g and mm) of *Myotis myotis macrocephalicus*

| | Body | | Tail | Ear | Foot | Forearm |
	Mass	Length				
			Males (3)			
		n = 2	n = 2	n = 2	n = 2	
Mean		70	56.5	27	14.5	64
Range		69–71	56–57	24–30	13–16	64.0–64.5
			Females (7)			
Mean		60	58	23	13	61
SD		4.5	3.3	1.0	1.9	0.9
Range		51–65	54–62	21.5–24.5	11.0–16.0	58.9–62.0

Forearm length of 61–74 mm and body mass of 28–28.5 g were recorded (B. Shalmon, pers. comm.).

Table 41
Skull measurements (mm) of *Myotis myotis macrocephalicus* (Figure 45)

| | Males (2) | | Females (1) |
	Mean	Range	Mean
GTL (M n = 1)	25.5		21.9
CBL	23.7	23.6–23.8	20.4
ZB	15.75	15.6–15.9	14.0
BB	11.2	10.8–11.6	10.4
IC	5.25	5.2–5.3	5.0
UT	10.5	10.4–10.6	8.9
LT	11.3	11.2–11.4	9.7
M	19.5	19.2–19.8	17.2

Karyotype: 2N = 44, FN = 50, based on specimens from Europe (Bovey, 1949), but needs confirmation (Stebbings, 1977).

Myotis myotis macrocephalicus Harrison and Lewis, 1961 is the subspecies present in Israel (Harrison and Bates, 1991).

Distribution: It occurs in south and central Europe from England to the Ukraine, and in Asia Minor, Syria and Lebanon.

In Israel it was found in the Ḥula Valley, Ẓefat, Herẓliyya, near Haifa (M. Dor, pers. comm.), the Lower Galilee (Makin, 1977) and the Upper Galilee (16 specimens in one cave in August; Muskin, 1993) and the Golan (B. Shalmon, pers. comm.), but it is rare (Figure 46). Due to the difficulty of distnguishing between this species and *M. blythii*, some of the reports of the occurrence of these species in Israel are confused, and cannot be resolved here.

Habits: In Lebanon (Lewis and Harrison, 1962) it roosts in colonies numbering hundreds, in caves and buildings. It hangs by its feet, singly, in pairs or small groups from cracks in the walls (Atallah, 1977). Maximum intensity of its calls is 35 kHz (Shalmon, 1994).

Food: It feeds on insects, usually large moths, crickets, grasshoppers and beetles, as well as spiders which are consumed in flight (Stebbings, 1977).

Reproduction: Males with large testes were caught in mid-June on Mount Ḥermon (B. Shalmon, pers. comm.), but no other data on its reproduction in Israel are available.

Parasites: The following parasites were found on this bat: *Nycteribia pedicularia, N. latreillii, N. schmidlii, N. vexata, Stylidia biarticulata, Basilia nana, Penicillidia dufourii* and *P. conspicua* (Diptera Pupipara), *Ixodes vespertilionis* (Ixodidae), *Spinturnix myoti* (mites), as well as endoparasites: Protozoa — *Polychromophilus murinus* and *Trypanosoma* sp. (Theodor and Costa, 1967).

Myotis blythii (Tomes, 1857)   Lesser Mouse-eared Bat

Figs 47–48   נשפון מצוי

Vespertilio blythii Tomes, 1857. *Proceedings zool. Soc. Lond.*, (1857): 53 Nasirabad, Rajaputana, India.

Synonymy in Harrison and Bates (1991).

Similar to *Myotis myotis*, and was considered conspecific. Forearm length ranges from 54–64 mm. The tail is relatively longer than in *M. myotis* and the ears are shorter.

Myotis blythii omari Thomas, 1906 is the subspecies present in Israel (Harrison and Bates, 1991).

Distribution: It occurs in the Mediterranean region of Europe and north-west Africa, Asia Minor and east to China (Corbet, 1978).

In Israel it is occasionally found in the Galilee, the Golan, Mount Ḥermon and the Yizre'el Valley (Figure 48), but is very rare (see remark in the description of *M. myotis*). B. Shalmon (pers. comm.) estimated their present (1996) number in Israel at 50 individuals.

Table 42
Body measurements (g and mm) of *Myotis blythii omari*

	Body		Tail	Ear	Foot	Forearm
	Mass	Length				
	Males (15)					
	n = 4	**n = 7**	**n = 8**	**n = 10**		**n = 14**
Mean	18	69	58	21	13.5	58
SD	1.5	5.2	3.3	1.9	1.5	2.2
Range	17–20	61–75	52–63	17–24	11–17	54–63
	Females (2)					
	n = 1	**n = 1**				
Mean		60	63	22.5	14.5	60
Range				22–23	14–15	59.5–60

Forearm length of up to 61 mm and body mass of up to 27.5 g were recorded (B. Shalmon, pers. comm.).

Table 43
Skull measurements (mm) of *Myotis blythii omari* (Figure 47)

	Males (6)			Sex unknown (1)	Females (1)
	Mean	Range	SD		
GTL (M n = 5)	22.6	21.8–23.0	0.5	22.3	22.5
CBL (M n = 5)	21.0	19.9–21.6	0.7	20.4	21.5
ZB	14.3	14.0–14.5	0.2	14.0	14.6
BB	9.9	9.6–10.3	0.3	10.0	10.5
IC	5.5	5.3–5.7	0.2	5.4	5.1
UT (M n = 5)	9.4	8.9–9.5	0.3	8.9	9.1
LT	9.9	9.4–10.1	0.35	9.8	9.7
M	17.2	16.7–17.4	0.3	17.3	17.2

Karyotype: 2N = 44, FN = 50, based on specimens from Europe (Zima and Kral, 1984).

Habits: Its habits strongly resemble those of *M. myotis*. It roosts in colonies which number hundreds, in dark, deep caves, but Lewis and Harrison (1962) found several specimens in small holes under a bridge in Lebanon. It emerges at late dusk or after dark. Maximum intensity of its calls is at 45 kHz (range 32–145) (Shalmon, 1994).
Food: It feeds on insects, chiefly moths and beetles, taken in flight or from the ground.
Reproduction: Males collected in July in Lebanon had enlarged testes (Lewis and Harrison, 1962).
Parasites: The following ectoparasites were found on this bat: *Nycteribia latreillii, N. vexata* and *Penicillidia dufourii* (Diptera) (Theodor and Costa, 1967).

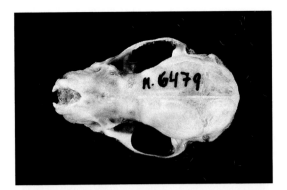

Figure 47: *Myotis blythii omari.* Dorsal, ventral and lateral views of the cranium, and a view of the mandible.

1 cm

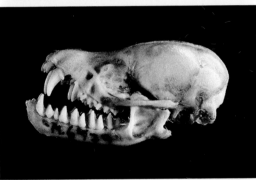

Figure 48: Distribution map of *Myotis blythii omari.*

● – museum records, ○ – verified field records.

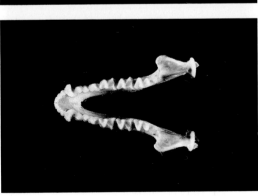

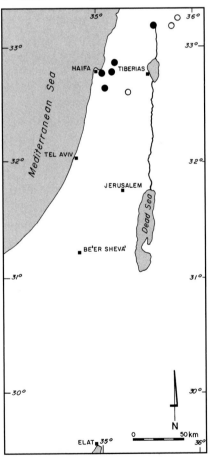

115

Myotis emarginatus Geoffroy, 1806   Notched-ear (Geoffroy's) Bat

Figs 49–50   נשפון פגום־אוזן

Myotis emarginatus E. Geoffroy, 1806. *Annales Mus. Hist. Nat. Paris*, 8:198. Charlemont, Givet, Ardennes, France.

Synonymy in Harrison and Bates (1991).

This is a small bat, similar in form to *M. mystacinus*, but closer in size to *M. nattereri*. The posterior margin of the ear has distinct angular emargination slightly above the middle. The tragus is pointed with straight anterior and posterior edges. It has a distinctive yellowish-red colour dorsally.

Myotis emarginatus emarginatus is the subspecies present in Israel (Harrison and Bates, 1991).

Table 44
Body measurements (g and mm) of *Myotis emarginatus emarginatus*

| | Body | | Tail | Ear | Foot | Forearm |
	Mass	Length				
	Males (1)					
	8	40	40	13	10	42
	Females (11)					
	n = 9				**n = 9**	**n = 10**
Mean	6.1	46	38	14	8	40
SD	0.5	3	4	2	2	2
Range	5–6.7	40–52	30–42	12–16	5–10	36–42

Forearm length of 38–40 mm and body mass of 6 g were recorded (B. Shalmon, pers. comm.).

Table 45
Skull measurements (mm) of *Myotis emarginatus emarginatus* (Figure 49)

| | Males (1) | Females (8) | | |
	Mean	Mean	Range	SD
GTL	16.5	16.3	16.0–16.6	0.3
CBL	15.2	15.0	14.6–15.5	0.3
ZB	9.6	9.8	9.5–10.0	0.2
BB	7.2	7.4	7.0–7.9	0.3
IC	2.8	3.7	3.6–3.8	0.1
UT	6.4	6.5	6.3–6.6	0.1
LT	6.9	6.8	6.3–7.2	0.3
M	12.0	12.1	11.8–12.5	0.2

Karyotype: 2N = 40, FN = 50, based on specimens from Europe (Zima and Kral, 1984).

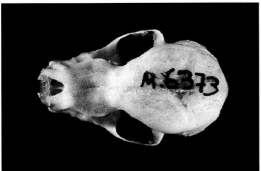

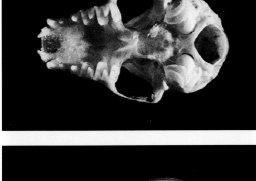

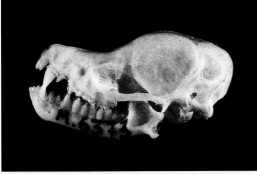

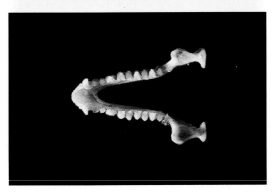

Figure 49: *Myotis emarginatus emarginatus.* Dorsal, ventral and lateral views of the cranium, and a view of the mandible.

1 cm

Figure 50: Distribution map of *Myotis emarginatus emarginatus.*

● – museum records, ○ – verified field records.

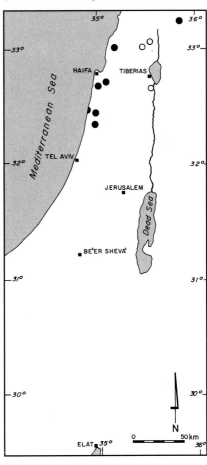

117

Distribution: It occurs in southern Europe (north to the Netherlands and south Poland), and east to Iran and south to Morocco (Corbet, 1978).

In Israel it formerly occurred in the Carmel and was also found near the Yarmouq, as well as in a cave in Bitan Aharon (Makin, 1977); and a dead specimen was found in 1990 in the western Galilee; it is very rare nowadays but has been recorded in the northern Galilee and Mount Ḥermon (B. Shalmon, pers. comm.; Figure 50). B. Shalmon (pers. comm.) estimated their present (1996) number in Israel at 50 individuals.

Habits: It roosts in colonies of hundreds in caves, tunnels and cellars. It generally hangs singly from the roof but occasionally hangs in small clusters. This species is currently rare in Israel and only a few individuals were found in a recent survey of the Carmel (Shalmon, 1994). In the Carmel, two of the caves where it was found were maternal colonies of mothers and young (Makin, 1977). Maximum intensity of calls is at 50–60 kHz (range 40–60. Shalmon, 1994).

Reproduction: Pregnant females, each with one foetus, were found in the Carmel on 6 May 1962 (Harrison and Bates, 1991) and a colony with young was found at the end of May at Bet Ḥananya. Hence, it is conceivable that parturition takes place in May. Twenty two specimens collected in Naḥal Oren (Mount Carmel) on 6 May 1962 were all females, probably from a nursery colony.

Myotis capaccinii (Bonaparte, 1837)   Long-fingered Bat
Figs 51–52   נשפון גדות

Vespertilio capaccinii Bonaparte, 1837. *Iconografia della Fauna Italica*, 1:fas. 20. Sicily. Synonymy in Harrison and Bates (1991).

This is a small, grey bat with relatively large feet, three-quarters of tibial length. The outer edge of the tibia is hairy. The tip of the tail protrudes about 2–4 mm from the uropatagium. The ears are relatively small and are narrow across their bases. The tragus attains more than half the length of the ear.

Myotis capaccinii bureschi Heinrich, 1936 is the subspecies present in Israel (Harrison and Bates, 1991).

Distribution: This bat occurs in the Mediterranean zone of Europe and north-west Africa, Asia Minor and the Middle East from Israel and Lebanon to Iran and Uzbekistan (Corbet, 1978).

In Israel it is found in the Upper and Lower Galilee, Mount Ḥermon, Carmel, Yizre'el and Ḥula Valleys (M. Dor and B. Shalmon, pers. comm.). It is the most common species of its genus in Israel, but populations are now (1996) small, estimated at 5000 by B. Shalmon (pers. comm.; Figure 52).

Habits: It roosts in colonies which can number hundreds in caves, some of which are used only for wintering (Makin, 1977). In the Upper Galilee it is known to hibernate in one particular cave, where up to 2700 specimens occur in July (Muskin, 1993). It also occupies several other caves in the region and moves between them during

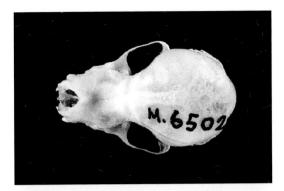

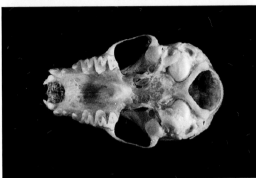

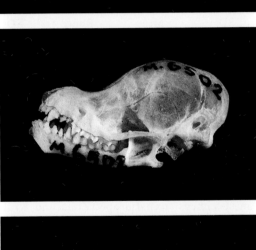

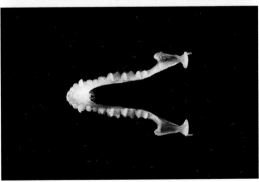

Figure 51: *Myotis capaccinii bureschi.* Dorsal, ventral and lateral views of the cranium, and a view of the mandible.

1 cm

Figure 52: Distribution map of *Myotis capaccinii bureschi* .

● – museum records, ○ – verified field records.

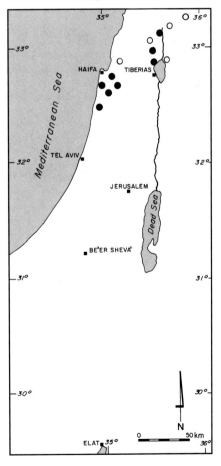

119

Table 46
Body measurements (g and mm) of *Myotis capaccinii bureschi*

| | Body | | Tail | Ear | Foot | Forearm |
	Mass	Length				
			Males (10)			
					n=8	n=8
Mean	6.5	51	35	12	11	39.5
SD	0.7	5.1	4.5	1.2	0.6	0.9
Range	5.7–8	42–60	28–41	10–14	10–12	38–41
			Females (10)			
	n=2	n=7				
Mean	8	46	37	12	11	40
SD		2	3	1	1	1
Range	5.7–10	42–50	30–42	10–14	9–12	39–41

Forearm length ranging 37.4–40.9 (males) and 38.4–43.6 (females) mm and body mass of up to 9.9 g were recorded (B. Shalmon, pers. comm.).

Table 47
Skull measurements (mm) of *Myotis capaccinii bureschi* (Figure 51)

| | Males (9) | | | Females (3) | | |
	Mean	Range	SD	Mean	Range	SD
GTL	15.3	14.9–15.8	0.3	15.3	15.2–15.4	
CBL	14.0	13.4–14.4	0.3	13.8	13.7–13.9	
ZB	9.1	8.8–9.3	(n=3)	9.2	9.1–9.2	(n=2)
BB	7.4	6.9–7.7	0.3	7.3	7.1–7.7	
IC	3.8	3.7–4.0	0.1	3.8	3.8–3.9	
UT (M n=7)	5.5	5.4–5.7	0.1	5.4	5.3–5.4	
LT (M n=8)	5.8	5.6–6.0	0.2	5.8	5.6–5.9	
M	10.5	10.0–11.0	0.35	10.5	10.4–10.6	

Karyotype: 2N=44, FN=50, based on specimens from Europe (Zima and Kral, 1984).

summer (Muskin, 1993). It often shares its roosts with other species (*M. nattereri*). A cave may be used only by members of one sex at one time of the year, and at other times by both sexes (Makin, 1977). They form dense clusters in the cave and emerge from the roost soon after sunset. It often hunts above water and near dense vegetation (Muskin, 1993). In Israel calls are produced with maximum energy at 50 kHz (Muskin, 1993).

Food: It preys on flying insects.

Reproduction: In Israel, pregnant females were found during April and young at the end of April, suggesting that parturition takes place at the end of April. Females bear one young.

Parasites: The following ectoparasites were found on this bat: *Ixodes vesoertilionis* (Ixodidae) and *Spinturnix psi* (mites) (Theodor and Costa, 1967).

Myotis nattereri (Kuhl, 1818)   Natterer's Bat

Figs 53–54   נשפון דק אוזן

Vespertilio nattereri Kuhl, 1818. *Annalen Wetterau Ges. Naturk.* 4:33. Hanau, Hessen, Germany.
Synonymy in Harrison and Bates (1991).

A small bat. A clear line of demarcation exists between the dark back and the light underparts from shoulder to ear. A line of small, inflected bristles occurs on the rear edge of the uropatagium. The ears are narrow with the tragus tall and longer than those of *M. capaccinii* and *M. emarginatus*. The tragus length is two-thirds of the ear length. The wings are broad.

Myotis nattereri hoveli Harrison, 1964 is the subspecies present in Israel (Harrison and Bates, 1991).

Distribution: This bat occurs in most of Europe, Morocco, and in Asia from the Caucasus to Japan (Corbet, 1978).

In Israel it is found in the Mediterranean zone where it was reported from the Galilee, near Jerusalem, near Haifa, Herzliyya and in the Hula and Yizre'el Valleys and it was formerly quite common (M. Dor, pers. comm.; Makin, 1977). It has currently been reported by Muskin (1993) and Shalmon (pers. comm.) from several caves in the Upper Galilee and the foothills of Judea, where several dozen to 200 were observed in each cave (Figure 54). B. Shalmon (pers. comm.) estimated their present (1996) number in Israel at 1500 individuals.

Habits: In Israel FM signals are at 35–78 kHz, with highest pulse intensity at 35–45 kHz (Muskin, 1993, Shalmon, 1994), for a duration of two msec.

Food: Two pellets examined from Zoarim Cave in the Upper Galilee contained remains of Lygaeidae, Coleoptera, Diptera and a spider (Whitaker *et al.*, 1994)

Reproduction: Males with large testes were captured during October, indicating that copulation takes place in October (B. Shalmon, pers. comm.). Four females, each

Table 48
Body measurements (g and mm) of *Myotis nattereri hoveli*

	Body		Tail	Ear	Foot	Forearm
	Mass	Length				
			Males (8)			
	n = 3	n = 7	n = 7			
Mean	5	44	43	16	8	40
SD	0.8	4.9	2.1	1.6	0.9	1.2
Range	4.6–6.0	38–50	40–46	13–17	7–9	38–42
			Females (10)			
Mean		42	46	16	7	40
SD		2	2	1	1	1
Range		41–47	42–48	14–17	6–8	39–41

Forearm length ranging up to 41 mm and masses of 5.7–8 g (males) and 9–11 g (pregnant females) were recorded (B. Shalmon, pers. comm.).

Table 49
Skull measurements (mm) of *Myotis nattereri hoveli* (Figure 53).

	Males (2)		Females (5)		
	Mean	Range	Mean	Range	SD
GTL	16.6	16.6	16.1	15.6–16.3	0.3
CBL	15.15	14.8–15.5	15.1	14.8–15.4	0.2
ZB	9.45	9.2–9.7	10.1	10.0–10.5	0.2
BB	7.6	7.5–7.7	7.9	7.7–8.2	0.2
IC	3.85	3.7–4.0	3.9	3.8–4.0	0.1
UT	6.3	6.2–6.4	6.0	5.7–6.4	0.3
LT	6.6	6.5–6.7	6.5	6.2–6.8	0.3
M	11.8	11.6–12.0	11.8	11.3–12.1	0.3

Karyotype: 2N = 44, FN = 50, based on specimens from Europe (Zima and Kral, 1984).

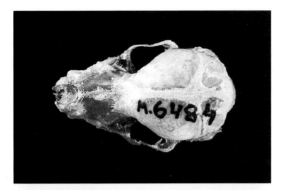

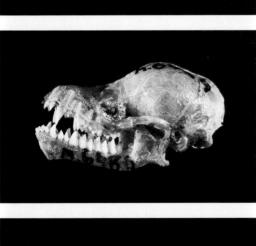

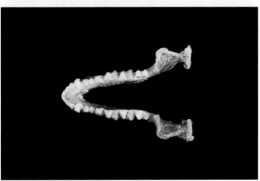

Figure 53: *Myotis nattereri hoveli.* Dorsal, ventral and lateral views of the cranium, and a view of the mandible.

1 cm

Figure 54: Distribution map of *Myotis nattereri hoveli* .

● – museum records, ○ – verified field records.

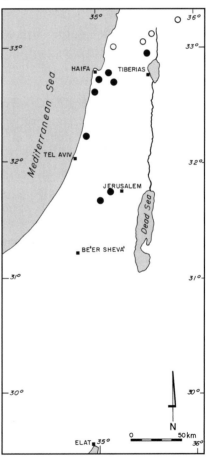

123

carrying one young, were found on 3 April 1955 in Givʻot Zaid, near Haifa (M. Dor, pers. comm.) and a young specimen with body length of about 4 cm was found on 10 May 1950 (presently in Tel Aviv University Museum). Hence, it seems that parturition takes place during March–May. In Israel post-partum females were found at the end of April 1975 in a cave in the Yizre'el Valley (Makin, 1977).

Parasites: The following parasites were found to infect this bat: *Nycteribia pedicularia* and *Basilia nana* (Diptera Pupipara), *Spinturnix myoti, Ichonoryssus graulosus* and *I. flavus* (mites) as well as *Polychromophilus murinus* and *Trypanosoma* sp. (Protozoa) (Theodor and Costa, 1967).

Myotis mystacinus (Kuhl, 1819)   Whiskered Bat
Figs 55*, 55   עטלפון משופם

Vespertilio mystacinus Kuhl, 1819. *Ann. Wetterau Ges. Naturk.*, 4(2), p. 202.

This is a small bat, possessing long, silky fur with golden, grey-brown tips and a light grey belly. The face and ears are black. The ears are relatively long and the tragus tall and narrow.

There is only a single specimen of this species preserved in an Israeli museum (TAU M.9456). The foot is short (6.1 mm) and its length is less than half of the tibia (16.8 mm). Its skull measurememts are (Figure 55*): GTL = 13.8 mm, CBL = 13.4 mm, ZB = 8.1 mm, BB = 7.0 mm, IC = 4.1 mm, UT = 5.3 mm, LT = 5.9 mm, M 10.2 = mm. The following measurements were provided by B. Shalmon (pers. comm.) for individuals caught in the Golan Heights and Mount Ḥermon during the summer of 1995. Mean forearm length of 13 males and 4 females was 38.0 mm (range 34.8–40.0) and 39.5 mm (range 37.6–41.1), respectively, and their mean body mass was 7.5 g (range 6.0–8.6) and 7.2 g (range 6.0–8.5).

Distribution: This species occurs throughout the Palaearctic region from Ireland to Japan and from southern Scandinavia to Turkey and Iran (Corbet, 1978). Tristram (1888) reported its existance from southern Lebanon.

In Israel it was found in summer 1995 in the Golan Heights and Mount Ḥermon (B. Shalmon, pers. comm., Figure 55).

Habits: It inhabits open woodland, and often hunts near and above water. Maximum intensity of calls at 50 kHz (range 40–60; Shalmon, 1994).

Reproduction: Males with developed testes were found between June–September. A maternal colony of several dozen females was found in the Golan on 22 June 1995, in which two females were still lactating (B. Shalmon, pers. comm.).

* Photographs added in proof [see Fig. 55*].

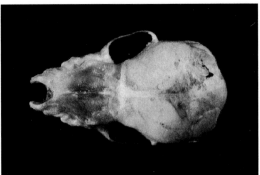

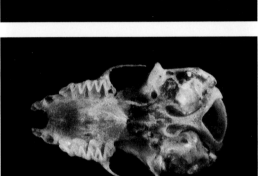

Figure 55*: *Myotis mystacinus* Dorsal, ventral and lateral views of the cranium, and a view of the mandible.

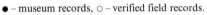

1 cm

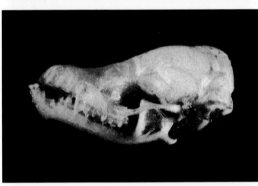

Figure 55: Distribution map of *Myotis mystacinus.*

● – museum records, ○ – verified field records.

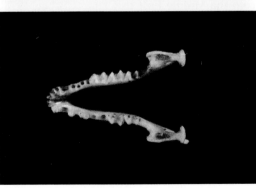

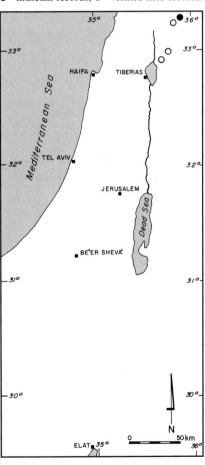

125

EPTESICUS Rafinesque, 1820

Eptesicus Rafinesque, 1820. *Annals of Nature*, p. 2.

A genus which is very similar to *Pipistrellus*. In Israel it is represented by two species: *E. serotinus* in the Mediterranean region and *E. bottae* in the desert.

Dental formula is $\dfrac{2.1.1.3}{3.1.2.3} = 32$.

Key to the Species of Eptesicus in Israel
(After Harrison and Bates, 1991)

1.   Large species; forearm exceeds 47 mm and greatest length of skull exceeds 20 mm. Pelage colour dark. **Eptesicus serotinus**

–   Smaller species; forearm less than 47 mm and greatest length of skull less than 18.5 min. Pelage colour pallid. Size medium; Second upper incisor relatively small, scarcely exceeding the cingulum of the first incisor in height. **Eptesicus bottae**

Eptesicus serotinus (Schreber, 1774)   Serotine Bat
Figs 56–57   אפלול מצוי

Vespertilio serotinus Schreber, 1774. *Die Saugetiere in Abbildungen nach der Natur.*, 1:167, pl. 53.

This is a medium-sized, dark brown bat with a black and nearly naked muzzle. The underpart is lighter in colour than the back. About 5 mm of the end of the tail protrudes out of the uropatagium.

Eptesicus serotinus serotinus is the subspecies present in Israel (Harrison and Bates, 1991).

Distribution: It has a wide distribution in the central and southern Palaearctic region in Europe, North Africa and Asia, from the Middle East to China (Corbet, 1978). In Israel it was collected only a few times: two females where found in Naḥal 'Ammud (Harrison and Bates, 1991), and one specimen was collected from a building in Tel Aviv on 14 July 1976. Several specimens were mist-netted in the Golan Heights and Mount Hermon on 19–20 June 1995 and a colony of several hundred was found in the Golan Heights in September 1995 (B. Shalmon, pers. comm.). Muskin (1993) reported a cave in the Upper Galilee where 20–30 specimens were observed during June 1990 (Figure 57). B. Shalmon (pers. comm.) estimated their present (1996) number in Israel at 1000 individuals.

Habits: In Israel when searching it produces FM signals of highest intensity at 30 kHz in the Galilee (Muskin, 1993).

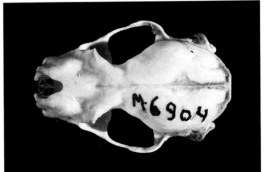

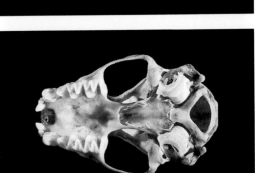

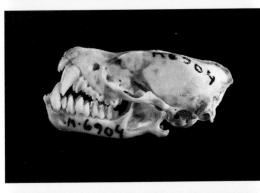

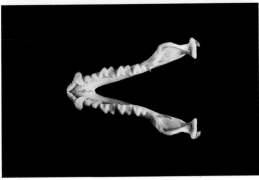

Figure 56: *Eptesicus serotinus serotinus.* Dorsal, ventral and lateral views of the cranium, and a view of the mandible.

1 cm

Figure 57: Distribution map of *Eptesicus serotinus serotinus.*

● – museum records, ○ – verified field records.

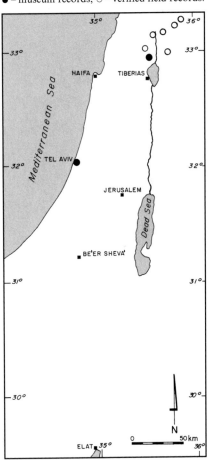

127

Table 50
Body measurements (g and mm) of *Eptesicus serotinus serotinus*

| | Body | | Tail | Ear | Foot | Forearm |
	Mass	Length				
			Males (2)			
			n = 1			
Mean	11.5	75	45.5	15	10	48
Range	10–13		40–51			47–49
			Females (1)			
	14.4	76	55			54.5

Forearm length of up to 53.5 mm and body mass of up to 25.5 g were recorded during June and September (B. Shalmon, pers. comm.).

Table 51
Skull measurements (mm) of *Eptesicus serotinus serotinus* (Figure 56)

| | Males (2) | | Females (1) |
	Mean	Range	
GTL	20.7	20.0–21.4	21.9
CBL	17.4	15.2–19.7	20.2
ZB	12.7	12.1–13.4	14.1
BB	9.25	9.2–9.3	10.4
IC	4.4	4.3–4.5	4.5
UT	7.2	6.7–7.7	7.4
LT	7.9	7.7–8.2	8.3
M	17.9	14.4–15.5	15.8

Karyotype: 2N = 50, FN = 46, based on specimens from Tunisia (Baker *et al.*, 1975).

Reproduction: Males with large testes were found on Mount Ḥermon in June 1995 and September 1994, indicating that mating takes place during autumn (B. Shalmon, pers. comm.). The two females collected in Naḥal 'Ammud in mid-April and early May were both pregnant with one foetus each. Two caves with colonies of nursing females were found in the Golan Heights on 22 June 1995 (B. Shalmon, pers. comm.).

Eptesicus bottae (Peters, 1869)   Botta's Serotine Bat

Figs 58–59   אפלול הנגב

Vesperus bottae Peters, 1869. *Monatsberichte K. preuss. Akad.wiss., Berlin*, p. 406. Yemen, Arabia.

Synonymy in Harrison and Bates (1991).

E. bottae is smaller and lighter in colour than *E. serotinus*. The underparts are white. The ears and muzzle are darker than the back. About 3 mm of the tail protrudes from the uropatagium.

Eptesicus bottae innesi Lataste, 1887 is the subspecies occurring in Israel (Harrison and Bates, 1991).

Distribution: It occurs in Egypt, Asia Minor, Arabia and east to Kazakhstan, Iran and Pakistan (Harrison and Bates, 1991).

In Israel it is not uncommon in the Negev and Judean Deserts, and along the 'Arava Valley from Elat to 'En Gedi (Yom-Tov *et al.*, 1992a; Figure 59).

Habits: It occurs in various habitats in the desert, including rocky slopes, open areas, farms and cultivated fields. It also forages near electric lights. In Iraq it roosts in buildings. Its flight is strong and quite high and it is noisy on the wing (Harrison, 1975). Calls are produced at frequencies ranging from 15–82 kHz, with maximum energy at 32–33 kHz (Shalmon, 1994).

Reproduction: Males with large testes were caught during May and September (B. Shalmon, pers. comm.). A female specimen caught by Harrison in Yotvata on 22 April 1962 was carrying two embryos. A female caught on 24 April 1988 near the Dead Sea was pregnant, and two other females caught on 24 May 1988 were post-lactating. Lactating females were caught on 2 May 1995 in the northern 'Arava (B. Shalmon, pers. comm.). Hence, parturition appears to take place in late April–early May.

Table 52
Body measurements (g and mm) of *Eptesicus bottae innesi*

	Body		Tail	Ear	Foot	Forearm
	Mass	Length				
	n = 2	n = 2	n = 2	n = 2	n = 2	n = 3
Mean	8.1	57.5 42.5	14	8	43	
Range	5.4–10.7	55–60	42–43	12–16	7.5–8.9	42–44

The sample size includes two males, and a specimen of unknown sex. In a sample of 22 specimens (13 males, 9 females) from the Negev, mean forearm length was 42.0 mm (SD = 1.5, range 39.7–44.9) and mean body mass was 10.6 g (SD = 1.5, range 8.5–13). There were no significant differences in forearm length and mass between males and females (B. Shalmon, pers. comm.).

Table 53
Skull measurements (mm) of *Eptesicus bottae innesi* (Figure 58)

	Males (2)		Sex unknown (1)
	Mean	Range	
GTL	16.25	15.9–16.6	16.0
CBL	15.65	15.4–15.9	15.3
ZB	10.6	(n = 1)	10.2
BB	8.25	8.1–8.4	8.0
IC	3.7	3.7	3.6
UT	5.65	5.6–5.7	5.6
LT	6.3	6.3	6.2
M (M n = 1)	12.4		11.7

Karyotype: 2N = 50, FN = 46, based on specimens from Jordan (Qumsiyeh, 1996).

130

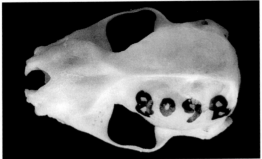

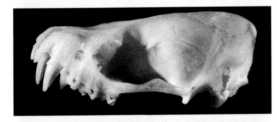

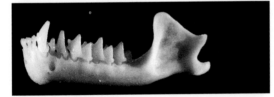

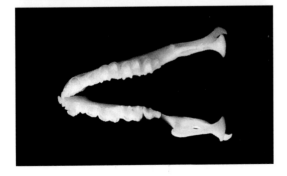

Figure 58: *Eptesicus bottae innesi*. Dorsal, ventral and lateral views of the cranium, and a view of the mandible.

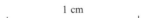

1 cm

Figure 59: Distribution map of *Eptesicus bottae innesi*.

● – museum records, ○ – verified field records.

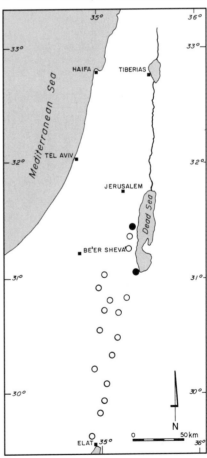

131

PIPISTRELLUS Kaup, 1829

Pipistrellus Kaup, 1829. *Skizzierte Entwicklungs-Gesch. Naturl. Syst. Europ. Thierwelt.*, 1:98

The genus *Pipistrellus* cannot be diagnosed globally by any universally applicable morphological characters. The species allocated to this genus, *sensu* Hill and Harrison (1987), can be separated from other Vespertilionid genera by their bacular morphology, but the retention of the genus is a matter of convenience (Harrison and Bates, 1991). It is very similar to *Eptesicus*. In Israel it is represented by five species (*P. pipistrellus, P. kuhlii, P. savii, P. rueppellii* and *P. bodenheimeri*), possibly six (if *P. ariel* is considered a separate species).

Dental formula of Israeli forms is $\dfrac{2.1.2.3}{3.1.2.3} = 34$.

Key to the Species of Pipistrellus in Israel
(After Harrison and Bates, 1991)

(*Pipistrellus ariel* is not included, as it is likely to be synonymous with *P. bodenheimeri*)

1. Small species; forearm 28.1–31.6 mm, greatest length of skull 10.8–12.0 mm. Both belly and dorsal parts whitish buff. **Pipistrellus bodenheimeri**
– Larger species 2
2. Forearm 30.0–34.0 mm, greatest length of skull 12.6–14.3 mm. Belly pure white to hair roots, back dark brown or sandy buff. There is a marked contrast between the colour of the belly and the back. **Pipistrellus rueppellii**
– No sharp contrast between dorsal and ventral sides 3
3. Forearm 29.0–37.0 mm, greatest length of skull 12.2–14.5 mm. Belly whitish with dusky bases, back brown or pale buff with dark bases. Wing membranes pallid with white border. **Pipistrellus kuhlii**
– Relatively small species. Forearm 29.0–31.1 mm, greatest length of skull 11.6–12.4 mm. Belly light brown with dusky grey bases, dorsal colour uniform dark brown. Dorsal profile elavated. **Pipistrellus pipistrellus**
– Relatively larger species. Forearm 31.5–36.5 mm, greatest length of skull 13.1–13.8 mm. Belly whitish buff with grey bases, dorsal colour yellowish buff with dark bases. Dorsal profile nearly strait. **Pipistrellus savii**

Pipistrellus pipistrellus (Schreber, 1774) Common Pipistrelle
Figs 60*, 60 עטלפון אירופי

Vespertilio pipistrellus Schreber, 1774. *Die Saugetiere in Abbildungen nach der Natur.*, 1:167, pl. 54. France.

This is a small pipistrelle with the dorsal parts uniform dark brown, ventral parts light brown and wing membranes blackish. Its ears are short, and a short, curved, blunt tragus and a post-calcarial lobe are characteristic to this species. There is only a single specimen (TAU M.9446) of this species available in Israeli museums. Its skull measurements (Figure 60*) are: GTL = 11.8 mm, CBL = 11.6 mm, BB = 6.1 mm, IC = 3.4 mm, UT = 4.0 mm, LT = 4.4 mm, M = 8.0 mm. Mean forearm length of seven specimens caught on Mount Ḥermon was 30.2 mm (range 29.0–31.1) and their mean body mass was 4.8 g (range 3.5–5.5; B. Shalmon, pers. comm.). *Pipistrellus pipistrellus pipistrellus* is apparently the subspecies occurring in our region (Harrison and Bates, 1991).

Karyotype 2N = 42, FN = 48, based on specimens from Europe (Bovey, 1949), but needs confirmation (Stebbings, 1977).

Distribution: It is widely distributed in the Palaearctic region from Britain to China (Corbet, 1978).

In Israel it is a rare species, collected only once in the Upper Galilee (Makin, 1977), but heard several times in this region (Muskin, 1993). On 19–20 June 1995 several specimens were caught on Mount Ḥermon and the Golan (B. Shalmon, pers. comm.; Figure 60).

Habits: In the Upper Galilee it occurs in a variety of habitats, but prefers to forage above open water and wet habitats (Muskin, 1993). In the Galilee maximum energy of its calls is at 48 kHz (Muskin, 1993), weaker than those of *P. kuhlii* (Muskin, 1993).

Reproduction: Two lactating females were caught on Mount Ḥermon on 20 June 1995 (B. Shalmon, pers. comm.).

---

* Photographs added in proof [see Fig. 60*].

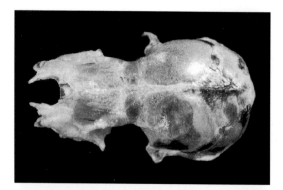

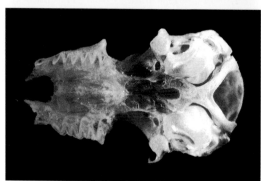

Figure 60*: *Pipistrellus pipistrellus pipistrellus*. Dorsal, ventral and lateral views of the cranium, and a view of the mandible.

1 cm

Figure 60: Distribution map of *Pipistrellus pipistrellus pipistrellus*.

● – museum records, ○ – verified field records.

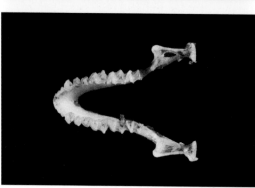

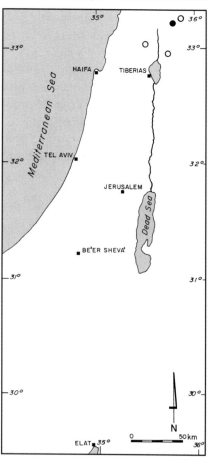

134

Pipistrellus kuhlii (Kuhl, 1819) Kuhl's Pipistrelle

Figs 61–62; Plate X:3 עטלפון לבן שוליים

Vespertilio kuhlii Kuhl, 1819. *Annalen Wetterau Ges. Naturk*, 4:199. Trieste (Italian/ Yugoslavian border).

Synonymy in Harrison and Bates (1991).

This species is quite similar to *P. pipistrellus* but more robust. The colour of the back is light brown and the underparts whitish. The border of the uropatagium, between the fifth digit and the tail, is white. The ears are tall and their tips narrower than in *P. pipistrellus* and the tragus is tall and narrow.

The subspecies present in Israel has not been defined (Harrison and Bates, 1991).

Distribution: It occurs in southern Europe and east to Turkestan and Pakistan, throughout south-west Asia and in much of Africa north, east and south of the Sahara. It is also found in west Africa (Corbet, 1978).

In Israel it is the most common insectivorous bat in the Mediterranean zone, but is also found in the Northern Negev and Judean Deserts and in the Dead Sea Area (Figure 62).

Habits: It roosts in colonies of hundreds in hollow trees or under bark, and very commonly in wooden buildings or buildings with wooden beamed-roofs where it can become troublesome due to the accumulation of faeces. It also inhabits wooden shacks formerly used as human habitation. Colonies become very noisy when disturbed, as individuals scramble deeper into the interior of their roosting site. This is the only species of insectivorous bat in Israel that does not roost in caves. Because of the ample availability of roosting facilities, it is extremely common. As it does not roost in caves, it was not affected by their fumigation and is still the most common insectivorous bat in Israel.

In Israel it emerges from its roost early in the evening, even before dark, and all colony members emerge within 15 min. During spring and summer it is active mainly during the first hours of the night, but from mid-May until the end of August activity is bi-modal (Barak and Yom-Tov, 1991). It flies straight and fast, often parallel to walls, and is capable of great manoeuvrability. It forages in uncluttered environments (i.e. fields) to which it flies along permanent routes, but also near street lights, where its hunting success increases in parallel with increase in group size from 1 to 5, probably because their calls cause the prey to disperse from the pole (Barak and Yom-Tov, 1989). There is an annual cycle of body mass, in which females and grouped males attain maximal mass in autumn (and may weigh up to 7.5 g) before hibernation, and minimal (about 5 g) in late winter (February). It does not show continuous hibernation, but is active, on and off, throughout the winter. In Israel the maximum energy of its calls is at 40–42 kHz (Muskin, 1993). The remains of 24 specimens were found in pellets of barn owls (*Tyto alba*) in Israel (Dor, 1947).

Food: It feeds on small insects, preferably Lepidoptera and Diptera, which are detected from about 1–1.5 m. The prey is caught in the mouth, and sometimes

Table 54
Body measurements (g and mm) of *Pipistrellus kuhlii*

| | Body | | Tail | Ear | Foot | Forearm |
	Mass	Length				
			Males (11)			
Mean	4.6	44	38	10	6	34 (n = 9)
SD	1	4	3	1	1	1
Range	3.0–6.2	39–52	31–40	8–11.5	5–8	33–36
			Females (12)			
Mean	5.1	46	38	10	7	34 (n = 9)
SD	1	10	2	4	1	1
Range	4–7	38–50	35–40	7–15	6–8	32–36

Specimens as heavy as 8.5 g were reported by Barak (1989).

Table 55
Skull measurements (mm) of *Pipistrellus kuhlii* (Figure 61)

| | Males (8) | | | Females (10) | | |
	Mean	Range	SD	Mean	Range	SD
GTL	13.7	13.3–13.9	0.2	13.6	13.4–14.4	0.3
CBL	12.7	12.1–13.1	0.3	12.6	12.3–13.3	0.3
ZB	8.6	8.4–8.8	0.2	8.7	8.6–8.7	(n = 2)
BB	6.5	6.3–6.8	0.2	6.6	6.3–7.2	0.3
IC	3.5	3.4–3.7	0.1	3.6	3.4–3.8	0.1
UT	5.0	4.7–5.4	0.2	4.9	4.8–5.1	0.1
LT	5.3	5.2–5.5	0.1	5.2	5.0–5.3	0.1
M	10.0	9.7–10.1	0.1	9.9	9.5–10.7	0.4

Karyotype: 2N = 44, FN = 50, based on two specimens from Jordan (Qumsiyeh, 1996).

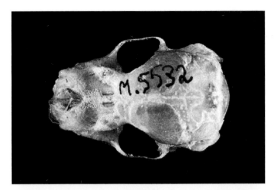

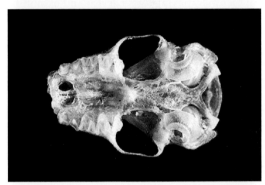

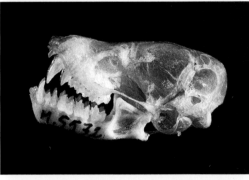

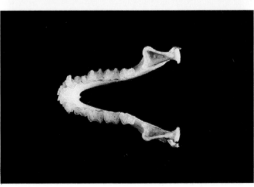

Figure 61: *Pipistrellus kuhlii.*
Dorsal, ventral and lateral views of the cranium, and a view of the mandible.

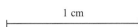

1 cm

Figure 62: Distribution map of *Pipistrellus kuhlii.*

● – museum records, ○ – verified field records.

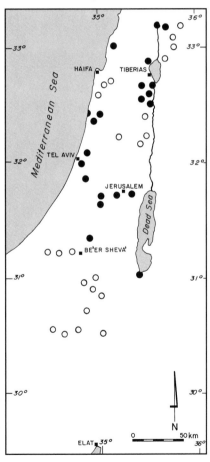

137

secured by folding the wing and the uropatagium. Five pellets examined from the Upper Galilee contained mainly remains (in decreasing order) of winged ants, Coleoptera, Chrysomelidae, Lepidoptera, and Cerambycidae (Whitaker *et al.*, 1994). Barak (1989) reported that specimens caught at the beginning of activity in the evening are at least 25% lighter in comparison to specimens caught in the morning, indicating a large mass increase due to extensive feeding.

Reproduction: This bat has been studied by Barak and Yom-Tov (1991) in the Bet She'an Valley, northern Israel. There are two types of males: those which roost singly all year round in permanent places along the females' flying routes, and those which roost in small groups (7–16 individuals) elsewhere. Single males are characterized by attaining maximal body mass in October, having larger testes and exhibiting song flight and false landing displays. In contrast, the body mass of grouped males fluctuates like that of the females; they have small testes and do not perform such displays. Single males court and copulate with females during September–October, and their testes are greatly enlarged during this season. From March, females form breeding colonies, and pregnant females were found between March–May. Births in the colony are not well synchronized, and lactating females were observed during May–July. Litter size is 1–2 (Harrison and Bates, 1991). Breeding colonies are deserted by October.

Parasites: The following parasites were found to infect this bat: *Basilia daganiiae* (Diptera Pupipara), *Cacodmus tunetanus* (Hemiptera), *Ischnopsyllus consimilis* (Siphonaptera), *Spinturnix acuminatus* and *Steatonyssus murinus* (mites) and *Trypanosoma* sp. (Protozoa) (Theodor and Costa, 1967).

Pipistrellus rueppellii (Fischer, 1829)   Rueppell's Pipistrelle
Figs 63–64; Plate X:2   עטלפון ריפל

Vespertilio rüppellii Fischer, 1829. *Synopsis Mamm.*, p. 109. Dongola, Sudan.
Synonymy in Harrison and Bates (1991).

This pipistrelle is distinguished from other species of its genus by its pure white underparts which sharply contrast with the dark brown back. The ears and muzzle are black.

Pipistrellus rueppellii rueppellii is the subspecies occurring in Israel (Harrison and Bates, 1991).

Distribution: This bat occurs throughout most of Africa (although apparently not in the Sahara) and in Iraq (Corbet, 1978).

In Israel it was found several times in the Dead Sea Area, once near Elat and once near Haifa (Figure 64). It is possible that it performs local movements.

Habits: In Egypt it was collected from under rocks (Qumsiyeh, 1985). In Israel it was caught mainly near water pools. Association with riverine conditions is reported by Smithers (1983). Calls are produced at frequencies ranging from 37–90 kHz, with maximum energy at 50–60 kHz (Shalmon, 1994).

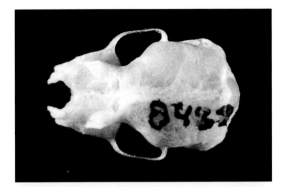

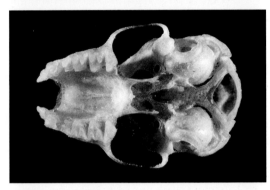

Figure 63: *Pipistrellus rueppellii rueppellii.* Dorsal, ventral and lateral views of the cranium, and a view of the mandible.

1 cm

Figure 64: Distribution map of *Pipistrellus rueppellii rueppellii.*

● – museum records.

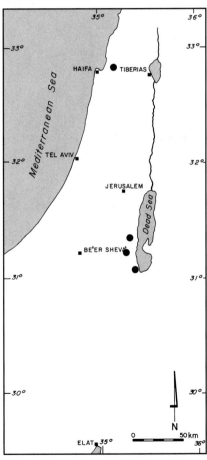

139

Table 56
Body measurements (g and mm) of *Pipistrellus rueppellii rueppellii*

| | Body | | Tail | Ear | Foot | Forearm |
	Mass	Length				
			Males (3)			
Mean	4.9	51	32	7	6	31
Range	4.5–5.7	50–52	31–34	7–8	6–7	30–32
			Females (3)			
Mean		46	33	12	7	33
Range		41–54	32–34	12.1–12.3	7.0–7.6	33.0–33.7

The heaviest male caught so far weighed 6.5 g (Yom-Tov).

Table 57
Skull measurements (mm) of *Pipistrellus rueppellii rueppellii* (Figure 63)

| | Males (3) | | Female (3) | |
	Mean	Range	Mean	Range
GTL	13.2	13.0–13.5	13.8	13.8–13.9
CBL	12.2	12.1–12.4	12.9	12.6–12.9
ZB (F n = 2)	8.5	8.2–8.7	8.6	8.4–8.7
BB	7.0	6.7–7.3	7.1	6.8–7.3
IC	3.7	3.6–3.7	3.9	3.8–4.0
UT	4.4	4.4–4.5	4.8	4.6–4.9
LT	4.8	4.8–4.9	4.9	4.8–5.0
M	9.3	9.1–9.6	9.8	9.6–10.1

Food: Of seven pellets examined from Nawit pools near the Dead Sea, six contained only Lepidoptera, and the seventh also some unidentified insects (Whitaker *et al.*, 1994).

Reproduction: A male with large testes was caught on 4 July 1995 (R. Feldman, pers. comm.). Otherwise nothing is known about its reproduction in Israel.

Pipistrellus savii (Bonaparte, 1837)   Savi's Pipistrelle
Figs 65–66   עטלפון סאבי

Vespertilio savii Bonaparte, 1837. *Iconografia della Fauna Italica*, 1: fasc. 20. Pisa, Italy.
Synonymy in Harrison and Bates (1991).

This pipistrelle is similar to *P. kuhlii* but has several distinctive characters. About
3 mm of the tail tip protrude from the uropatagium, the tail is shorter than the
length of the head and the body, and the wings, including the uropatagium, are
black. The muzzle, ears and cheeks are darker than the golden back.
Pipistrellus savii caucasicus Satunin, 1901 is the subspecies occurring in Israel
(Harrison and Bates, 1991).
Distribution: It occurs in the Mediterranean zone of Europe and North Africa,
and throughout Asia east to Japan (Corbet, 1978).
In Israel it is found in the Upper Galilee (Harrison and Makin, 1988), and Muskin
(1993) reports that it was heard several times in the Upper Galilee. During 18–20
June 1995 several specimens were mist-netted in the Golan and Mount Hermon
(B. Shalmon, pers. comm.; Figure 66).

Table 58
Body measurements (g and mm) of *Pipistrellus savii caucasicus*

| | Body | | | | | |
	Mass	Length	Tail	Ear	Foot	Forearm
	(n = 2)					
Mean	5.3	38	30.5	8	4.5	33
Range	3.7–6.8	38.0	29–32	6–10	4–5	31.5–34

The sample size includes 1 male, and one individual of unknown sex. In a sample of
17 specimens (14 males, 3 females) from Mount Hermon, mean body mass and forearm length were 7.4 g (range 5.6–9.0; SD = 1.1) and 33.6 mm (range 31.9–36.5;
SD = 1.3) (B. Shalmon, pers. comm.).

Table 59
Skull measurements (mm) of *Pipistrellus savii caucasicus* (Figure 65)

| | (2) | |
	Mean	Range
GTL	13.8	(n = 1)
CBL	13.1	(n = 1)
ZB	7.95	7.7–8.2
BB	6.75	6.2–7.3
IC	3.3	3.1–3.5
UT	4.5	4.2–4.8
LT	4.7	4.6–4.8
M	9.5	9.0–10.0

The sample size includes 1 male, and one individual of unknown sex
Karyotype: 2N = 44, FN = 50, based on specimens from Europe (Zima, 1982b).

141

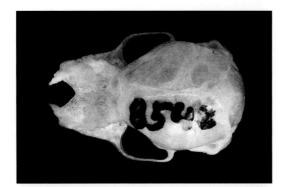

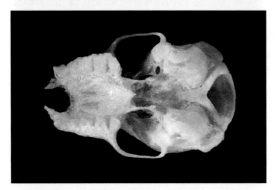

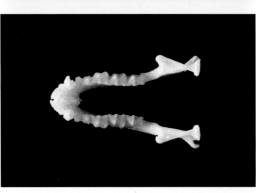

Figure 65: *Pipistrellus savii caucasicus.* Dorsal, ventral and lateral views of the cranium, and a view of the mandible.

1 cm

Figure 66: Distribution map of *Pipistrellus savii caucasicus.*

● – museum records, ○ – verified field records.

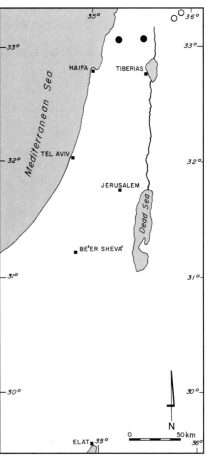

Habits: The call of a specimen caught in Mount Ḥermon was at maximum energy 45–50 kHz (range 20–75. Shalmon, 1994).

Reproduction: Two lactating females were caught near Mount Ḥermon, on June 19 and 20, 1995 (B. Shalmon, pers. comm.).

Pipistrellus bodenheimeri Harrison, 1960   Bodenheimer's Pipistrelle
Figs 67–68   עטלפון בודנהיימר

Pipistrellus bodenheimeri Harrison, 1960. *Durban Mus. Novit.*, 5:261. Yotvata, Wadi Araba, 40 km north of Eilat, Israel.

This is the smallest Israeli bat. Females are significantly larger and heavier than males (Yom-Tov *et al.*, 1992b). Its colour is light buff, but young specimens are light grey. Only the tip of the last vertebra protrudes from the uropatagium. The antitragus is well developed and is narrow and high.

Distribution: It occurs in Arabia, Israel and Sinai.

In Israel it is a resident, and is the most common bat in the Dead Sea Area (Yom-Tov *et al.*, 1992a), and very common elsewhere along the 'Arava Valley (Figure 68).

Habits: It roosts in crevices. In the Dead Sea Area it was observed foraging near vegetation, rocky, slopes, street lights and above water. In the Dead Sea Area it hibernates between October and April and capture rates of this bat during winter were 5% of the summer rate. (Yom-Tov *et al.*, 1992b). Calls are produced at frequencies ranging from 17–71 kHz, with maximum energy at 44–45 kHz (Shalmon, 1993).

Food: It feeds on small insects which are hunted and consumed in flight. Fourteen pellets examined from three localities in the Dead Sea Area and the 'Arava contained remains of Lepidoptera, Coleoptera and Diptera (Whitaker *et al.*, 1994). It feeds intensively in the evening and gains about 15% of body mass within two hours after starting its activity in the evening (Yom-Tov *et al.*, 1992b).

Reproduction: Males with large testes were observed from May until October, and it seems that mating takes place in autumn. Some information from the Dead Sea Area was gathered by Yom-Tov *et al.* (1992b). Lactating females were caught between May and early July, but females with protruding nipples were observed until early November, and post-lactating females in mid-October. Hence, parturition takes place from the end of April to May. It is estimated that 90% of females give birth every year. The first volant young was caught in mid-June. A female carrying two embryos was caught on 22 April 1962 by Harrison (1964).

Table 60
Body measurements (g and mm) of *Pipistrellus bodenheimeri bodenheimeri*

	Body		Tail	Ear	Foot	Forearm
	Mass	Length				
			Males (11)			
	n = 5	**n = 7**	**n = 10**	**n = 8**	**n = 9**	
Mean	2.5	34	31	10.5	5	29
SD	0.35	1.7	1.6	0.9	0.35	1.0
Range	2.0–2.9	32–37	29–34	8.5–11	4.8–6.0	28–31
			Females (6)			
	n = 4	n = 2		n = 5		
Mean	4.0	36.5	33	11	5	30
SD	1.5		3.2	0.6	1.0	0.45
Range	2.9–6.0	36–37	29–37	10–11.4	3–6	29–30.2

Forearm length of up to 32.8 mm was recorded (B. Shalmon, pers. comm.).

Table 61
Skull measurements (mm) of *Pipistrellus bodenheimeri bodenheimeri* (Figure 67)

	Males (6)			Females (6)		
	Mean	Range	SD	Mean	Range	SD
GTL(n = 5)	11.3	11.1–11.5	0.2	11.4	11.2–11.7	0.2
CBL(n = 5)	10.7	10.5–10.8	0.1	10.9	10.6–11.2	0.2
ZB (F n = 5)	6.8	6.2–7.4	0.4	6.9	6.7–7.0	0.1
BB (F n = 5)	6.0	5.7–6.2	0.2	5.9	5.7–6.3	0.3
IC	2.9	2.7–3.1	0.1	2.8	2.7–3.1	0.2
UT (F n = 5)	3.7	3.6–3.9	0.1	3.7	3.6–3.8	0.1
LT	4.1	3.9–4.5	0.2	4.0	3.8–4.1	0.1
M	7.9	7.8–8.3	0.2	7.9	7.7–8.2	0.2
TB (n = 5)	2.8	2.5–3.6	0.45	2.7	2.4–3.0	0.2

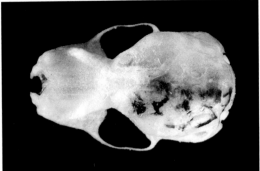

Figure 67: *Pipistrellus bodenheimeri boden-heimeri*. Dorsal, ventral and lateral views of the cranium, and a view of the mandible.

1 cm

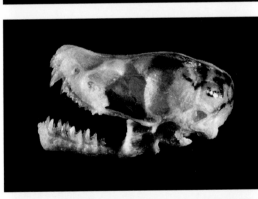

Figure 68: Distribution map of *Pipistrellus bodenheimeri bodenheimeri*.

● – museum records, ○ – verified field records.

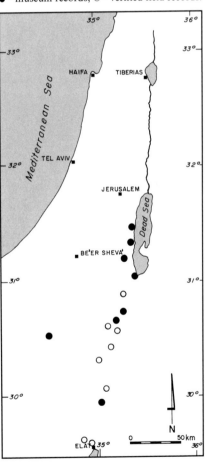

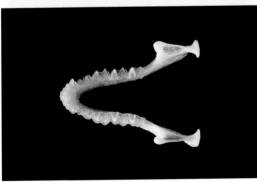

145

Pipistrellus ariel Thomas, 1904   Egyptian Desert Pipistrelle
Fig. 69   עטלפון אריאל

Pipistrellus ariel Thomas, 1904. *Annals Mag. Nat. Hist. Lond.*, 14:157. Wadi Alagi, north-eastern Sudan, 22°N 35°E, 2000 feet.

This species is probably conspecific with *P. bodenheimeri* (Yom-Tov *et al.*, 1992a). However, Harrison and Bates (1991) consider it as a separate species, distinguished by its narrower and more pointed muzzle, taller and narrower ears, developed tragus and poorly developed antitragus.

Table 62
Body measurements (mm) of *Pipistrellus ariel*

Body		Tail	Ear	Foot	Forearm
		Males (1)			
		33	9	5	29

Table 63
Skull measurements (mm) of *Pipistrellus ariel*

	Male (1) Mean
GTL	11.7
CBL	11.1
ZB	6.7
BB	5.6
IC	2.8
UT	3.8
LT	4.05
M	7.5

Distribution: Only three specimens of this species are known — the type from Sudan, another from Kordofan and a third from Naḥal Ẓe'elim in the Dead Sea Area, Israel (Figure 69).

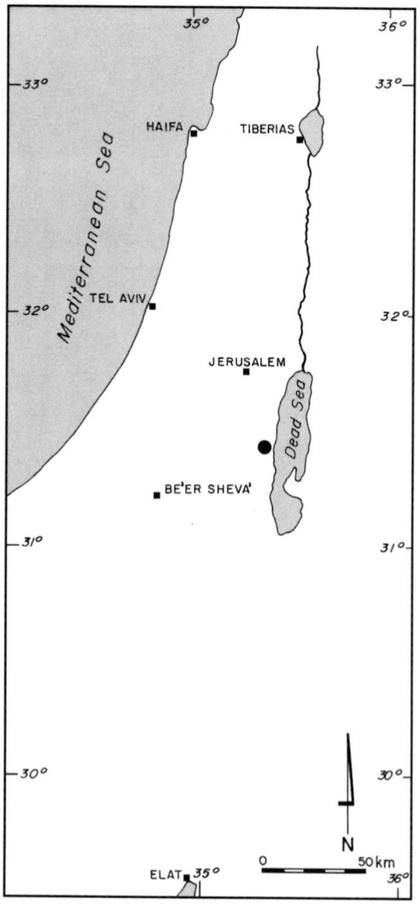

Figure 69: Distribution map of *Pipistrellus ariel ariel*.
● – museum records.

OTONYCTERIS Peters, 1859

Otonycteris Peters, 1859. *Monatsberichte K. preuss. Akad. Wiss.*, p. 223.

This genus is monospecific.

Otonycteris hemprichii Peters, 1859   Hemprich's Long-eared Bat
Figs 70–71; Plate X:4   אודן

Otonycteris hemprichii Peters, 1859. *Monatsberichte K. preuss. Akad. Wiss.*, p. 223. No local-
    ity; restricted by Kock (1969) to the Nile valley between north of Aswan, Egypt and Chon-
    dek, Northern Sudan.
Synonymy in Harrison and Bates (1991).

This is a medium-sized bat with very large ears which are almost twice the length of
the head and are not connected at their base by a membrane. The eyes are relatively
large. The tail is shorter than the forearm, and included in the flight membrane. It
has two pairs of pectoral mammae. The dorsal parts are grey, contrasting with the
paler underparts.
Otonycteris hemprichii jin Cheesman and Hinton, 1924 is the subspecies present in
Israel (Harrison and Bates, 1991).
Distribution: It is widely distributed in the desert zone from Morocco through Egypt
and east to Afghanistan and Kashmir (Harrison and Bates, 1991).
In Israel it was recorded from the Negev and Judean Deserts, as far west as Nabi
Musa, near Jericho. A female was collected in Rosh Ha'Ayin, about 10 km east
of Tel Aviv (Figure 71).
Habits: It normally occurs in extreme desert areas, fluttering like a butterfly over
water and near vegetation very close to the ground (Yom-Tov, 1993). It is a "whis-
pering" bat, whose calls are of low intensity. Calls are produced at frequencies ran-
ging from 20–37 kHz, with maximum energy at 34 kHz (Shalmon, 1993). Heim de
Balsac (1965) reported this species from pellets of barn owls (*Tyto alba*), in Algeria.
Food: It feeds on insects and arachnids, including scorpions, solifuges and spiders.
Eleven pellets examined near a freshwater pool near Sapir in the 'Arava contained
only remains of Scarabaeidae (Whitaker*et al.*, 1994).
Reproduction: Atallah (1977) reported three pregnant females, each with one embryo
on 2 May 1966, found in Azraq Oasis, Jordan, and estimated that parturition takes
place in early June. One female at the end of lactation was netted on 4 July 1994 in
the Negev (B. Shalmon, pers. comm.).

Table 64
Body measurements (g and mm) of *Otonycteris hemprichii jin*

| | Body | | Tail | Ear | Foot | Forearm |
	Mass	Length				
	Males (3)					
	n = 2					
Mean	18	69	56.5	36	13	63
Range	18.0–18.5	60–78	55–60	35–38	12–14	60–66
	Females (4)					
	n = 2					
Mean	22.5	67	52.5	36.5	12	63
SD	3.9	4.8	3.8	1.7	0.4	
Range	22–23	63–72	48–55	34–42	10–14	60–66
	Unknown Sex (2)					
Mean		23.8	71.5	53.5	35.5	
Range		17–30	70–73	52–55	33–38	

Table 65
Skull measurements (mm) of *Otonycteris hemprichii jin* (Figure 70)

| | Males (2) | | Females (3) | | Unknown sex (1) |
	Mean	Range	Mean	Range	
GTL	24.2	24.1–24.3	24.2	23.6–25.0	
CBL	22.1	(n = 1)	21.9	21.7–22.3	
ZB	14.6	14.3–15.0	15.0	14.7–15.4	13.7
BB	9.7	9.6–9.8	10.6	10.0–11.2	10.8
IC	4.4	(n = 1)	4.3	4.3–4.4	4.1
UT	8.2	8.0–8.4	8.2	8.1–8.4	8.1
LT	9.3	9.2–9.5	9.2	9.0–9.5	9.1
M	16.5	(n = 1)	16.8	16.3–17.0	16.3
TB	5.95	5.8–6.1	5.75	5.7–5.8 (n = 2)	5.6

Dental formula is $\frac{1.1.1.3}{3.1.2.3} = 30$.

Karyotype: 2N = 28, FN = 46, based on a specimens from Jordan (Qumsiyeh and Bickman, 1993).

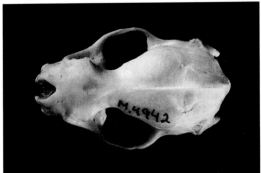

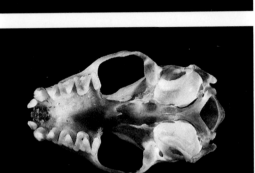

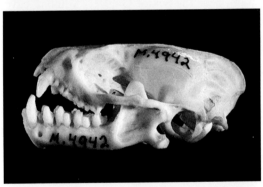

Figure 70: *Otonycteris hemprichii jin*. Dorsal, ventral and lateral views of the cranium, and a view of the mandible.

├─────── 1 cm ───────┤

Figure 71: Distribution map of *Otonycteris hemprichii jin*.

● – museum records, ○ – verified field records.

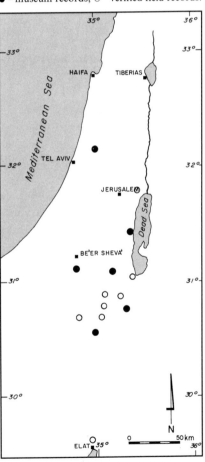

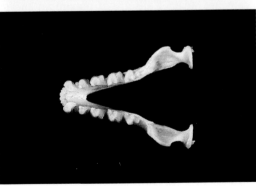

PLECOTUS E. Geoffroy, 1818

Plecotus E. Geoffroy, 1818. *Description de l'Egypte*, 2:112.

This genus is represented in Israel by one species, *P. austriacus*.

Plecotus austriacus (Fischer, 1829)   Grey Long-eared Bat
Figs 72–73; Plate XI:1   אזנן

Vespertilio auritus austriacus Fischer, 1829. *Synopsis Mamm.*, p. 117. Vienna, Austria.
Synonymy in Harrison and Bates (1991).

This is a small bat with large ears, connected at their base by a membrane. The ears are almost as long as the body or the forearm. The tragus is large and pointed. The tail is longer than the forearm and its end is not included in the uropatagium. The wings are short and rounded.

Plecotus austriacus christii Gray, 1838 is the subspecies present in Israel (Harrison and Bates, 1991).

Distribution: It is widely distributed in the Palaearctic region from Japan to Cape Verde islands and south to Senegal, including Arabia and Sinai (Harrison and Bates, 1991).

In Israel it was found on Mount Ḥermon, near the Sea of Galilee, 'Adullam cave, near Jerusalem and the 'Arava Valley from Elat to the Dead Sea Area, and in 'Avedat. On 19 June 1995 a female was caught on Mount Ḥermon (B. Shalmon, pers. comm.). Hence, it seems that this species formerly occurred all over Israel and is presently found along the Rift Valley and the Negev (Figure 73).

Habits: It roosts in small groups or singly in caves and ruins. When sleeping the big ears are folded back, below the wings. It flies low, slow and fluttering, and sometimes hovers low above ground. In Egypt it roosts in dark areas of the pyramids, old monuments and old houses (Qumsiyeh, 1985). A group of 30 was found hibernating in the 'Arava Valley in March , eight of which were checked and all were females (Makin, 1977). Hence, sexual segregation apparently exists during hibernation. It is a "whispering" bat, whose calls are weak. Calls are at maximum energy between 35–45 kHz (range 20–50. Shalmon, 1993).

Food: It feeds on moths, Diptera, small beetles and other insects and spiders which are caught in flight or picked from the ground and vegetation. Prey is located by either echolocation or sight, and often eaten at feeding sites where remains are found.

Reproduction: A male with large testes was found on Mount Ḥermon on 11 September 1994 (B. Shalmon, pers. comm.), probably indicating that mating occurs in autumn. Of 48 females found in a cave near 'Arad on 13 April 1989, 25 were lactating, as well as two females caught in Ne'ot HaKikar on 2 May and 14 June 1989. Two other females caught on 20 August and 17 September 1989 in Ne'ot HaKikar

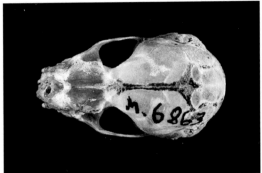

Figure 72: *Plecotus austriacus christii.* Dorsal, ventral and lateral views of the cranium, and a view of the mandible.

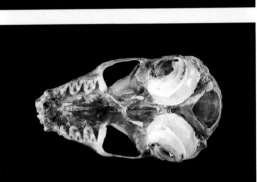

1 cm

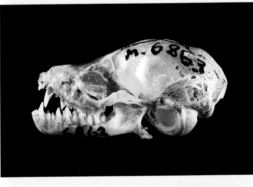

Figure 73: Distribution map of *Plecotus austriacus christii .*

● – museum records, ○ – verified field records.

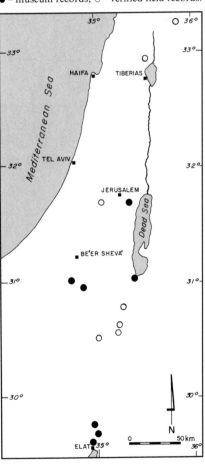

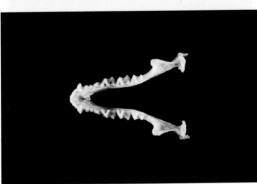

151

Table 66
Body measurements (g and mm) of *Plecotus austriacus christii*

| | Body | | Tail | Ear | Foot | Forearm |
	Mass	Length				
			Males (1)			
	n = 2					n = 2
Mean	6.5	52	43	36	5	39.5
Range	6.2–6.8					
			Females (10)			
	n = 5			n = 13		n = 11
Mean	6.8	50	47	36	8	41
SD	1.0	6.4	4.2	1.6	1.4	1.2
Range	5.5–7.8	41.0–58.0	43.0–54.8	33.0–38.4	5.0–9.6	38.6–42.1

Body mass of up to 11.5 g (a lactating female) and forearm length ranging from 37.4–42.8 mm were recorded. Field data indicate that there are no significant differences in either body mass or forearm length between the sexes or between samples from May and September (B. Shalmon, pers. comm.).

Table 67
Skull measurements (mm) of *Plecotus austriacus christii* (Figure 72)

| | Males (2) | | Females (3) | |
	Mean	Range	Mean	Range
GTL	17.05	17.0–17.1	17.3	17.0–17.5
CBL	15.75	15.7–15.8	15.6	15.2–16.1
ZB	9.00	9.0	8.7	8.6–8.8
BB	8.00	8.0	8.1	8.1–8.2
IC	3.35	3.3–3.4	3.3	3.1–3.4
UT	5.6	5.5–5.7	5.6	5.3–5.7
LT	6.1	6.1	5.9	5.9–6.0
M	10.6	10.5–10.7	11.1	10.9–11.5
TB	4.5	4.5	4.7	4.5–4.8

Dental formula is $\dfrac{2.1.2.3}{3.1.3.3} = 36$.

Karyotype: 2N = 32, FN = 50, based on a specimens from Jordan (Qumsiyeh and Bickman, 1993).

were post-lactating (Yom-Tov *et al.*, 1992a). Lactating females and young were caught on 1 May 1995 in the northern 'Arava and on 19 June 1995 on Mount Hermon (B. Shalmon, pers. comm.). Apparently parturition in the Negev takes place in April.

Parasites: The following ectoparasite was found on this bat: *Argas transgariepinus* (Argasidae) (Theodor and Costa, 1967).

BARBASTELLA Gray, 1821

Barbastella Gray, 1821. *London Medical Repository*, 15:300.

This genus is represented in Israel by one species, *B. leucomelas.*

Barbastella leucomelas (Cretzschmar, 1826)　　　　Arabian Barbastelle
Figs 74–75　　　　　　　　　　　　　　　　　　　　　　בלומף שחור

Vespertilio leucomelas Cretzschmar, 1826. In: Rüppell, *Atlas Reise Nordl. Africa, Saugetiere*: 73. pl. 28b. Arabia Petraea (= Sinai).

This is a small bat with long and almost square ears which are joined together at their base by a membrane. The tragus is triangular and the nostril openings point

Table 68
Body measurements (mm) of *Barbastella leucomelas leucomelas*

	Body				
	Length	Tail	Ear	Foot	Forearm
		Females			
	n = 1	n = 1	n = 1	n = 2	n = 2
Mean	60	42	15	7	39
Range				7	39.2–39.4

Body mass of four specimens caught in the 'Arava Valley ranged from 7.0–10.7 g and forearm length ranged from 36.6–38.8 mm (B. Shalmon, pers. comm.).

Table 69
Skull measurements (mm) of *Barbastella leucomelas leucomelas* (Figure 74)

	Females (2)	
	Mean	Range
GTL	14.4	14.3–14.5
CBL	13.4	13.4
ZB	7.6	7.5–7.6
BB	7.1	7.0–7.1
IC	3.6	3.5–3.7
UT	4.5	4.3–4.6
LT	4.9	(n = 1)
M	9.5	(n = 1)

Dental formula is $\dfrac{2.1.2.3}{3.1.2.3} = 34$.

Karyotype: 2N = 32, FN = 50, based on specimens from Europe (Zima, 1978).

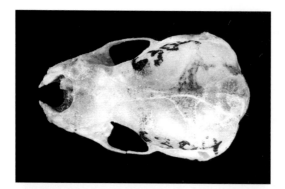

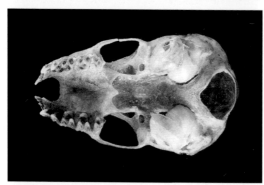

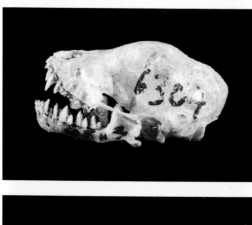

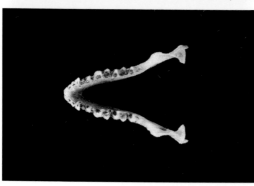

Figure 74: *Barbastella leucomelas leucome-las*. Dorsal, ventral and lateral views of the cranium, and a view of the mandible.

1 cm

Figure 75: Distribution map of *Barbastella leucomelas leucomelas*.

● – museum records, ○ – verified field records.

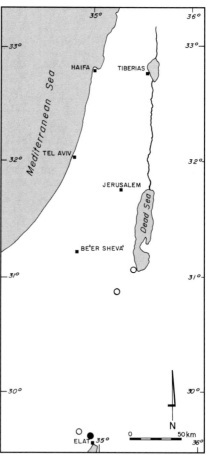

upwards and outwards. The tip of the tail protrudes from the uropatagium. The colour of the back is blackish and the underparts are lighter, fading to white in the inguinal region; this is reflected in its scientific (Greek) name — "leuco" — white, "melas" — black. The hair is very long.

Barbastella leucomelas leucomelas is the subspecies occurring in Israel (Harrison and Bates, 1991).

Distribution: It occurs in Asia from China and Japan to the Caucasus and Iran, as well as in Sinai.

Five specimens were observed in Israel: two specimens were collected by Harrison and Makin (1988) near Elot; two females were captured on two separate occasions later in the 'Arava Valley (4 April 1970 and 18 March 1995; B. Shalmon, pers. comm.) and one specimen was caught near Ne'ot HaKikar in July 1995 (R. Feldman, pers. comm.; Figure 75).

Habits: The specimens collected in Israel were caught foraging above an agricultural area and near water. Calls recorded near Elot had a highest pulse intensity at 45 kHz (B. Shalmon, pers. comm.).

Reproduction: A pregnant female, collected at Elot on 4 April 1970, gave birth three days later and a lactating female was caught at Ne'ot HaKikar in July 1995 (R. Feldman, pers. comm.).

MINIOPTERUS Bonaparte, 1837

Miniopterus Bonaparte, 1837. *Iconografia della Fauna Italica*, 1: pt 21.

This genus is represented in Israel by one species, *M. schreibersii*.

Miniopterus schreibersii (Kuhl, 1819)   Schreiber's Bat or Long-winged Bat
Figs 76–77   כנפן

Vespertilio schreibersii Kuhl, 1819. *Annalen Wetterau Ges. Naturk.*, 4:185. Kulmbazer Cave, Mountains of Southern Banat, Hungary.
Synonymy in Harrison and Bates (1991).

This is a bat with long, narrow wings; the tail is as long as the body and included in the uropatagium. The ears are relatively small, and the tragus slender. The second phalanx of the third finger is about three times as long as the first.

Miniopterus schreibersii pallidus Harrison, 1956 is the subspecies present in Israel (Harrison and Bates, 1991).

Distribution: This bat is widely distributed in the Old World, from southern Europe, Morocco through Iran and China to Australia, and in much of subSaharan Africa and Madagascar (Harrison and Bates, 1991).

In Israel it was one of the most common bats of the Mediterranean zone until the 1950s, and was recorded from the Ḥula Valley, Żefat, Kurdani (near Haifa) and

155

Table 70
Body measurements (g and mm) of *Miniopterus schreibersii pallidus*

	Body		Tail	Ear	Foot	Forearm
	Mass	Length				
	Males (7)					
	n = 4	**n = 4**	**n = 4**	**n = 4**		
Mean	9.8	49	53.5	8	10	45.5
SD	1.3	2.0	1.7	0.5	1.0	1.0
Range	8.0–11.0	46.0–50.0	52.0–56.0	7.0–8.0	9.0–11.3	44.0–47.0
	Females (2)					
	n = 1	**n = 1**	**n = 1**	**n = 1**		
Mean	9.0	48	56	8	10.2	45.7
Range					10.0–10.5	45–46.4

Body mass of up to 16.0 g and forearm length of up to 49.0 mm were recorded (B. Shalmon, pers. comm.).

Table 71
Skull measurements (mm) of *Miniopterus schreibersii pallidus* (Figure 76)

	Males (3)			Females (2)		
	Mean	Range	SD	Mean	Range	SD
GTL	15.1	14.9–15.3		15.3	15.3–15.4	
CBL	14.6	14.5–14.6		14.6	14.5–14.7	
ZB	8.8	8.6–9.2		8.3	8.3–8.3	
BB	7.6	7.5–7.7		7.6	7.5–7.7	
IC	3.7	3.7–3.8		3.7	3.7–3.7	
UT	5.8	5.8–5.9		5.7	5.6–5.8	
LT	6.2	6.2–6.3		6.3	6.1–6.4	
M	11.5	11.5–11.5		11.2	11.0–11.5	

Dental formula is $\dfrac{2.1.2.3}{3.1.3.3} = 36$.

Karyotype: 2N = 46, FN = 50, based on specimens from Tunisia (Baker *et al.*, 1975).

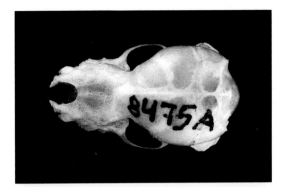

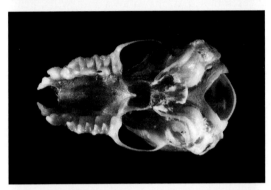

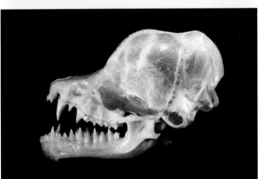

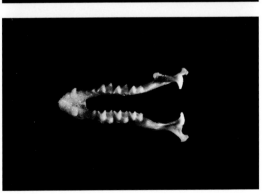

Figure 76: *Miniopterus schreibersii pallidus.* Dorsal, ventral and lateral views of the cranium, and a view of the mandible.

1 cm

Figure 77: Distribution map of *Miniopterus schreibersii pallidus.*

● – museum records, ○ – verified field records.

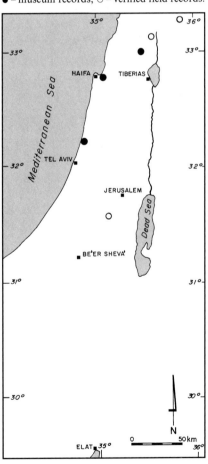

157

Herẓliyya (M. Dor, pers. comm.). It seems that this species was especially susceptible to cave fumigation, for today it is known from only one cave in Mount Meron, where several thousand roost during winter, and from Mount Ḥermon. A maximum number (3600) of emerging bats of this species was counted during September, and minimum (several hundred) during June (Muskin, 1993). Several males were mist-netted on Mount Ḥermon on 20 June and 11 September 1995 (B. Shalmon, pers. comm., Figure 77).

Habits: It roosts in large colonies, numbering thousands, in dark caves and mines. Sexual segregation was recorded. Foraging starts early in the evening, and is carried out at 10–20 m above ground. It is one of the fastest fliers among bats, with an average flight speed of 50 km/hr. Its flight resembles that of swallows and swifts, being fast with sharp turns. Migrations and movements between summer and winter roosts are well known in Europe, and in Australia a 1000 km migration was recorded (Fenton, 1983). In Israel M. Dor (pers. comm.) found roosts with only females near Ẓefat in September 1946, and with only males in Herẓliyya in May 1947, but mixed roosts were found in January, February and March in several localities (M. Dor, pers. comm.). In the Galilee calls are produced at frequencies ranging from 45–88 kHz (Muskin, 1993), with maximum energy at 55–60 kHz.

Food: Nothing is known from Israel regarding its diet.

Reproduction: A lactating female was caught on Mount Ḥermon on 2 May 1995 (B. Shalmon, pers. comm.).

Parasites: The following parasites were found to infect this bat: *Nycteribosca kollari, Nycteribia pedicularia, N. schmidlii, N. vexata, Penicillidia dufourii* and *P. conspicua* (Diptera Pupipara), *Ixodes vespertilionis* and *I. simplex* (Ixodidae), *Spinturnix psi* (mites) and *Polychromophilus melanipherus* (Protozoa) (Theodor and Costa, 1967).

CARNIVORA

This order is represented in Israel by five families with 17 genera: Mustelidae, Viverridae, Canidae, Hyaenidae and Felidae. The Syrian bear (*Ursus arctos*, Ursidae) inhabited the Galilee until the end of the 19th century.

MUSTELIDAE
Otters, Martens, Badgers, Polecats סמורײם

This family is represented in Israel by five genera, each with one species: *Lutra, Mellivora, Meles, Vormela* and *Martes*.

Key to the Species of Mustelidae in Israel

1.	Small species, greatest length of skull less than 88 mm.	2
–	Large species, greatest length of skull more than 100 mm.	3
2.	Head and body length in excess of 580 mm. Four upper and four lower premolars present.	**Martes foina**
–	Head and body length less than 500 mm. Three upper and three lower premolars present. Forehead with a prominent white band; pelage on the back a variegated yellow and brown. Skull length 48–56 mm; lower carnassial with a distinct metaconid.	**Vormela peregusna**
3.	Feet fossorial, plantigrade, with long claws on the fore feet. Tail short, less than half the body length.	4
–	Feet adapted for swimming, toes webbed, with short claws on the fore feet. Tail more than half the body length and thickened at base.	**Lutra lutra**
4.	Face with a median white stripe and two lateral black stripes, enclosing eyes. m1 enlarged and nearly rectangular in outline.	**Meles meles**
–	Crown of head and back white; the face, limbs and belly are black. m1 small, transversely elongated with a medial constriction.	**Mellivora capensis**

LUTRA Brunnich 1780

Lutra Brunnich, 1780. *Zoologiae Fundamenta*. p. 34.

Lutra lutra (Linnaeus, 1758)
Figs 78–79; Plate XI:3–4

Common Otter
לוטרה

Mustela lutra Linnaeus, 1758. *Systema Naturae*, 10th ed., 1:45. Uppsala, Sweden.
Synonymy in Harrison and Bates (1991).

The body is elongated and flexible and the long tail is strong and somewhat flattened dorso-ventrally. The legs are short and strong and the digits are connected with strong webs. On the hind foot the fourth digit is elongated, equalling the third one in length. The eyes, ears and nostrils are located on the dorsal part of the head, enabling the submerged otter to expose only a small part of the head yet still being able to see, hear and smell while swimming. Its long, strong vibrasse form a front-pointed fan, thus enabling it to sense even weak water currents caused by its prey. Males are significantly larger and heavier than females.

Lutra lutra seistanica Birula, 1912 is the subspecies present in Israel (Harrison and Bates, 1991).

Table 72
Body measurements (g and mm) of *Lutra lutra seistanica*

| | Body | | | | |
	Mass	Length	Tail	Ear	Foot
	Males (7)				
Mean	10583	758	459	24	123
SD	1258	55	41	6	10
Range	9120–12240	704–832	428–550	15–35	102–133
	Females (5)				
Mean	7000	681	375	25	121
SD	2000	33	19	5	21
Range	5000–10000	640–720	360–400	21–30	100–143

Table 73
Skull measurements (mm) of *Lutra lutra seistanica* (Figure 78)

| | Males (8) | | | Females (8) | | |
	Mean	Range	SD	Mean	Range	SD
GTL	120.8	114.2–128.7	4.6	113.6	108.4–123.5	4.7
CBL (M n = 7)	120.3	114.2–128.7	4.3	113.3	108.6–123.5	4.5
ZB	71.3	63.8–74.7	3.6	68.1 (F n = 7) 64.6–76.5		3.9
BB	54.6	51.3–55.2	2.0	52.2	48.8–58.0	2.7
IC	15.3	13.6–17.5	1.3	14.5	13.1–16.2	1.2
UT	37.4	36.6–38.2	0.6	35.6	34.3–38.9	1.5
LT (M n = 7)	45.1	43.8–46.6	1.1	42.6	41.0–46.6	1.7
M (M n = 7)	75.3	71.1–82.7	4.2	71.4	67.0–78.0	3.4

Dental formula is $\frac{3.1.4.1}{3.1.3.2} = 36$.

Karyotype: 2N = 38, FN = 60, based on a specimen from Germany (Günther and Gebauer, 1982).

160

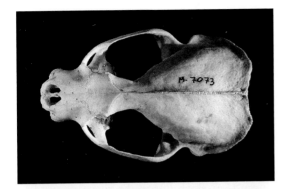

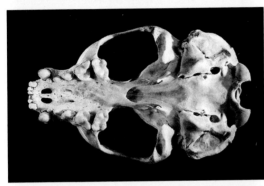

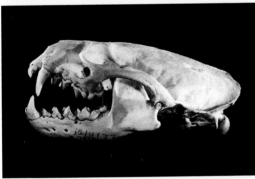

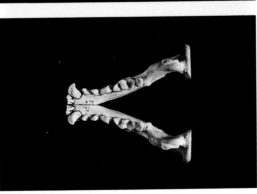

Figure 78: *Lutra lutra seistanica*. Dorsal, ventral and lateral views of the cranium, and a view of the mandible.

5 cm

Figure 79: Distribution map of *Lutra lutra seistanica*.

● – museum records.

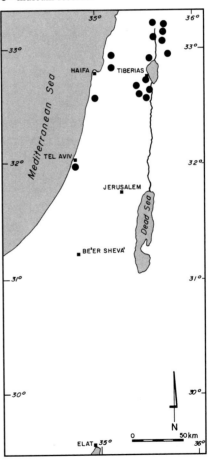

Distribution: This species is widely distributed in the Palaearctic region, except for the Siberian tundra and desert areas in the Sahara, Arabia and Iran (Corbet, 1978). In Israel it was widely found along rivers and streams in the Mediterranean region. In the Coastal Plain it occurred from the Lebanese border to Naḥal Soreq in the southern Coastal Plain, and all along the Jordan Valley. However, hunting by fish breeders, water pollution and the draining of streams and wadies in most of Israel caused a dramatic decline in its populations. Along the Coastal Plain it now occurs only north of 'Akko, if at all and in the Jordan Valley it occurs in the Ḥula, around Lake Kinneret (Sea of Galilee) and Bet She'an Valley and possibly south of it towards the Dead Sea (Figure 79). We estimate its present population size in Israel at about 100 individuals, most of which live in the rivers descending from the Golan.

Habits: This species is mainly nocturnal, but is also sometimes active during the day. It lives near fresh water streams and rivers, but at times wanders several kilometres inland. It climbs well and is an excellent swimmer, able to dive for up to eight minutes (Harrison and Bates, 1991). During fast swimming the tail and rear part of the body move up and down to propel the animal forward, while the front legs are pressed backwards to the body and move only when the otter performs turns, and the hind legs operate together as an oar. In slow swimming front and rear legs are used alternately. It is capable of sharp twisting in the water, and often rolls on its back and sides while swimming. In the water it appears shiny, due to the air bubbles trapped in the fur. Upon emerging from the water it shakes its body vigorously, and later rolls on the ground, preferably on grass or sand, to squeeze out the water from its fur. It is capable of jumping from the water up to 60 cm in the air. When inactive it hides below boulders, rocks and other cover, in dense vegetation of bramble (*Rubus sanctus*), and reeds (*Arundo donax* and *Phragmites australis*) and in burrows which it digs in the bank of a water course, with one opening below the surface of the water and another one on the shore. The burrow contains a dry resting chamber. Otters are territorial, and mark their territories with faeces and secretions of the anal glands on the shore on rocks, tree trunks and other hard objects. Territory size is about 1 km^2, and often constitutes a long strip along a water course. Its marking spots are located mainly along these water courses. Each territory also features permanent feeding sites where food remains are found.

The otter is very vocal and produces many calls, including whistling cries and a sharp alarm scream.

Food: In the water it catches fish, water snakes (*Natrix tessellatus*) (in the stomach of one 9 kg male caught at Ma'ayan Ẓevi many water snakes were found, weighing a total of 600 g), amphibians, crustaceans and molluscs, while on land it feeds on small birds and mammals as well as on invertebrates. It eats small fish holding its head above the water with its front feet manipulating the prey, but carries larger fish to the shore. It occasionally eats sweet fruit.

Reproduction: Gestation lasts 59–63 days, and litter size ranges from 1–3. In Israel otters give birth throughout the year, and young are found all year round. The female gives birth in its hiding place. The young open their eyes at 35 days, and at about that time start to feed on solid food. They are weaned when two months old, become independent when 6–9 months old and become sexually mature and attain adult size when about one year old. Otters can live up to 15 years in the wild and 24 years in captivity (Harrison and Bates, 1991).

Relations with man: Fish breeders often claim that otters cause considerable damage to fish in fishponds. However, this claim is exaggerated, because of the otters' large territories and their diverse diet. By eating dead and easily caught sick fish, water snakes and crustaceans, they actually help fishbreeders.

Until the forties of this century, otters were common around Lake Kinneret (Sea of Galilee), where they did not cause any damage, but were hunted to extinction because of their fur, which was sold in the Old City of Jerusalem.

Otters make excellent pets. Those raised in captivity from a young age become very tame, and interact well with man.

MELLIVORA Storr, 1780

Mellivora Storr, 1780. *Prodromus methodi mammalium.*, p. 34, tab. A.

Mellivora capensis (Schreber, 1776) Ratel, Honey Badger
Figs 80–81; Plate XII:1–2 גירית דבש

Viverra capensis Schreber, 1776. *Die Saugethiere in Abbildungen nach der Natur.*, pl. 125, also
    1777., 3:450, 588. Cape of Good Hope.
Synonymy in Harrison and Bates (1991).

The body is heavily built and very powerful, especially the very muscular front legs. Under the thick skin is a layer of loose elastic connective tissue that, together with the subcutaneous fat layer, makes for very loose connection between the strong body and the thick, strong skin. If the honey badger is seized by a large predator, it can turn around in its skin and defend itself with teeth and claws. Of the five claws on the forefoot, the three central ones are very long and strong and are the main digging tools. It has short limbs and very small ears. It has a very distinctive coloration: the entire ventral part, including the limbs, muzzle, face and cheeks and distal part of the tail are black, and these black areas are sharply defined from the upper white mantle. The fur is short and there is no real winter fur; in winter the fur becomes more dense, but remains short and without underwool. Males are significantly larger and heavier than females.

The name honey badger is hardly justified. It is not related at all to the common badger (*Meles*) and any similarity is very superficial. Captive specimens did not show any interest in honey, neither pure honey nor honeycombs. It is not omnivorous, unlike *Meles*, but decidedly carnivorous. Specimens that raid bee-hives, as they did in Israel, are probably more interested in the larvae and pupae than in the honey.

However, its scientific name reflects the belief that it eats honey: "voro" — to devour in Latin, "mel" — honey in Latin.

Mellivora capensis wilsoni Cheesman, 1920 is the subspecies present in Israel (Harrison and Bates, 1991).

Table 74
Body measurements (g and mm) of *Mellivora capensis wilsoni*

| | Body | | | | |
	Mass	Length	Tail	Ear	Foot
	Males (5)				
Mean	8952	708	168	10 (n = 1)	112
SD	2253	22	17		4
Range	5000–10500	670–720	140–180		107–115
	Females (2)				
Mean	6175	640	130	10 (n = 1)	109
Range	4850–7500	620–660	110–150		108–110

Table 75
Skull measurements (mm) of *Mellivora capensis wilsoni* (Figure 80)

| | Males (4) | | | Females (1) |
	Mean	Range	SD	Mean
GTL	134.4	132.8–135.4	1.1	117.7
CBL	134.2	131.9–136.7	2.0	114.4
ZB	78.8	77.7–80.0	1.0	68.0
BB	64.4	54–65.2	0.6	57.5
IC	33.9	33.3–34.6	0.6	33.0
UT	39.4	38.6–41.0	1.1	33.3
LT	44.2	42.5–45.7	1.3	39.0
M	87.7	84.6–89.2	2.7	72.3

Dental formula is $\frac{3.1.3.1}{3.1.3.1} = 32$.

Distribution: This species is widely distributed in Africa and in south-western Asia from the Mediterranean to India (Corbet, 1978).

In Israel it is rare but distributed throughout the country. Individuals have been captured or observed in the 'Arava, but also in the Galilee and the Judean Hills and even in the Coastal Plain (Figure 81).

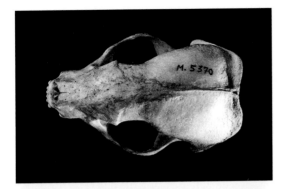

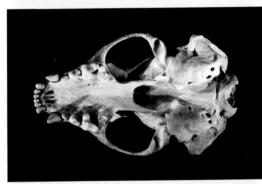

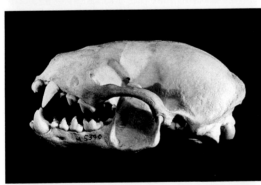

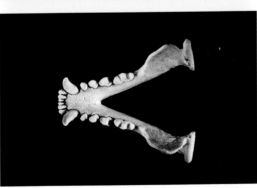

Figure 80: *Mellivora capensis wilsoni.* Dorsal, ventral and lateral views of the cranium, and a view of the mandible.

5 cm

Figure 81: Distribution map of *Mellivora capensis wilsoni.*

● – museum records.

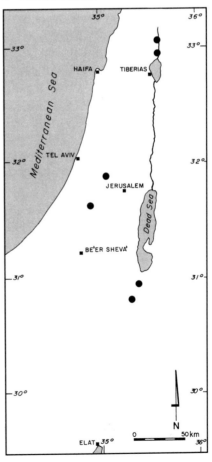

165

Habits: Under the thick hide, its body is heavily built and very powerful. It is very agile on the ground and a climber on rocks. In captivity it shows extraordinary mobility, agility and curiosity. In its cage it will pile up various objects and climb on them in order to get at out-of-reach food. It is very playful, and animals in two adjacent cages were observed playing by pushing various objects from one cage to the other under the metal dividing door. During play it twists and turns easily, demonstrating the looseness of its hide on its body. It is able to manipulate objects with its front paws like with hands in a primate-like fashion. One pair, kept in a room with a sink, learned to push their sleeping boxes below the sink, climbed up, opened the tap and played in the water.

It is predominantly nocturnal, but has been seen active during the day (Harrison and Bates, 1991). When inactive it hides in burrows it digs in the ground, under boulders and in natural caves. When attacked it fights with great courage, biting savagely. It marks its territory with secretion from the anal glands, and the male also marks by dragging its penis on the ground for several tens of centimetres while urinating.

The honey badger produces a great variety of calls, including wails, rattles, groans and roars.

Food: It is mainly carnivorous, obtaining most of its food by digging in the ground with its long, strong claws. It feeds on invertebrates and small vertebrates, mostly reptiles but also on mammals and birds. In Israel it has been observed raiding wooden beehives. It also eats carrion, and the Bedouin claim it digs up and eats human corpses (Harrison and Bates, 1991). They also claim that this species, as well as the common badger, attacks people, going mainly for the genital parts.

Reproduction: It is apparently monogamous, as in captivity members of a pair live in close cohabitation. At the Tel Aviv University Research Zoo it breeds all year long, provided that suitable conditions (an opportunity to dig and hide in the ground) are provided. Pregnancy there lasted 62–74 days and there was no delayed implantation of the ovum. Litter size was 1–2. The female gives birth once a year, but if the cubs are lost it can become pregnant again the same year. The newborn develops slowly: eyes open at 2.5 months, and it starts reacting to noises at three months, at which time it first emerges from its burrow. A captive animal achieved adult size when 17 months old. Longevity in captivity recorded to date is up to 24 years, but could be longer.

Parasites: This species is infected by fleas (*Pulex irritans*) and ticks (*Rhipicephalus sanguineus*) (Theodor and Costa, 1967).

MELES Boddaert, 1785

Meles Boddaert, 1785. *Elenchus animalium*, 1:45.

Meles meles (Linnaeus, 1758)　　　　　　　　　　　　　Badger
Figs 82–83; Plate XII:3–4　　　　　　　　　　　　　גירית מצויה

Ursus meles Linnaeus, 1758. *Systema Naturae*, 10th ed., 1:48. Uppsala, Sweden.
Synonymy in Harrison and Bates (1991).

The body is heavily built and powerful. The fur is grey; pure grey when fresh and brownish-grey when worn. The legs and underparts are black. The head is white, with a black band extending on each side from the snout over eye and ear to the neck. The black ears have white margins. The hair is long, and in winter a dense, woolly underfur develops. Males are significantly larger and heavier than females. *Meles meles canescens* Blanford, 1875 is the subspecies present in Israel (Harrison and Bates, 1991).

Distribution: It is widely distributed in the Palaearctic region from western Europe to Japan, but not in arid areas (Corbet, 1978).

In Israel it is confined to the Mediterranean region, with some penetration to semi-desert areas in the central Negev (for example, in Sede Boqer; Figure 83).

Fossil record: The earliest remains of this species found in Israel are from the Upper Palaeolithic period (Tchernov, 1988).

Habits: It is nocturnal. During the day it hides in elaborate burrow systems that it digs in the ground with its strong front paws which are equipped with four long claws. In Israel these burrow systems are not as complex as they are in Europe, and are normally inhabited by fewer individuals, mostly by one pair and during and after the breeding season also by their offspring.

When attacked, the badger erects the stiff hairs on its flanks, looking much larger than it actually is, but if the attacker is a larger animal than itself it rolls its body up and hides the head and legs below the body. Predators find it difficult to bite into this ball, but the badger, time and again, suddenly extends its head for a second and bites into the feet or snout of the attacker. It is capable of running fast, and is a good swimmer and climber.

Food: It is omnivorous and its diet includes earthworms, snails, insects, reptiles and rodents as well as fruit, bulbs, acorns, green leaves and carrion. It digs up much of its food from the soil.

Reproduction: It is monogamous, and members of the pair maintain close contact. Groups of badgers are apparently composed of parents with their offspring. In Europe there is delayed implantation of the embryo. This phenomenon was also found in Israel. Captive badgers copulate in April–May. Copulation lasts about half an hour and is repeated over several days, even weeks. Both pair mates are quite noisy before and during copulation. The two to four young are born in February

167

Table 76
Body measurements (g and mm) of *Meles meles canescens*

	Body				
	Mass	Length	Tail	Ear	Foot
Males (17)					
Mean	8500	696	142	42	107
SD	2200	68	19	5	7
Range	6500–14000	560–815	105–180	32–50	95–120
Females (10)					
Mean	7747	663	147	43	103
SD	1699	36	34	10	5.4
Range	5270–70840	620–750	83–200	32–70	92–110

Table 77
Skull measurements (mm) of *Meles meles canescens* (Figure 82)

	Males (10)			Females (9)		
	Mean	Range	SD	Mean	Range	SD
GTL	131.9	127.6–137.2	4.7	124.8	115.8–129.7	4.2
CBL	123.4	116.2–132.8	4.4	117.6	111.0–122.7	3.4
ZB	79.3	76.2–83.6	2.4	74.6	71.0–77.4	2.2
BB	60.9	58.8–63.6	1.7	56.1	55.7–59.5	3.1
IC	24.8	23.2–27.0	1.9	24.3	22.1–27.1	1.5
UT	41.3	38.6–47.2	2.5	39.4	37.8–41.5	1.2
LT	47.5	41.0–50.2	2.6	46.1	44.4–48.0	1.2
M	87.2	83.0–94.7	3.2	82.0	77.7–84.7	2.4

Dental formula is $\frac{3.1.4.1}{3.1.4.2} = 38$.

Karyotype: 2N = 44, based on specimens from Europe (Zima and Kral, 1984).

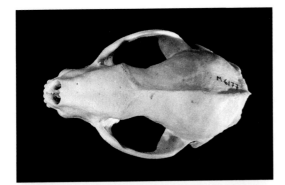

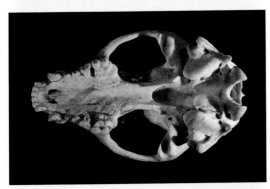

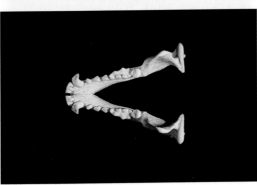

Figure 82: *Meles meles canescens*. Dorsal, ventral and lateral views of the cranium, and a view of the mandible.

5 cm

Figure 83: Distribution map of *Meles meles canescens*.

● – museum records.

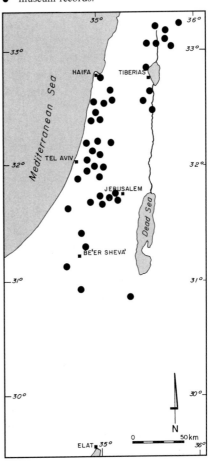

169

in one of the chambers in the burrow system, which is padded by the mother with soft, dry plant material. In Israel, the delayed implantation lasts 9–10 months, as compared to Europe where it lasts about seven months (in Lebanon, 2–4 young are born in March–April; Lewis *et al.*, 1968). The young open their eyes when about a month old, and when two months old they start to wander outside the burrow together with their parents. They are weaned when three months old, but often stay with their parents until the following autumn. Sexual maturity is achieved at two years. Maximum longevity in captivity is 15 years.

Parasites: This species is infected by fleas (*Pulex irritans* and *Echidnophaga popovi*), Mallophaga (*Trichdectes melis*), Diptera Pupipara (*Hippobosca longipennis*), ticks (Ixodidae — *Rhipicephalus sanguineus, Haemaphysalis erinacei, Ixodes kaiseri;* Argasidae — *Ornithodoros tholozani*) (Theodor and Costa, 1967).

Relations with man: In contrast to Britain, Israeli badgers are not considered to be an important vector in spreading disease to domestic animals, neither are they an agricultural pest, but are protected by law like other mammals.

VORMELA Blasius, 1884

Vormela Blasius, 1884. *Berichte naturf. Ges. Bamberg.* 13:9.

Vormela peregusna (Guldenstaedt, 1770)   Marbled Polecat
Figs 84–85; Plate XIII:1   סמור

Mustela peregusna Guldenstaedt, 1770. Novi comment. *Acad. Sci. Imp. Petropol.*, 14(1):441.
 Banks of the River Don, Southern Russia.
Synonymy in Harrison and Bates (1991).

This is the smallest of Israeli carnivores. The body is elongated, with short legs, enabling the animal to penetrate into rodent burrows. The body is covered with a mixture of yellow and grey spots and the throat, legs, underparts and the tip of the tail are black. The head is white with two broad, black, transverse bands, that cover the eyes and the ears, the tips of which are white. With adult reproductive males, the yellow spots turn a beautiful orange colour and the grey spots turn black. Males are significantly larger and heavier than females.

Vormela peregusna syriaca Pocock, 1936 is the subspecies present in Israel (Harrison and Bates, 1991).

Distribution: It occurs in the Palaearctic region from south-east Europe through central Asia to Mongolia and China (Corbet, 1978).

In Israel it occurs throughout the Mediterranean region, north of a line between Gaza and 'Arad (Figure 85). It mainly occupies open habitats with low (*Poterium spinosum*) or medium height (*Calicotome villosa*) bushes as well as fields and orchards.

Fossil record: The earliest remains of this species found in Israel are from the Upper Palaeolithic period (Tchernov, 1988).

Habits: Its social behaviour was studied by Ben David (1988). The size of its home range is about 0.5 km² and the range travelled a night varies from several dozen metres up to a kilometre. Home ranges of several individuals may overlap.

It is mainly nocturnal but individuals, particularly young animals in spring, were seen active during the day. When inactive it hides in rodent burrows (for example of *Meriones tristrami* and *Spalax leucodon*), under rocks and also in other cover (it often uses pipes) and burrows it digs by itself. A hiding place (or burrow) is usually used only once. In some years it is much more common than in others, and this phenomenon seems to be due to fluctuation in food availability (rodent cycles).

Marbled polecats have well developed anal glands whose fetid smelling secretion is used for territorial marking on stones and trees roots. When endangered and unable to flee or hide, the marbled polecat arches its back, lifts its tail and releases a foul-smelling secretion from the anal glands. The smell can be described as extremely awful and presents an effective protection. Together with the secretion the animal emits a loud shriek. The striking coloration, with the black and white head and brown and yellow back probably serves as a warning signal.

When caught it often feigns death, laying motionless in the hand and then jumps suddenly while leaving a strong smell from its anal gland secretion. It screams when in danger.

Food: It is carnivorous, feeding on rodents, reptiles, birds and invertebrates. *Gryllotalpa* is a favourite prey item, particularly during the summer months (Ben David *et al.*, 1991). However, its scientific name reflects the incorrect belief that it eats honey: "voro" — to devour in Latin, "mel" — honey in Latin.

Reproduction: During April the males moult and on their new fur the yellow spots are replaced by orange coloured ones. Mating occurs in April–May. The male holds the female by the nape with its teeth and copulates for about 15 minutes. There appears to be delayed implantation of the ovum, as active pregnancy starts in December–January, and lasts 6–8 weeks. The female gives birth once a year in January–March, so the delayed implantation and pregnancy in this species continue for up to ten months. In captivity, females gave birth also in March and July. Litter size is 1–8. Sex ratio at birth is 1:1. The young are raised in a lined nest in their mother's burrow and only the mother takes care of the young. They start feeding on solid food when about one month old, open their eyes when six weeks old, and start to replace their milk teeth when two months old, at which time they start leaving the maternal nest by themselves. In April–May they disperse, and at this time of the year, the young marbled polecats are also active during the day. Sexual maturity of females is achieved when three months old, but they will only give birth when one year old. Males only become sexually mature when one year old. In captivity they live up to 10 years.

Table 78
Body measurements (g and mm) of *Vormela peregusna syriaca*

| | Body | | | | |
	Mass	Length	Tail	Ear	Foot
			Males (16)		
	n = 14				
Mean	364	295	178	26	41
SD	78	19	18	4	2
Range	280–475	270–330	170–210	20–40	40–45
			Females (9)		
	n = 7				
Mean	250	287	174	25	38
SD	77	42	12	7.2	2.1
Range	160–398	255–390	160–200	14–35	35–40

Table 79
Skull measurements (mm) of *Vormela peregusna syriaca* (Figure 84)

| | Males (10) | | | Females (10) | | |
	Mean	Range	SD	Mean	Range	SD
GTL (F n = 9)	54.2	52.7–55.6	1.0	50.5	48.3–52.0	1.01
CBL	51.7	49.6–53.5	1.1	48.7	46.5–51.1	1.34
ZB	31.0	28.5–33.0	1.6	29.2	28.1–30.3	0.68
BB	29.6	27.8–31.0	0.9	26.5	25.0–28.4	1.11
IC	10.2	9.0–11.3	0.6	10.1	9.4–11.2	0.56
UT (F n = 9)	17.8	16.2–18.7	0.7	16.7	15.9–17.2	0.44
LT (F n = 9)	20.7	19.2–22.0	0.8	19.2	18.3–19.8	0.52
M (F n = 8)	33.0	32.7–34.5	0.8	30.6	29.3–31.8	0.70

Dental formula is $\dfrac{3.1.3.1}{3.1.3.2} = 34$.

Karyotype: 2N = 38, based on two males from Syria (Peshev and Al-Hossein, 1989).

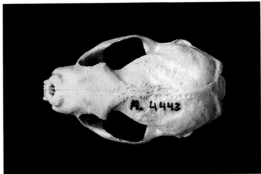

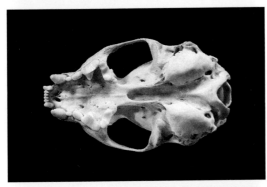

Figure 84: *Vormela peregusna syriaca.*
Dorsal, ventral and lateral views of the cranium, and a view of the mandible.

5 cm

Figure 85: Distribution map of *Vormela peregusna syriaca.*

● – museum records.

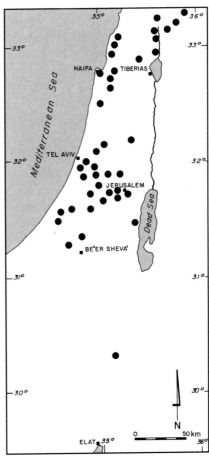

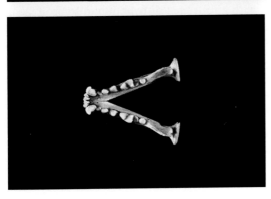

173

Parasites: This species is infected by ticks (Ixodidae — *Rhipicephalus sanguineus* and *Haemaphysalis adleri*) (Theodor and Costa, 1967).

Relations with man: Hand reared marbled polecats make very tame and fascinating pets, that, similar to tame skunks (*Mephitis*) use their anal glands only when extremely stressed.

MARTES Pinel, 1792

Martes Pinel, 1792. *Actes Soc. Hist. Nat. Paris*, 1:55.

Martes foina Erxleben, 1777   Beech Marten, Stone Marten
Figs 86–87; Plate XIII:2   דלק

Martes foina Erxleben, 1777. *Systema regni animales, Classis 1, Mammalia*, p. 458. Germany. Synonymy in Harrison and Bates (1991).

This is a typical mustelid with an elongated, lithe body and short legs with five digits and sharp claws. The general colour is blackish-brown and the white throat patch in this subspecies is divided by a median blackish-brown band. Males tend to be larger and heavier than females.

Martes foina syriaca Nehring, 1902 is the subspecies present in Israel (Harrison and Bates, 1991).

Distribution: Widely spread in the Palaearctic region from Europe throughout Asia to Mongolia and northern India (Corbet, 1978).

In Israel it occurs in the mountainous part of the Mediterranean region from the Galilee south to the Hebron area, with some penetration to semi-desert areas along the Jordan Valley and the Dead Sea Area. One specimen was caught in 'En Gedi, near the Dead Sea, where a suitable habitat exists in the cliffs, trees and dense vegetation of the oasis. It occurs mostly in areas of oak forest (*Quercus calliprinos* and *Pistacia palaestina*), but also in rocky areas where this vegetation has been eliminated (Figure 87).

It was formerly considered to be rare, with a restricted distribution to some areas in the Galilee. It began to expand its distribution about 20 years ago and is now found in all the above-mentioned areas. The extension of its distribution has been documented by specimens run over on roads. As in Europe, in Israel too this species can live in human settlements. People living in a village in Upper Galilee reported stone martens living in the lofts of their houses.

Fossil record: The earliest remains of this species found in Israel are from the Upper Palaeolithic period (Tchernov, 1988).

Habits: The stone marten is nocturnal. It is a very agile climber, equally at home on the ground and on trees. It is a fast runner, typically running a short distance then stopping to survey the surroundings before continuing. It screams when in danger.

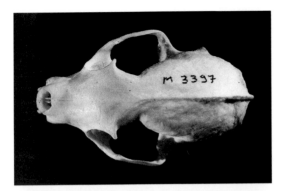

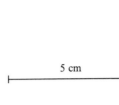

Figure 86: *Martes foina syriaca*. Dorsal, ventral and lateral views of the cranium, and a view of the mandible.

5 cm

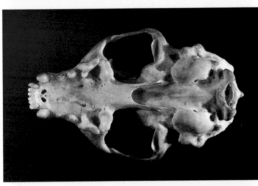

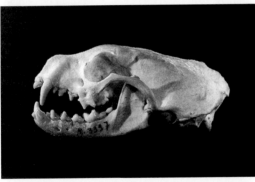

Figure 87: Distribution map of *Martes foina syriaca*.

● – museum records, ○ – verified field records.

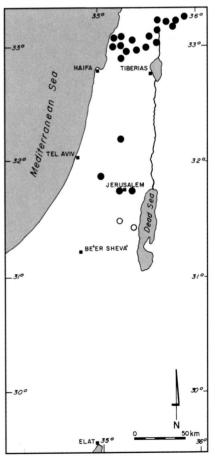

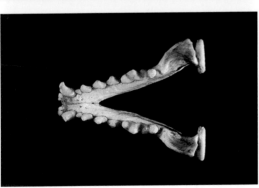

175

Table 80
Body measurements (g and mm) of *Martes foina syriaca*

	Body		Tail	Ear	Foot
	Mass	Length			
Males (10)					
Mean	1283	432	216	37	85
SD	334	39	25	6	7
Range	700–1075	345–496	160–254	28–48	75–92
Females (8)					
Mean	1051	411	237	30	77
SD	85	23	39	4	7
Range	925–1150	390–465	183–296	24–37	68–90

Table 81
Skull measurements (mm) of *Martes foina syriaca* (Figure 86)

	Males (10)			Females (4)		
	Mean	Range	SD	Mean	Range	SD
GTL	84.6	82.0–87.3	1.62	78.7	77.8–79.5	0.75
CBL	78.1	77.0–80.3	1.30	74.5	73.4–75.2	0.75
ZB (F n = 3)	50.1	46.9–52.1	1.65	44.6	43.4–45.3	1.02
BB	36.2	35.0–38.0	0.88	33.9	33.0–34.9	0.80
IC	18.0	16.0–20.1	1.18	17.0	15.8–17.7	0.85
UT (F n = 3)	29.4	27.7–31.2	0.88	27.0	26.6–27.5	0.47
LT (F n = 2)	33.8	32.4–35.2	0.84	31.2	30.7–31.6	0.64
M (F n = 3)	54.2	53.5–56.0	0.72	50.1	49.2–51.5	1.25

Dental formula is $\frac{3.1.4.1}{3.1.4.2} = 38$.

Karyotype: 2N = 38, based on specimens from Europe (Renzoni, 1970).

Food: It is omnivorous, feeding on various small vertebrates, invertebrates and fruits, particularly sweet ones.
Reproduction: No information from Israel.

CANIDAE Jackals, Wolves and Foxes כלביים

This family is represented in Israel by two genera — *Canis* and *Vulpes*.

Dental formula of all Israeli canids is $\frac{3.1.4.2}{3.1.4.3} = 42$.

Key to the Species of Canidae in Israel
(After Harrison and Bates, 1991)

1. Tail relatively short, not reaching the ground when standing erect. Frontal region of skull elevated, swollen with air cells. **2**
- Tail relatively long, reaching the ground when standing erect. Frontal region of skull flattened, not swollen with air cells. **3**
2. Size very large, greatest length of skull more than 180 mm. Upper molars lacking a well-defined cingulum. **Canis lupus**
- Size medium to large, greatest length of skull less than 180 mm. Upper molars with a well-defined cingulum. **Canis aureus**
3. Rear side of ears black, not greatly enlarged. Greatest length of skull more than 117 mm; tympanic bullae relatively small. **Vulpes vulpes**
- Ears without black or dusky tips, relatively large. Greatest length of skull less than 117 mm; tympanic bullae relatively large. **4**
4. Tail tip white. C-M3 is more than 49 mm. **Vulpes rueppellii**
- Tail tip black. C-M3 is less than 48 mm. **Vulpes cana**

CANIS Linnaeus, 1758

Canis Linnaeus, 1758. *Systema Naturae*, 10th ed., 1:38.

This genus is represented in Israel by two species — *C. aureus* (Golden Jackal) and *C. lupus* (Wolf).

Canis aureus Linnaeus, 1758   Asiatic Jackal
Figs 88–89; Plate XIV:1   תן

Canis aureus Linnaeus, 1758. *Systema Naturae*, 10th ed., 1:40. Province of Lar, Persia. Synonymy in Harrison and Bates (1991).

The Asian jackals to which the Israeli population of *Canis aureus syriacus* Hemprich and Ehrenberg, 1833 belongs (Harrison and Bates, 1991), are quite different from the African representatives of this species. They are more clumsily built, their legs, tail, ears and snout are shorter, and specimens in Israel are much smaller

Table 82
Body measurements (g and mm) of *Canis aureus syriacus*

| | Body | | | | |
	Mass	Length	Tail	Ear	Foot
	Males (13)				
	n = 11				
Mean	9218	742	237	82.5	161
SD	2430	102	24	5	11
Range	6100–12980	600–900	200–280	78–90	150–185
	Females (10)				
Mean	9159	759	244	82	161
SD	2007	59	17	7	12
Range	5000–12000	610–820	200–260	70–95	130–170

The measurements shown here are from individuals from the Mediterranean region. Samples of six males and five females from the Mediterranean area were compared to five males and five females from desert areas, and no significant difference in size was found between them. In addition there is no significant difference between males and females. A population of smaller jackals, that may have represented a different subspecies, inhabited the desert oasis of 'Ein Fashkha at the north-west corner of the Dead Sea. Unfortunately this entire population was poisoned before any specimens could be collected.

Table 83
Skull measurements (mm) of *Canis aureus syriacus* (Figure 88)

	Males (11)			Females (10)		
	Mean	Range	SD	Mean	Range	SD
GTL	168.3	160.6–178.5	5.7	161.5	148.9–171.0	6.0
(male n = 8)				(female n = 9)		
CBL	158.9	151.7–167.7	5.7	152.8	143.1–166.5	6.1
ZB	89.6	78.4–95.8	5.5	84.3	77.5–93.8	4.8
BB (m n = 10)	54.7	50.4–56.7	1.9	52.5	50.8–56.7	2.2
IC (m n = 10)	27.1	24.3–33.2	3.0	25.6	22.6–31.0	2.6
UT	72.0	67.3–74.9	2.3	68.0	65.6–72.0	1.8
LT	79.9	74.3–83.7	2.8	76.7	74.5–79.8	2.0
M	126.0	118.2–132.7	5.0	120.6	114.9–128.1	3.8
TB	25.8	23.7–29.2	1.5	25.1	21.0–28.5	2.0

Karyotype: 2N = 74, based on a male, apparently from North Africa (Matthey, 1954).

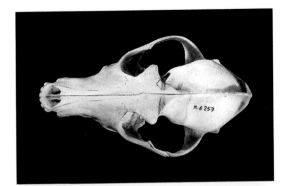

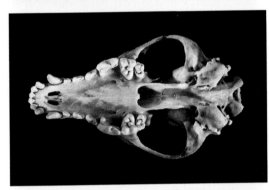

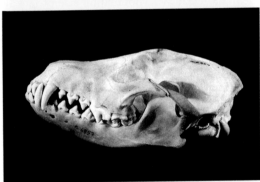

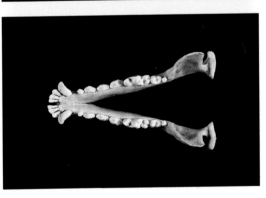

Figure 88: *Canis aureus syriacus*. Dorsal, ventral and lateral views of the cranium, and a view of the mandible.

5 cm

Figure 89: Distribution map of *Canis aureus syriacus*.

● – museum records.

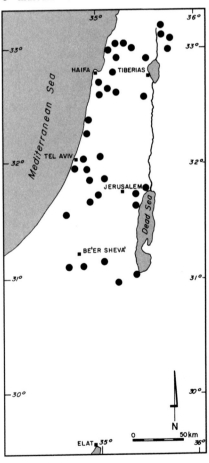

than the Egyptian subspecies *C. a. lupaster*. The colour is brownish-grey, the tail tip is black, and the lower part of the sides, the legs, ears and snout are yellowish-brown. In some specimens this "golden" colour extends higher up on the sides. This is reflected in the scientific name of the species; "aurum", meaning gold in Latin. In winter a dense, woollen underfur develops.

Distribution: This species occurs from south-eastern Europe through Turkey, Syria and Lebanon and central Asia to India and Indo-China, as well as in Africa, from Kenya north to Egypt and west to Senegal (Corbet, 1978).

In Israel it is very common in the Mediterranean region and penetrates into the northern and central Negev (Sede Boqer area), as well as along the Dead Sea and the northern 'Arava Valley, where it occurs mainly in and near oases and agricultural settlements (Figure 89). Jackals are not desert animals and in desert areas they occur only near human settlements.

Habits: Their social behaviour was studied by Golani and Keller (1974) and Golani and Mendelssohn (1971). Jackals are mainly nocturnal, but also crepuscular. They spend the day in thickets, burrows and under other cover. They are social animals, living in pairs during the breeding season and in family groups composed of parents and their offspring during the rest of the year. One year old animals may assist their parents in feeding the younger siblings. Jackals often spend many hours in allogrooming. Home ranges are marked with urine, in a fashion similar to dogs. Each pair keeps to its own home range, but young animals may transverse several territories. In 'En Fashkha, an oasis near the Dead Sea, population size of jackals during the 1970s increased due to high food availability at a feeding station run by the Nature Reserves Authority and food remains left by tourists. In this area jackals lived in two territorial groups numbering 10 and 20 individuals, respectively. Territorial limits were delineated by faeces arranged in piles or middens. In one of the groups several females gave birth and raised young simultaneously (Macdonald, 1979). This population was later exterminated by a poisoning campaign.

Jackals have a rich repertoire of vocal, facial and body expressions. Choruses of jackals are well known in Israel and other places in the Near East, and constitute one of the most characteristic night calls in the areas where they occur (Lewis *et al.*, 1967). A chorus starts with one individual calling, and within minutes many others of its own family and other families join in. The typical call has several components: cries, howls and trills. The call of young jackals is normally a short wailing. The jackal also uses barks (when attacked by other jackals), groans (during fights), and hums (during mating).

This species is commensal with man, and near certain human settlements in Israel the typical population density is one family (of up to six individuals) per km^2.

Food: It is omnivorous, feeding on small to medium size vertebrates (hares, rodents, birds, reptiles), carrion and vegetable matter and also catches fish in shallow water. Near human settlements it relies to a large extent on garbage. When hunting it relies primarily on hearing and sight, and uses smell only when close to its prey. In pursuit, if the prey disappears from sight the jackal will jump high on its hind legs while running.

Reproduction: Courtship behaviour starts in October and lasts until February, when copulation takes place. Mating is stereotypic: a bachelor male will appear in an area and bark, and several females may gather near it. When only one female remains, the male and female run towards and circle one another. They travel together, urinate in the same places and sniff one another's urine. During this period the female urinates like a male, i.e. while lifting a hind leg. After urinating the two scrape the ground with all four legs, and it is conceivable that while doing so they also leave a secretion from glands on the soles. During the next stage the male licks the oestrous female's vulva. Following copulation, the female presses its mouth to the side of the male's mouth, and the male regurgitates food for her. Jackals probably mate for life. Gestation last 60–63 days, and the young are born in spring (April–May). Litter size is 3–8 (in the Tel Aviv University Research Zoo), and delivery occurs in a burrow several metres long dug by both parents, but mostly by the female. The female remains with the pups for several days, living on and producing milk from the fat reserves accumulated earlier. The pups open their eyes when 10–14 days old, and when three weeks old they also start to feed on solid food which the parents regurgitate for them. They are weaned when three months old, and at that time the young start to wander outside the natal burrow and to hunt by themselves. They generally reach sexual maturity when two years old, but some do so when only one year old. In captivity they may live 16 years, but most become senile when 10–12 years old.

Parasites: This species is infected by fleas (*Pulex irritans, Ctenocephalides canis* and *C. felis*), and ticks (Ixodidae — *Rhipicephalus sanguineus, Haemaphysalis adleri, H. otophila* and *Ixodes redikorzevi*) (Theodor and Costa, 1967).

Relations with man: During this century, the population size of jackals in Israel has depended to a large extent on human activity. A poisoning campaign carried out by the Plant Protection Department of the Ministry of Agriculture in 1964 caused an almost complete extermination of the jackal in Israel (Mendelssohn, 1972). The rationale for this campaign was the mostly imaginary damage caused by jackals to plastic covers of various crops. This population reduction was followed by a population explosion of rodents, hares and chukar partridges, causing far greater damage to agriculture than that caused by jackals. The rise in human standard of living, accompanied by an increase in garbage dumps with many food remains, and changes in agricultural practices, including the common practice of disposing carcasses of turkeys and chickens on garbage dumps, enabled a recovery in the population size of jackals everywhere during the 1980s and 1990s. In the Golan, food availability due to farmers disposing of carcasses caused a population density increase from 0.2 jackals/km^2 during the 1970s (Ilani, 1979) to 2.5 jackals/km^2 in 1985 (a total of about 4000 individuals; Frankenberg and Pevzner, 1988). In this region jackals cause considerable damage by attacking calves, mostly during parturition: about 1.5–1.9% of calves born are preyed upon, mostly by jackals, causing an annual damage of about US$ 42,000. Most (75.5%) jackal attacks on calves occur during the first two days after birth. The face and tongues of calves attacked during parturition are eaten while still not completely born, but death is mainly caused by opening of the posterior part of the calves' abdomen after delivery. In some cases the mother's vaginal area is also damaged, and several cows had to be destroyed because of serious wounds of this kind (Yom-Tov *et al.*, 1995). Jackals also cause damage by feeding on melons, grapes, watermelons and other crops, and by being a vector for rabies. However, they are also beneficial to agriculture by feeding on voles, hares, young porcupines and other agricultural pests. During a vole mass increase in the Golan in 1985, jackals fed extensively on voles, as evidenced by their scats being composed solely of vole hairs.

There is a controversy over whether the jackal is the fox mentioned in the Bible. The story of Samson and the 300 foxes (Judges 15) makes it somewhat likely that the Biblical fox is indeed the jackal. However, that jackals have shorter tails than foxes, which would make it difficult to tie them together, is seen as evidence for the fox being the animal used by Samson.

Canis lupus Linnaeus, 1758   Wolf
Figs 90–92; Plate XIII:3–4   זאב

Canis lupus Linnaeus, 1758. *Systema Naturae*, 10th ed., 1:39. Sweden.
Synonymy in Harrison and Bates (1991).

This species is highly polymorphic in size and pelage. In Israel size variation is very large, and wolves from the Golan are much heavier (one adult male had a weight of 32.3 kg) than those in the 'Arava (Males there weigh 18–22 kg, and females 16–18 kg; R. Hefner, pers. comm.), and larger (greatest length of the skull is 8.7% and 13.4% larger in males and females, respectively, in the Golan and north of Israel in comparison with the specimens from the 'Arava (Figure 90) with a gradient from north to south in accordance with Bergmann's rule (Figure 2, see p. 34). Northern wolves are darker than those of the Negev and 'Arava. However, the body extremities (tail, ears and feet) of northern specimens are not proportionally shorter than those from the 'Arava, contrary to what is expected from Allen's rule, and are 38% (tail), 11% (ears) and 22% (feet) of body length in both populations.

The wolves from central and south-central Israel are identical in size and coloration with *C. l. pallipes* Sykes, 1831 from India; wolves from the extreme south of Israel (Yotvata and Elat) are, according to size, *C. l. arabs* Pocock, 1934, but in coloration they are similar to *C. l. pallipes* and differ from a specimen of *C. l. arabs* from southern Saudi Arabia. A very small female from southern Sinai, according to size undoubtedly a *C. l. arabs*, also had the *pallipes* coloration. Specimens from the Golan are not only larger and darker than *C. l. pallipes*, but their skull also has different proportions. Specimens from the Galilee (now extinct) are intermediate between the Golan wolf and *C. l. pallipes*.

182

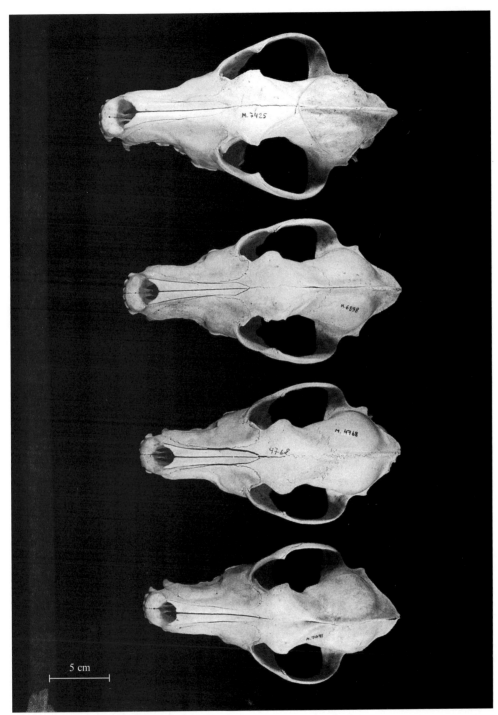

Figure 90: Skulls of male wolves (*Canis lupus*) from the Golan Heights (largest), ʿEin Fashkha, Sede Boqer and ʿEn Yahav (ʿArava Valley, smallest).

Table 84
Body measurements (g and mm) of *Canis lupus*
from the Golan and northern Israel

	Body		Tail	Ear	Foot
	Mass	Length			
Males (7)					
	n = 6			**n = 6**	
Mean	27058	1072	402	122	240
SD	5732	83	15	14	13
Range	17500–32300	995–1200	390–430	105–125	220–255
Females (1)					
Mean	24,200	970	370	120	220

Table 85
Body measurements (g and mm) of *Canis lupus pallipes*
from the 'Arava (Figure 91)

	Body		Tail	Ear	Foot
	Mass	Length			
Males (9)					
Mean	20407	992	345	110	219
SD	3927	91	30	11	14
Range	16200–28320	845–1130	300–395	92–124	195–246
Females (8)					
Mean	16447	997	306	107	205
SD	4631	119	20	10	12
Range	14000–21200	760–1120	270–332	90–120	190–230

Distribution: This species was widely distributed in the Holarctic region, in North America south to 20°N, all over Europe and most of Asia, but its occurrence in much of its former range, particularly in Europe and North America is currently restricted (Corbet, 1978).

In Israel it occurred throughout the country, but in the Mediterranean region its populations suffer from continuous persecution by man in the form of poisoning and shooting, and this population has become greatly reduced. Only a few dozen individuals survive in the Mediterranean region, mainly in the Golan and the Judean Hills. However, in the Negev, the Dead Sea Area and the Rift Valley there are still viable populations of this species (Figure 92). Some of these wolves were shot by Bedouin after crossing the Jordanian border, as happened to four radio-collared wolves studied by R. Hefner in the southern 'Arava. We estimate that currently there are about a hundred wolves in the desert areas, and less than 50 in the Mediterranean region of Israel.

Table 86
Skull measurements (mm) of *Canis lupus* from the Golan and northern Israel

| | Males (9) | | | Females (2) | |
	Mean	Range	SD	Mean	Range
GTL	238.8	226.5–244.6	6.2	234.4	230.3–238.4
CBL	219.9	212.5–224.7	4.5	217.7	211.2–224.1
ZB	123.9	115.1–134.8	6.4	123.1	116.5–129.6
BB	67.9	64.4–70.6	2.3	67.95	66.0–69.9
IC	41.2	36.7–44.5	2.7	40.7	38.0–43.3
UT	98.1	91.5–101.9	3.5	97.2	94.5–99.0
LT	111.5	107.1–115.3	2.5	109.4	107.8–111.0
M	177.3	171.6–181.8	3.4	176.2	171.3–181.1
TB	29.6	25.5–31.4	1.8	29.6	28.1–31.0

Table 87
Skull measurements (mm) of *Canis lupus pallipes* from the 'Arava

| | Males (12) | | | Females (5) | | |
	Mean	Range	SD	Mean	Range	SD
GTL	219.6	206.3–242.5	10.1	206.6	193.4–214.1	8.4
CBL	203.2	192.9–225.3	8.8	189.5	179.8–194.6	6.9
ZB	113.1	103.9–122.7	6.4	105.4	98.4–108.3	4.3
BB	64.7	58.9–71.6	3.2	62.2	59.1–64.1	1.9
IC	40.4	35.9–48.9	3.8	37.1	33.9–39.2	2.3
UT	91.7	86.4–100.1	3.8	86.9	82.0–90.9	3.9
LT	103.0	96.9–110.8	3.8	98.7	93.2–100.8	3.2
M	163.3	146.4–180.5	8.8	154.5	146.8–160.7	5.2
TB	27.7	24.2–30.9	1.8	27.0	24.7–28.4	1.4

Karyotype: 2N = 78, FN = 76 (Zima and Kral, 1984)

Fossil record: The earliest remains of this species found in Israel date to about 150,000 years BCE (Tchernov, 1988).

Habits: This species was studied in the southern 'Arava by R. Hefner, a ranger of the Nature Reserve Authority. The wolf is mainly nocturnal, but is often seen active during the day, particularly during winter and in areas where it is not persecuted by man. Wolves in Israel can be solitary, or live in pairs or small family groups numbering 3– 6, but sometimes up to 12 wolves are seen together. Such large groups are seen mainly during the end of summer and autumn, when young forage with their parents. They travel within a home range which they mark with urine and by defecating in prominent places. A female marked with a radio collar near Sede Boqer in the central Negev had a home range of 23 km^2 in summer and 60 km^2 in winter, and travelled a maximum of 10.5 km per day (Afik and Alkon, 1983). During the seven months of the study this female was accompanied by an adult male. In the southern 'Arava wolves travel on average 13 km per night but may walk up to 40 km per night. Mean walking speed is 1.3 km/hour. Some individuals in the 'Arava study travelled up to 150 km north of the place where they were first caught. The average size of home range was 12–42 km^2, depending on the way it was calculated (R. Hefner, pers. comm.).

Wolves have a rich repertoire of vocal, facial and body posture expressions.

Food: Wolves are mainly carnivorous, hunting gazelles, hares, rodents, partridges and other vertebrates. They also feed near human settlements, scavenge on garbage and on carcasses of domestic animals, as well as sweet fruits, such as dates. In the 'Arava they sometimes specialize in hunting feral domestic cats in kibbutzim (R. Hefner, pers. comm.). Despite wild pigs being very common in Israel, preying on this species has not been recorded. They often pick up animals that have been run over on roads, and are then often hit themselves by cars. Being killed on roads seems to be the most important mortality factor for wolves in Israel.

Reproduction: In Israel wolves mate between December and February, in the south earlier than in the north. Gestation lasts 63 days. In northern Israel young are born at the end of April, in the Negev in March and the beginning of April, and in the 'Arava several weeks earlier. Litter size is 2–6. Both parents dig a burrow where the young are born. One burrow near Ḥazeva had a tunnel about 60 cm in diameter and was eight metres long with a 1.5 by one metre room at its end, where four young were found. The pups open their eyes when 10–12 days old, and when a month old they also start to feed on solid food brought by the parents and regurgitated for them. During the first weeks of their life, the fur of the pups is soft, woolly and dark grey. When about two months old they leave the maternal burrow and wait outside for their parents to bring food. When three months old the young start to forage together with their parents, by eight months old they attain adult size, and remain with their parents until about one year old. If they remain with the family after the birth of a subsequent litter, they participate in food delivery to the new young. In the 'Arava, some young were observed with their parents until 18 months old. In one case two females together raised 12 cubs (there were six very small and six larger cubs) in one burrow without any apparent animosity between the females (R. Hefner, pers. comm.). Wolves reach sexual maturity when 2–3 years old. In captivity they live up to 16 years.

Parasites: This species is infected by ticks (Ixodidae — *Rhipicephalus sanguineus* and *Haemaphysalis otophila*) (Theodor and Costa, 1967).

Relations with man: The conflict between wolves and man continues to this day, in spite of the reduced populations of the former. In the Mediterranean region their populations suffer from continuous persecution by man in the form of poisoning and shooting, and these populations have become greatly reduced and only a few dozen individuals now survive. In the Golan wolves sometimes attack cattle, particularly calves and in the 'Arava they prey occasionally upon calves. However, some wolves in the 'Arava are very accustomed to humans and can be approached to a

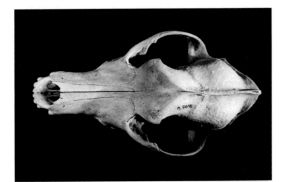

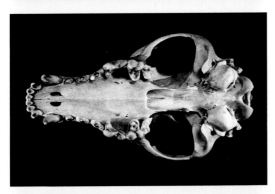

Figure 91: *Canis lupus*. Dorsal, ventral and lateral views of the cranium, and a view of the mandible.

5 cm

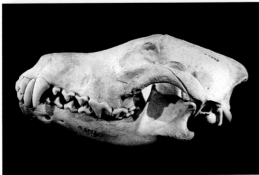

Figure 92: Distribution map of *Canis lupus*. ● – museum records.

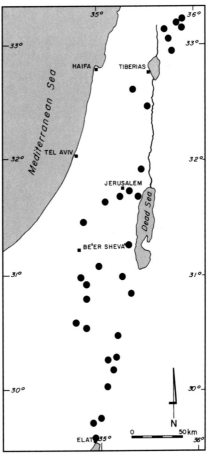

187

distance of 20–50 m. In the 'Arava and the Negev wolves are scavengers at garbage dumps, and the existence of these dumps near every army camp and many agricultural settlements has enabled their continued existence.

The wolf is the ancestor of the dog, and to this day they interbreed. Cases of mating of feral or domestic dogs with wolves are known from the Golan (during the early 1990s) and Yizre'el Valley (in 1963/4). It is not easy to discriminate between some dogs and wolves, but in general wolves have longer legs and larger paws (particularly the front legs), their thorax is narrower and the abdomen is seen as a continuation of the thorax, while in the dog it is contracted. Wolves carry their tail straight downwards, while with hybrids the distal part is turned upwards. The tympanic bullae and the teeth, particularly the carnassials, are larger in wolves. The area of the mouth is white or light grey in wolves, whereas this area is black in most dogs.

VULPES Frisch, 1775

Vulpes Frisch, 1775. *Natursystem der Vierfussingen Thiere*, p. 15.

These are medium-sized or small canids with long tail and ears. A distinctive mid-dorsal dark band extends from the nape of the neck caudally, becoming a mid-dorsal crest throughout the length of the tail. A distinctive dorsal spot is present at one third from the base of the tail, indicating the position of the caudal gland. This genus is represented in Israel by three species — *V. vulpes* which occurs throughout the country, *V. rueppellii* which inhabits sandy areas and wide wadis in the desert, and *V. cana* which inhabits steep, rocky cliffs in the desert (Figure 93).

Vulpes vulpes (Linnaeus, 1758) Red Fox
Figs 93–95; Plate XIV:2–4 שועל מצוי

Canis vulpes Linnaeus, 1758. *Systema Naturae*, 10th ed., 1:40. Sweden.
Synonymy in Harrison and Bates (1991).

This species is very polymorphic in its body size and pelage colour, and many subspecies have been described. In Israel, the fox pelage corresponds to the description of two subspecies: *V. v. arabica* Thomas, 1902 and *V. v. palaestina* Thomas, 1920. *V. v. palaestina* has a brown-reddish-yellow back, grey to dark grey flanks and dark grey belly and breast. *V.v. arabica* has a yellow back, light-grey or white flanks, and light grey or white underparts. Summer pelage is much thinner without much of the fine, woolly underfur, and of lighter colour (Lewis *et al.*, 1967). The *arabica* type foxes are mostly found in the desert area, but occasionally specimens have also been found on Mount Carmel and in other areas of central and even northern Israel, where the common type is the *palaestina* form, but specimens of this type are also occasionally found in the 'Arava.

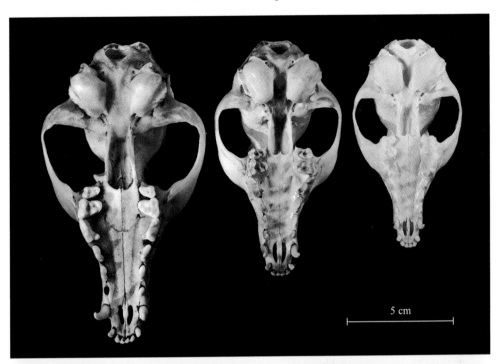

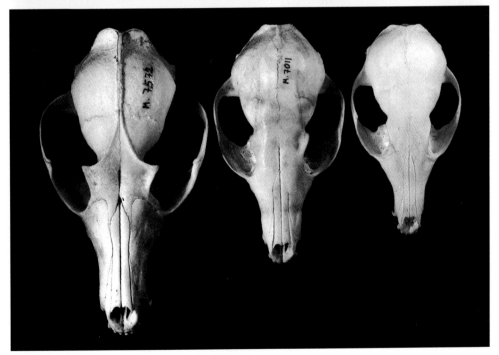

Figure 93: Skulls of *Vulpes vulpes palaestina* (largest), *V. rueppellii* and *V. cana* (smallest) from Israel.

Table 88
Body measurements (g and mm) of *Vulpes vulpes*

| | Body | | Tail | Ear | Foot |
	Mass	Length			
	Males (25)				
	n = 24				
Mean	3100	586	376	96	138
SD	600	34	36	12	11
Range	2150–4150	520–660	293–442	80–140	140–155
	Females (15)				
	n = 13			n = 14	n = 14
Mean	3047	572	373	91.5	130
SD	437	29	27	6	8
Range	2500–4000	530–620	330–432	79–100	114–140

In the 'Arava, mean body mass of males and females was 3557 and 3069 g, respectively (Assa, 1990).

Table 89
Skull measurements (mm) of *Vulpes vulpes* (Figure 94)

| | Males (12) | | | Females (11) | | |
	Mean	Range	SD	Mean	Range	SD
GTL	132.7	123.0–144.0	7.1	124.3	117.4–129.4	38.0
CBL	124.6	115.0–136.3	1.3	116.7	111.0–122.6	3.8
ZB	70.1	63.0–76.5	3.8	65.6	61.5–68.8	2.2
BB	45.1	41.9–47.5	1.6	44.4	42.6–47.2	1.2
IC	24.9	22.0–29.3	2.3	22.9	19.3–24.7	1.8
UT	59.0	53.9–65.0	3.7	55.6	52.6–57.6	1.6
LT	66.0	59.7–72.9	4.1	61.6	59.4–64.5	1.6
M	100.1	90.0–109.6	6.3	93.6	89.3–99.4	2.9
TB	22.8	22.0–24.0	0.8	23.6	22.7–25.0	0.7

Karyotype: 2N ranges from 34 to 42 in Europe (Zima and Kral, 1984).

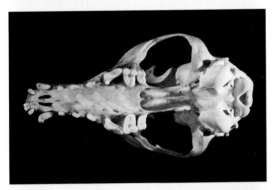

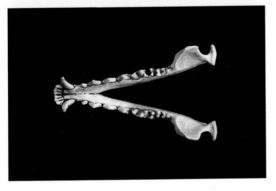

Figure 94: *Vulpes vulpes*. Dorsal, ventral and lateral views of the cranium, and a view of the mandible.

5 cm

Figure 95: Distribution map of *Vulpes vulpes*.
● – museum records.

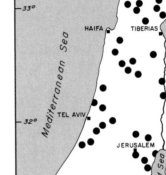

Distribution: This species is widely distributed in the Holarctic region in North America, Europe and most of Asia (Corbet, 1978). It is probably the most common carnivore in the Middle East (Lewis *et al.*, 1967), including Israel.

In Israel it is abundant throughout the country, where its population has increased due to food availability in garbage dumps (Figure 95).

Fossil record: The earliest remains of this species found in Israel date to about 250,000 years BCE (Tchernov, 1988).

Habits: It is mainly nocturnal, but at times is known to be active during the day. Its social behaviour was studied in the northern 'Arava by Assa (1990). In this area they are active from about half an hour after sunset and for the whole night, for 10–13 hours during winter. In the summer they are also crepuscular, probably due to the need to feed the young. There is no correlation between the moon phase and time of activity.

The fox spends its inactive periods in burrows dug by itself or in those of other animals, such as porcupines. In the 'Arava the burrow openings frequently face north, apparently to avoid overheating. Each individual has up to 10 burrows in its territory, and it frequently moves from one to another. During summer foxes often hide under bushes during the day.

It is territorial, marking its territory with faeces, with a strongly odorous excretion of the anal glands and with urine. Territory size depends on food availability. In the 'Arava, territory size of males ranged from 7–15 km^2, and that of a female was 6.5 km^2. The foxes travelled 10–20 km nightly, but most of their activity was carried out in a small area where food was abundant, such as a garbage dump. In an area of 80 km^2 in the 'Arava, eight adults and eight young were caught, but other foxes also resided there.

Foxes have a great variety of calls, including a coarse bark which males often utter during the mating season, but they lack the typical howling of the jackals and wolves.

Food: It is omnivorous, feeding on small (rodents, birds, reptiles) to medium sized (hares) vertebrates, invertebrates and succulent fruit, such as grapes and figs. Insects (beetles, grasshoppers etc.) are often an important component of their diet. In the Northern Negev (Revivim) the stomachs of several foxes contained large numbers of a large, subterranean cricket (*Lucinia*). In Ras Muhammad, southern Sinai, foxes were seen to dig up and eat eggs of sea turtles. They are frequently seen travelling along roads, apparently seeking carcasses of dead, run-over animals. They frequently scavenge on garbage dumps.

Remains of gazelle fawns have quite often been found near burrows with pups, in the 'Arava as well as in the Golan. It seems that foxes, when hard pressed to provide food for their growing cubs, will prey on this relatively large animal (1.5–2 kg).

Reproduction: In the 'Arava oestrus occurs in December and in northern Israel in February. Gestation lasts 51–56 days, and young are born between February (in the 'Arava) and April (in northern Israel). In the 'Arava delivery occurs in a burrow system with several openings. Before giving birth the female lines the nesting burrow with soft hair it plucks from its belly. Litter size ranges from 2–5 (mostly three). The parents move the young from one burrow system to another every 11–12 days, apparently in order to avoid high parasite load, or when disturbed by potential predators (i.e. humans). The young open their eyes when about two weeks old, start eating solid food when a month old, and by that time they start leaving the burrow to play near its opening. At this stage their fur is dark grey, and the black colour of the outer side of the ear appears when four months old. They are weaned when three months old. Both parents take care of the young and bring them food carried in the mouth. The young leave their parents when five months old, attain adult size when six months old and become sexually mature when 10 months old. In captivity maximum longevity was 12 years.

Parasites: This species is infected by fleas (*Pulex irritans* and *Ctenocephalides canis*), Mallophaga (*Felicola vulpis*), Diptera Pupipara (*Hippobosca longipenis*), ticks (Ixodidae — *Rhipicephalus sanguineus*, *Haemaphysalis adleri*, *H. otophila* and *Ixodes kaiseri*) (Theodor and Costa, 1967).

Relations with man: In Israel this species is the main vector of rabies among wildlife. They also cause some damage to agriculture by chewing at plastic irrigation pipes and eating grapes and other crops. They make delightful, though very smelly (because of the secretion of the anal glands) pets, recognizing their owners after several years of separation.

Vulpes rueppellii Schinz, 1825
Figs 96–97; Plate XV:1–2

Rueppell's Sand Fox
שועל החולות

Vulpes rueppelii (sic) Schinz, 1825. In: Cuvier, *Thierreich*, 4:508. Dongola, Sudan.
Synonymy in Harrison and Bates (1991).

The coloration is light brownish-yellow with some light grey on the sides. The tip of the tail is white. A dark brown band extends from the eye to the mouth. Specimens from Israel are more similar to the Egyptian *Vulpes rueppellii rueppellii* Schinz, 1825 than to the Arabian *V.v. sabaea* Pocock, 1934. Unlike the red fox, the back of the ear is not black, but yellow. The ears are relatively much longer than those of the red fox.

Distribution: This species occurs in the Saharo-Sindian deserts from Morocco to Pakistan (Corbet, 1978).

In Israel it occurs in wide wadis and on sand dunes in the Negev and the 'Arava (Figure 97), south and east of the 100 mm isohyet, as far north as Jericho. The recent population reduction in the 'Arava appears to be due to competition with the red fox, whose populations have increased and invaded the habitat of the sand fox.

Habits: The sand fox has been studied in Oman by Lindsay and Macdonald (1986). It is a monogamous, mainly nocturnal animal. When inactive it lives in burrows, with 1–5 openings. Its home range averages 30.4 km^2.

Table 90
Body measurements (g and mm) of *Vulpes rueppellii*

| | | Body | | | |
	Mass	Length	Tail	Ear	Foot
		Males (4)			
	n = 1		**n = 3**		**n = 3**
Mean	956	409	333	79	79
SD		33	11	6	22
Range		365–438	324–345	70–85	54–92
		Females (2)			
	n = 1	**n = 1**	**n = 1**	**n = 1**	
Mean	962	362	302	84	83
Range	725–1200				

Table 91
Skull measurements (mm) of *Vulpes rueppellii* (Figure 96)

| | Males (2) | | Females (3) | |
	Mean	Range	Mean	Range
GTL	115.5	114.4–116.6	101.5	99.5–103.8
CBL	109.9	108.4–111.5	97.3	94.8–99.2
ZB	60.5	60.1–61.0	52.0	51.3–52.9
BB	41.0	40.9–41.2	37.5	36.1–40.1
IC	18.6	18.6–18.7	18.3	17.4–19.3
UT	52.2	52.1–52.2	45.0	42.4–47.2
LT	58.3	57.6–59.0	48.3	46.5–49.4
M	85.6	84.7–86.5	74.9	73.7–76.2
TB	22.8	22.6–23.0	22.0	20.1–24.8

Karyotype: 2N = 40, based on a male, apparently from Algeria (Matthey, 1954).

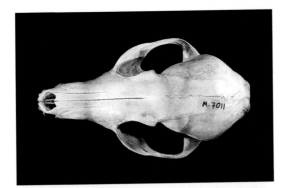

Figure 96: *Vulpes rueppellii.* Dorsal, ventral and lateral views of the cranium, and a view of the mandible.

5 cm

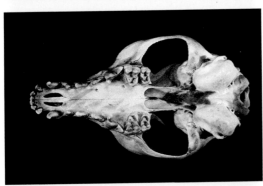

Figure 97: Distribution map of *Vulpes rueppellii.*

● – museum recordss.

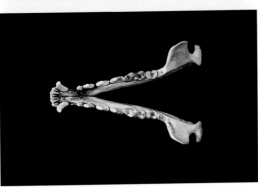

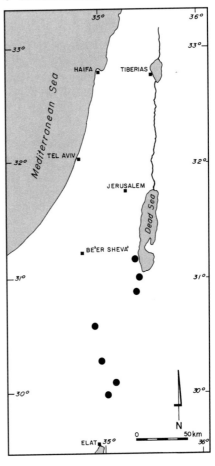

195

These foxes have a great variety of calls, including a coarse bark which males often make during the mating season, but they lack the typical howl of the jackals and wolves.

Food: It is mainly carnivorous, feeding on rodents, reptiles, many insects and birds, but also on fruit and vegetables.

Reproduction: Mating occurs in February. Gestation lasts 56 days and the young are born in April. Litter size is 1–4. They open their eyes when 12 days old, and are weaned when two months old. At this age their fur is dark-grey. Both parents care for the young, which are raised in a burrow dug by their parents. The young become independent when five months old, and reach sexual maturity at 10 months. In captivity maximum observed longevity is 12 years.

Parasites: This species is infected by fleas (*Synosternus pallidus*) and ticks (Ixodidae — *Rhipicephalus sanguineus*) (Theodor and Costa, 1967).

Vulpes cana (Blanford, 1877)   Blanford's Fox
Figs 98–99; Plate XV:3   שועל צוקים

Vulpes canus Blanford, 1877. *Journal Asiat. Soc. Bengal*, 46(2):321. Gwadar, Baluchistan.

The following description is from Geffen (1994). Blanford's fox is a small fox with a long, bushy tail that almost equals its body in length and breadth. The pads of the feet and digits are small and hairless and the claws are curved, sharp and semi-retractile. The body is brownish-grey ("can" — means grey in Latin), fading to pale yellow on the belly. The winter coat is soft and woolly with a dense, black underwool. Its dorsal region is sprinkled with white-tipped hair. The summer coat is less dense, the fur is paler coloured, and the white-tipped hairs are less apparent. The tail is similar in colour to the body. The tail tip is usually black, although in some individuals it is white. The forefeet and hindfeet are dorsally pale yellowish-white, while posteriorly they are dark grey. Unlike the other fox species in the Arabian deserts, the blackish pads of the feet and digits are hairless and the claws are cat-like, curved, sharp, and semi-retractile. The head is orange buff in colour, especially in the winter coat. The face is slender with a dark band extending from the upper part of the sharply pointed muzzle to the internal angle of the eyes. The ears are pale brown on both sides with long white hairs along the antero-medial border. Sexual dimorphism in this species is minimal (up to 6%, depending on character).

Distribution: It occurs in Afghanistan, northwestern Pakistan, southern Turkmenia, Oman, United Arab Emirates, Saudi-Arabia, Sinai, Jordan, eastern Egypt and Israel (Chilcott *et al.*, 1995, Harrison and Bates, 1991, Peters and Rodel, 1994).

In Israel it occurs in the mountainous ranges of the Negev and Judean Deserts, where it inhabits steep, rocky slopes, canyons, and cliffs, in the driest and hottest regions of the country (Figure 99).

Habits: Daily energy expenditure of free-ranging Blanford's foxes near the Dead Sea was studied by Geffen *et al.* (1992d). They concluded that foxes maintained their water and energy balances on a diet of invertebrates and fruits without drinking. In Israel they consume more fruit during the hot summer. Their metabolism during activity was 8.4 times higher than at rest.

In Israel, Blanford's fox appears to be fairly common in the Judean and Negev Deserts. Their density in 'En Gedi was estimated at two individuals/km^2 and at Elat it was 0.5 individuals/km^2 (Geffen, 1990).

In Israel these foxes are strictly nocturnal. Their onset of activity is governed largely by light conditions, and closely follows sunset. Foxes were active about 8–9 h/night, independent of the duration of darkness or normal climatic conditions. Average distance travelled per night was 9.3 km, and size of nightly home range averaged 1.1 km^2 (Geffen and Macdonald, 1992). Dry creek beds were the most frequently visited habitats in all home ranges.

Blanford's foxes are almost always solitary foragers, only occasionally foraging in pairs. Three types of foraging behaviour were observed: 1) unhurried movements back and forth between rocks in a small area (0.01–0.03 km^2), accompanied by sniffing and looking under large stones and occasionally digging a shallow scrape; 2) standing near a bush for a few seconds, alert with ears erect, prior to circling the bush or pouncing upon prey within, and then walking to another bush to repeat the sequence; and 3) short, fast sprint after small terrestrial or low-flying prey (Geffen *et al.*, 1992b).

They are organized in strictly monogamous pairs in territories of about 1.6 km^2 which overlap minimally (Geffen and Macdonald, 1992). Three out of five territories contained one, non-breeding yearling female during the mating season, but there was no evidence of polygyny (Geffen and Macdonald, 1992).

Dens used by these foxes in Israel were usually on a mountain slope and consisted of large rock and boulder piles or screes. They appeared to use only available natural cavities, and never dug burrows. Dens were used both for rearing young during spring and for day-time rest throughout the year. During winter and spring, both members of a pair frequently occupied the same den, or adjacent dens at the same site, while during summer and autumn they often denned in separate locations. Frequent changes in location of den from day to day were more common in summer and autumn (Geffen and Macdonald, 1992).

Its very bushy and long tail is probably an important counter balance during climbing. Among canids, Blanford's foxes have an astonishing jumping ability; captive individuals bounced from one wall to another or jumped to the highest ledges (2–3 m) in their cage with remarkable ease, and did so as part of their routine movements (Mendelssohn *et al.*, 1987). In the field, these foxes were observed climbing vertical, crumbling cliffs by a series of jumps up the vertical sections (Geffen *et al.*, 1992a). Their naked pads and sharp, curved claws enhance traction on vertical ascents (Mendelssohn *et al.*, 1987; Geffen *et al.*, 1992a).

Food: Blanford's foxes are insectivorous and frugivorous, feeding mostly on beetles, grasshoppers, ants and termites (Geffen *et al.*, 1992b). Plant foods consisted mainly of fruit of two caperbush species, *Capparis cartilaginea* and *Capparis spinosa*, fruits

Table 92
Body measurements (g and mm) of *Vulpes cana*

	Mass	Length	Tail	Ear	Foot
		Body			
		Males (3)			
	n = 1		**n = 2**		
Mean	956	424	334.5	77	73
Range		403–438	324–345	70–81	54–92
		Females (1)			
	n = 2				
Mean	962.5	362	302	84	83
Range	725–1200				

Table 93
Skull measurements (mm) of *Vulpes cana* (Figure 98)

	Males (5)			Females (3)		
	Mean	Range	SD	Mean	Range	
GTL	95.9	92.4–98.4	2.5	93.9	91.5–96.1	
CBL	91.7	82.7–95.3	5.3	89.1	83.5–93.0	
ZB	50.8	48.8–52.6	1.4	49.6	47.8–51.4	
BB	34.7	33.4–36.9	1.3	34.6	33.8–35.1	
IC	17.3	16.5–18.6	0.9	17.1	16.8–17.7	
UT	41.7	41.0–42.8	0.7	41.3	40.7–42.0	
LT	45.5	44.0–46.7	1.0	43.9	43.2–44.5	n = 2
M	71.5	67.5–73.3	2.4	69.5	69.3–69.7	n = 2

Data from a much larger sample (Geffen *et al.*, 1992b) show that specimens from 'En Gedi are about 12% heavier than those from Elat. This difference may be due to a difference in food availability. Mean body mass of males was 1.18 and 1.05 kg for 'En Gedi and Elat, respectively, and 1.11 and 0.99 kg for females from these localities.

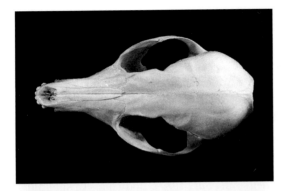

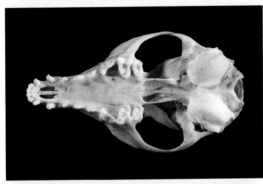

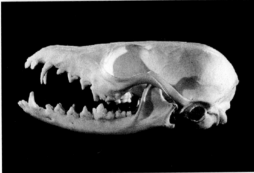

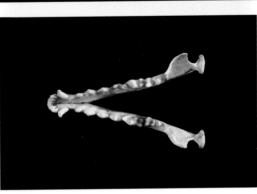

Figure 98: *Vulpes cana cana*. Dorsal, ventral and lateral views of the cranium, and a view of the mandible.

5 cm

Figure 99: Distribution map of *Vulpes cana cana*.

● – museum records.

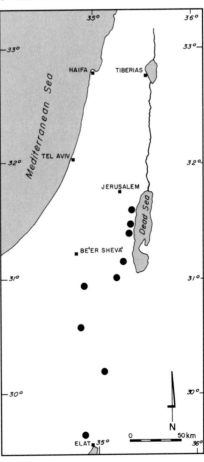

and plant material of *Phoenix dactylifera, Ochradenus baccatus* and *Fagonia mollis*, and various species of Gramineae were also eaten. Remains of vertebrates were present in ca. 10% of faecal samples analyzed. The seasonal fluctuation in body mass observed in this species, where foxes were heaviest in winter and lightest in summer (Geffen *et al.*, 1992b) may be correlated with food and water availability.

Reproduction: Blanford's foxes live in monogamous pairs (Geffen and Macdonald, 1992). Females are monoestrous and come into heat during January-February (in Israel). Gestation lasts 50–60 days, and litter size is 1–3 pups. Females have two to six active teats, and the lactation period is 30–45 days. Neonates are born with soft, black fur. From the body mass of three young born in captivity, a neonate body mass of 29 g was estimated (Mendelssohn *et al.*, 1987). Body mass of subadults is reached by ca. 3–4 months (700–900 g). At about two months of age the young start to forage, accompanied by one of the parents, and at three months of age they start to forage alone. Juveniles have similar markings to the adults, but their coat is darker and more greyish. Sexual maturity is reached at 10–12 months of age (Geffen, 1990).

Young are entirely dependent upon their mother's milk for food until they begin to forage for themselves. Adult Blanford's foxes have never been observed to carry food to the young and only one den was found with remains of prey at the entrance (Geffen and Macdonald, 1992). Observations of these foxes suggest that food is not regurgitated to the young, similar to other small canids. There is no indication that the male provides food either for the female or for the cubs, although males were observed grooming and accompanying 2–4 month old juveniles. Offspring often remain in their natal home range until autumn (October–November).

In two populations in Israel, old age or rabies were the primary causes of death (Geffen, 1990). There was one known case of mortality from predation, in which the predator was suspected to be a red fox. Life span of Blanford's foxes in these populations was estimated at 4–5 years. In captivity individuals reached six years of age.

VIVERRIDAE   Mongooses   גחניים

This family is represented in Israel by a single species, *Herpestes ichneumon*. Another species, *Genetta genetta*, was reported from Palestine by Tristram (1888) and Aharoni (1917). The type specimen of *Genetta genetta terrae-sanctae* in the British Museum of Natural History from Mount Carmel, described by Tristram, was found by Schlawe (1981) and H. Mendelssohn (unpublished data) to be identical with specimens from Algeria. The specimens that Aharoni (1917) claimed to have collected could not be located in any collection. Hence, there is currently no support for the existence of this species in Israel. *Genetta genetta* occurs in north-west Africa, Spain and southern France, but nowhere in the eastern Mediterranean countries.

HERPESTES Illiger, 1811

Herpestes Illiger, 1811. *Prodromus systematis mammalium et avium.*, p. 135.

Herpestes ichneumon (Linnaeus, 1758)   Egyptian Mongoose, Ichneumon
Figs 100–101; Plate XV:4; Plate XVI:1   נמיה

Viverra ichneumon Linnaeus, 1758. *Systema Naturae*, 10th ed., 1:43. "ad ripas Nili", Egypt.

The body of the Egyptian mongoose is elongated, the legs are short with five toes with blunt claws, the ears are short and the eyes are small with horizontal pupils. The grey fur is long, and when it bristles, for example during excitement, it causes the animal to appear much larger than it really is. In winter a yellowish, woolly underfur develops. The tip of the tail is black.

Herpestes ichneumon ichneumon is the subspecies present in Israel (Harrison and Bates, 1991).

Distribution: The Egyptian mongoose is widely distributed in Africa from Morocco to Egypt and south to the Cape Province, in the Iberian peninsula and in south-west Asia (Harrison and Bates, 1991).

In Israel it is common in the Mediterranean region, as well as in oases and agricultural settlements along the coast of the Dead Sea and the Jordan River, and in desert areas in the northern 'Arava and Northern Negev (Figure 101).

Fossil record: The earliest remains of this species found in Israel date to the Upper Palaeolithic period (Tchernov, 1988).

Habits: It has been studied by Ben-Yaakov and Yom-Tov (1983) and Hefetz *et al.* (1984). Mongooses are social animals, living in polygamous families comprising a male, 1–3 females and their cubs. They live in home ranges which they mark with an odorous secretion from the anal glands, smeared on stones, tree trunks and other objects. The anal region has a fold of bare skin around the anus, forming a pouch of about three cm in diameter, containing two glands, one on each side of the anus, which produce an odorous brownish secretion, and surrounded by two rows of sebaceous glands. Mongooses are capable of ejecting the secretion to a distance of about 50 cm, but use it mainly to mark their home ranges. The male secretion contains a specific compound not present in the female secretion. Mongooses create dung hills at permanent sites in their ranges which also serve as marking spots. All members of the family secrete and defecate in the marking spots.

Mongooses appear to use the sense of smell more than sight (when a specimen was offered a smelly, dead chick and a live, noisy one, it chose the dead one). However, when walking in high grass a mongoose will often stop and stand on its hind legs, supporting its body by the tail, in order to observe the area. They often creep ("herpestes" means a creeping thing in Greek).

Mongooses are diurnal and crepuscular and are not active under extreme climatic conditions such as strong winds, heavy rain or ambient temperature above 27°C

201

Table 94
Body measurements (g and mm) of *Herpestes ichneumon ichneumon*

| | Body | | Tail | Ear | Foot |
	Mass	Length			
			Males (11)		
Mean	3025	537	427	28	98
SD	368	22	30	8	5
Range	2500–3700	500–565	390–480	15–40	90–110
			Females (6)		
				n = 5	
Mean	2700	514	431	25	93
SD	489	73	31	4	6
Range	2150–3500	446–620	400–485	20–30	85–100

Table 95
Skull measurements (mm) of *Herpestes ichneumon ichneumon* (Figure 100)

| | Males (10) | | | Females (9) | | |
	Mean	Range	SD	Mean	Range	SD
GTL	102.5	96.5–106.1	2.9	95.9	87.3–104.0	6.5
CBL	100.6	96.0–102.7	2.0	94.8	87.6–102.1	5.5
ZB	50.8	48.7–53.6	1.7	47.5	42.2–53.5	3.8
BB	35.7	33.5–37.5	1.4	34.2	31.4–37.4	1.5
IC	17.9	16.1–19.5	1.2	17.4	15.0–20.1	1.7
UT	36.0	35.0–37.1	0.7	34.5	31.9–37.1	1.8
LT (F n = 4)	38.2	36.8–39.6	0.8	37.8	35.2–40.6	1.7
M (F n = 4)	67.1	62.2–69.6	2.4	64.9	59.6–71.5	4.2

Dental formula is $\frac{3.1.4.2}{3.1.4.2} = 40$.

Karyotype: 2N = 43 in males and 44 in females, FN = 64 (Zima and Kral, 1984).

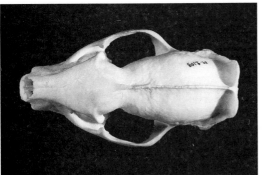

Figure 100: *Herpestes ichneumon ichneumon.* Dorsal, ventral and lateral views of the cranium, and a view of the mandible.

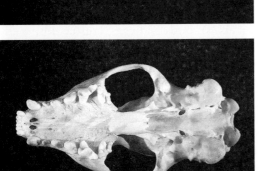

5 cm

Figure 101: Distribution map of *Herpestes ichneumon ichneumon.*

● – museum records.

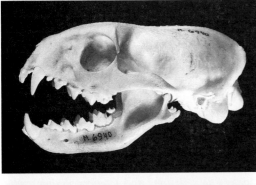

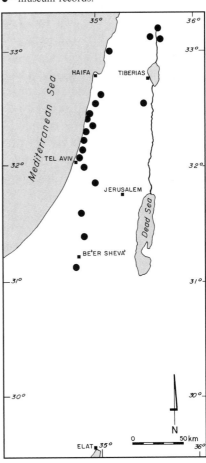

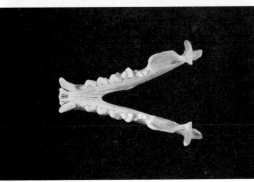

or below 11°C. They spend most of their time hiding in thickets and burrows excavated by other animals as well as in natural cavities, but also in open areas and use permanent trails in their home ranges. In populated areas they are sometimes nocturnal.

Mongooses have a variety of calls — a loud scream when attacked, accompanied by excretion from the anal glands which causes all members of the family to rush towards the attacked individual and to attack the foe; a low grumbling ("unk — unk") which serves as a contact call when the family is moving in dense vegetation, and other calls.

Food: They are omnivorous, feeding on a large variety of animals, including rodents, reptiles (skinks, lizards, snakes and terrapins), birds and eggs, crustaceans, snails, anuran amphibians and insects, particularly beetles, crickets and bugs. They are resistant to some extent to snake venom.

Reproduction: Reproductive behaviour was described in detail by Ben-Yaakov and Yom-Tov (1983). Mating takes place in February–March. The male sniffs the female's vulva, and both animals allogroom and lick one another before copulation takes place. There are several copulations, during and between which both animals emit pip sounds. Non-ejaculatory copulations last 30–60 seconds and ejaculatory ones 6–7 minutes. Gestation lasts 60 days, and mean litter size in captivity was 3.3 (range 1–4, n = 10). Females produce one litter annually between April–May, but if the cubs are lost they may have a replacement litter in the same year, and delivered as late as June. The cubs are covered with a thin, yellowish fur at birth, which changes to the adult fur when they are 3–4 weeks old. They open their eyes when 21 days old, start to react to noise when 25 days old, and appear outside the thicket in which they were born when 1.5 months old. They are always accompanied by an adult, although not necessarily a parent. They start to eat solid food when a month old, and are weaned at two months. The young are capable of hunting when four months old, but leave their family only when one year old, sometimes later. All the females in the family participate in rearing the cubs. The adults continue to partially feed the young even when they are a year old. The cubs have all their milk teeth at three months, and replace the first premolars when nine months old. Maximum longevity in captivity was 14 years.

Parasites: This species is infected by fleas (*Pulex irritans* and *Ctenocephalides felis*) and Mallophaga (*Felicola inequalis*) (Theodor and Costa, 1967).

Relations with man: Snakes, including venomous ones, are an important item in the mongoose diet. A poisoning campaign carried out in Israel by the Ministry of Agriculture against jackals in 1964 resulted in the population reduction of mongooses as well, and was followed by a considerable increase in the number of cases of viper bites. Recovery of the mongoose populations several years later was followed by a decrease in the number of bites (Mendelssohn *et al.*, 1971).

Mongooses can cause damage to poultry farms. Like many other carnivores, their killing urge is released by movement of prey. When they penetrate into an unprotected chicken house, they kill one chicken, but are stimulated by the movements

204

of the other frightened chickens, kill another one and so on, producing an overkill. They do not suck the blood, contrary to the belief of people who do not understand the reason for the overkill. However, they are very beneficial to man in other ways, killing rats, mice and venomous snakes.

Mongooses make excellent pets, when they are hand reared from a young age.

HYAENIDAE Hyaenas צבועיים

This family is represented in Israel by a single species, *Hyaena hyaena*.

HYAENA Brisson, 1762

Hyaena Brisson, 1762. *Le règne animal.*, ed. 2, pp. 13, 168.

Hyaena hyaena (Linnaeus, 1758)　　　　　　　　Striped Hyaena
Figs 102–103; Plate XVI:2–4; Plate XVII:1–2　　　　צבוע

Canis hyaena Linnaeus, 1758. *Systema Naturae*, 10th ed., 1:40. Benna Mountains, Laristan, Southern Iran.
Synonymy in Harrison and Bates (1991).

Hyaenas have strong, heavy jaws, suitable for cracking bones, one of the hyaena's main foods. The skull is correspondingly heavy. In order to support this heavy skull, the neck is strong and heavily muscled, and as broad as the head. The shoulders and forelegs are strong, to carry and move the heavy neck and head, resulting in a disproportionately large front part of the body in relation to the smaller rear part. This disproportion is also expressed in the size of the feet, with the front feet being over twice the size of the hind ones. The four toes are closely attached to each other and the tracks show no interstice between the toes, unlike canid tracks. Despite their build, hyaenas are good runners. The coarse fur is grey with black, vertical stripes. The throat has a large, black spot, very conspicuous when the hyaena lifts its head. Along the neck and back is a crest of long (20–30 cm), coarse, grey and blackish hair, that bristles in excitement and makes the hyaena look more formidable. In winter a light-grey, woolly underfur develops.

Hyaena hyaena syriaca Matschie, 1900 is the subspecies present in Israel (Harrison and Bates, 1991).

Distribution: It occurs in Asia from India to the Mediterranean, and in Africa from Morocco to Egypt and Kenya (Corbet, 1978).

In Israel it was once common throughout the country, but has currently almost entirely disappeared from the Mediterranean Coastal Plain, is rare in central and northern Israel, but still fairly common in desert areas of the Negev, Judea and the 'Arava (Figure 103). A survey carried out in the lower Jordan Valley estimated

Table 96
Body measurements (g and mm) of *Hyaena hyaena syriaca*

	Body		Tail	Ear	Foot
	Mass	Length			
			Males (10)		
Mean	32175	1092	295	145	223
SD	7732	96	19	13	7
Range	24800–45000	975–1240	260–320	128–165	210–232
			Females (10)		
Mean	32015	1085	302	149	210
SD	4786	61	41	10	8
Range	25000–39660	1020–1210	250–394	130–165	198–221

Table 97
Skull measurements (mm) of *Hyaena hyaena syriaca* (Figure 102)

	Males (10)			Females (11)		
	Mean	Range	SD	Mean	Range	SD
GTL	249.5	235.2–262.8	10.7	241.3	233.9–249.2	5.0
CBL	218.9	208.2–235.1	9.0	212.0	196.9–220.0	6.5
ZB	154.3	141.9–168.6	9.5	151	144.2–160.7	5.8
BB	70.0	67.4–73.4	2.2	70.9	65.7–72.3	2.4
IC	46.8	42.3–50.8	2.7	46.1	42.4–48.9	1.7
UT	92.1	86.9–99.1	4.2	88.8	84.9–93.8	2.9
LT	100.2	87.9–107.3	5.7	97.3	94–103.3	9.0
M	172.1	166.3–179.9	5.5	166.6	158.2–173.9	5.1

Dental formula is $\frac{3.1.4.1}{3.1.3.1} = 34$.

Karyotype: 2N = 40 for a specimens of unknown origin (Hsu and Arrighi, 1966).

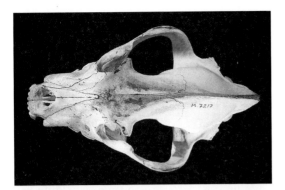

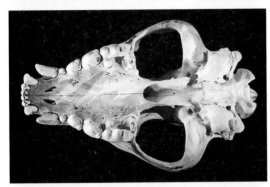

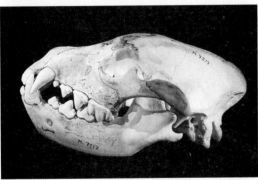

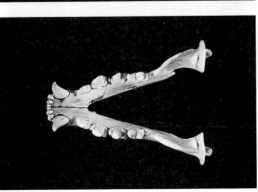

Figure 102: *Hyaena hyaena syriaca*. Dorsal, ventral and lateral views of the cranium, and a view of the mandible.

5 cm

Figure 103: Distribution map of *Hyaena hyaena syriaca*.

● – museum records.

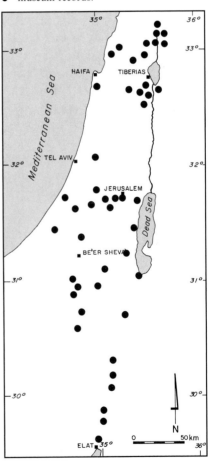

207

a population of between 19–36 hyaenas there, and at least 20 hyaenas were reported to have been killed by man (runover by cars or shot) between 1980–1985. The total population size in Israel is estimated to be 100–150.

Fossil record: The earliest remains of this species found in Israel date to about 150,000 years BCE (Tchernov, 1988).

Habits: It was studied in Israel by Bouskila (1983, 1984), Macdonald (1978), Skinner and Ilani (1980), D. Nissim (unpublished data) and Van Aarde *et al.* (1988). The hyaena is chiefly nocturnal, but at times is seen active by day. During the day it dwells in caves or burrows among rocks. It sometimes digs its own den, which it often takes over from a smaller animal, such as a porcupine. It is also known to share its den with porcupines. The cave may have several openings of different sizes. Hyaenas also use places with soft soil to bed down during the day or night, generally below overhanging rocks or bushes.

Hyaenas are solitary, and occupy home ranges which are sometimes shared by a male and a female inhabiting separate dens. In captivity they may form a long lasting monogamous relationship. The size of their home range varies, and appears to be larger in desert than in non-desert areas. In the Sede Boqer area hyaenas were observed 10 km from their den (Bouskila, 1984), which may suggest a home range of about 300 km^2. A radio-tracked female which was monitored for seven months had a home range of 60.9 km^2, and this range was partly overlapped by those of two other adults (Van Aarde *et al.*, 1988). They usually travel on permanent trails, and walk about 2–4 km/hr.

Home ranges are marked with a yellowish-white, fatty secretion from the anal glands. The secretion is smeared on stones, bushes and other objects along trails and in the territory. Faeces are deposited in piles, sometimes several times in the same place, suggesting that they also function as markers. Because the hyaenas eat much bone, their faeces are hard, dry, rounded and light-grey or off-white.

When hyaenas meet, for example near a carcass, they almost always sniff one another, mainly in the anal region. This is done either while standing or with one animal lying on the ground, raising its tail, extruding the rectum and so exposing the anal glands. The other animal sniffs the exposed anal region, and the two change positions to allow the first animal to sniff the second. This ceremony is also carried out between pair mates. When excited, hyaenas erect the very long hair of their dorsal crest. When threatening one another they fully expose their teeth, raise their tails and erect their mane while facing one another. When two hyaenas fight, they stand in an anti-parallel position, lower the front part of their body while folding the front legs at the carpal joints, fold the ears back and try to bite one another in the front legs and the ears. Such fighting is accompanied by loud, wailing vocalization, but the opponents circle each other, all the time on the carpal joints, and rarely cause severe damage. Most fights are not followed by biting, and one individual will either leave or lie on its back, exposing its belly and enabling the opponent to smell its anal glands. When encountering other carnivores near a carcass, hyaenas have priority over foxes, and generally also over wolves. In one case, however, a caracal chased a hyaena from a donkey carcass. In another case, a group of five wolves did not permit a hyaena to approach a carcass.

Hyaenas are not very vociferous, but continuous wailing calls are uttered when fighting or when cornered.

Food: They are scavengers, feeding on carrion of large animals and on bones, left over from carcasses cleaned by other scavengers such as vultures, but attacks on livestock have been reported, mostly small ones (goats, sheep, donkeys, calves), but these cases are decidedly rare (Harrison and Bates, 1991), as well as on medium to small size wild animals such as porcupines, hares, foxes, jungle cats, tortoises, rodents, reptiles and even chuckar partridges. The large head with its enormous muscles and strong jaws enable the hyaena to tear bones and heavy skin of large animals such as camels and to crack the shell of tortoises. They also eat fruit and vegetables, as well as refuse. The daily food consumption of an adult may be 7–8 kg (Skinner and Ilani, 1980), but they are able to consume up to 10 kg in one meal. Hyaenas often carry carcasses to their dens, where large accumulations of bones can be found. They are reported to have excavated shallow human graves, and to have consumed the body. An examination of the bones in a den in the Northern Negev Desert led Kerbis-Peterhans and Kolska Horwitz (1992) to state that hyaenas living in this den scavenged mainly on domestic stock that died naturally, including a high proportion of equids, camels and caprines.

Hyaenas travel large distances in order to find food, but usually not more than 20 km in one night. Several hyaenas may feed together on the same carcass.

Reproduction: Mating takes place mainly during winter and spring, but can occur all year long. Copulation is protracted, and may last for many minutes, up to half an hour. The gestation period is 90 days and litter size 1–4. Near Sede Boqer mean litter size was 2.5 (Bouskila, 1983). The young open their eyes when a week old, and start eating solid food when a month old. The young leave the den when two months old, and accompany their mother a month later. Lactation lasts until the young are six months old. Females give birth once annually. The young tend to play and fight each other near the entrance of their den, even during the day. They stay with their mother until she gives birth to her next litter.

Some females reach sexual maturity when a year old, and give birth when 15 months old, but others give birth when 27 months old. Males reach sexual maturity when about two years old.

Although mainly solitary, there have been observations from the Judean Desert of two females rearing their young together in the same den for five months, and separating only later to two dens 150 m apart. There are also observations of two and even three females walking their young together to food sources, and of adults which were not the parents, visiting the young (Bouskila, 1983). However, only the female appears to tend and bring food to the young.

In captivity they may live up to 30 years.

Parasites: This species is infected by fleas (*Pulex irritans*) and ticks (Ixodidae — *Haemaphysalis erinacei* and *Ixodes kaiseri*) (Theodor and Costa, 1967).

Relations with man: The hyaena's habit of walking on roads while seeking small road-killed animals results in high road mortality. The odour of carcass, even flattened by the traffic, remains and stimulates the hyaena's well developed sense of smell. During the 1970s, 15–30 hyaenas were killed on roads every year for six consecutive years (Ilani, 1979). It is hard to explain how the small hyaena population in Israel is able to withstand the mortality rate caused by road accidents, in addition to other mortality factors. The explanation may be the early maturity of females, which begin to breed when one year old. Some young females, therefore, may succeed in rearing a litter of cubs before they are run over. Road mortality is especially high for young, dispersing hyaenas. A ranger of the Nature Reserves Authority (Doron Nissim) carried out observations on three female hyaenas, with dens at a distance of several hundred metres from each other. He provided them with food, in order to enable them to remain close and not to risk crossing many roads. Each one reared 2–3 cubs. Nevertheless, all their cubs were killed on roads before they were one year old (D. Nissim, pers. comm.). One of us (H.M.) found on a stretch of one km of road three immature hyaenas which had been run over by cars within one week.

The hyaena has long been an object of superstition, having the reputation of a grave-robber. This may be the source of the custom of erecting tomb stones on graves, a relic of an earlier custom of heaping rocks on graves, in order to prevent the hyaena from exhuming the body. A tale, widely believed by Palestinian Arabs, runs as follows: If you meet a hyaena, it will rear on its hindlegs (hyaenas are actually not able to stand on their hind legs!), put its forelegs on your shoulders and breathe into your face. You will then be hypnotized and have to follow the hyaena into its den, where it will suck your brain. The spell can only be broken if, before you reach the den, somebody makes an incision in your skin and spills some drops of blood. The hyaena is in any case considered to be an awful, terrible and highly dangerous animal by the Arabs. However, there is no evidence that in Israel one has ever killed a human, even a child. There is also a belief that it can change its sex, mentioned also in the Talmud (Yerushalmi, Shabat, a, 3). This may be based on the similarity of the external sex organs in male and female *Crocuta*, that occurs in East Africa and was perhaps known also by the people of the Levant. Because of the infamous reputation of the hyaena, Palestinian Arabs used to keep hyaenas in small (about 80×60 cm) cages made of iron rods, on wheels. Before being introduced into this cage, the hyaena was blinded by holding a white-hot iron close to its eyes. The result was that the pupil of the eye did not contract and the green tapetum was visible even in full day light, giving the hyaena a more demonic appearance. The cage with the hyaena was wheeled to fairs and festivals and for paying a certain sum people were permitted to hit and to beat the hyaena. Therefore these hyaenas uttered their continued, weird, wailing fear calls as soon as they heard or smelled approaching people. When such a hyaena was saved from its ordeal by buying it, it continued to wail for weeks when approached. When released from its wheeled cage into a larger, more suitable cage, it remained on the same spot for weeks and turned around in the stereotyped movement to which it had adapted during the years spent in the small cage. Only gradually did it learn to walk and explore its new surroundings. The habit of displaying hyaenas in this way no longer exists among the Arabs living in Israel. Hyaenas sometimes cause damage by chewing plastic irrigation pipes, and to agricultural crops, such as melons, water melons, dates and other sweet fruit.

Hand-reared hyaenas become imprinted on their owners, and remain very tame and behave like dogs.

FELIDAE   Cats   חתוליים

This family is represented in Israel by three extant genera, *Felis*, *Caracal* and *Panthera*. The last cheetah, *Acinonyx jubatus*, was observed during the 1950s.

Key to the Species of Felidae in Israel
(After Harrison and Bates, 1991)

1.	Small sized cats. Greatest length of skull less than 105 mm.	2
–	Medium or large cats. Greatest length of skull more than 105 min.	3
2.	Fore legs without pronounced black elbow bars externally; soles of feet with short hairs only, not concealing the pads. Ears small, with only faint black tips. Tympanic bullae not greatly inflated, maximum diameter of each is less than 23 mm.	**Felis silvestris**
–	Fore legs with pronounced black elbow bars present externally; soles of feet with a mat of long hair concealing the pads. Ears very large, and broad with pronounced black tips. Tympanic bullae greatly inflated, maximum diameter of each is 25 mm or more.	**Felis margarita**
3.	Medium sized cats. Greatest length of skull less than 160 mm.	4
–	Large sized cats. Greatest length of skull more than 165 mm.	5
4.	Tufts on the ear apices long, about 50 mm in adults. Body and tail uniformly coloured, without a pattern.	**Caracal caracal**
–	Tufts on ear apices short, about 15 mm at most in adults. Pelage pattern well developed on limbs only. Tail ringed and tipped with black.	**Felis chaus**
5.	Face without prominent black stripe from each eye to the mouth; spots on the body mostly rosettes. Claws with complete claw sheaths. Skull rather flat, not highly domed above.	**Panthera pardus**

FELIS Linnaeus, 1758

Felis Linnaeus, 1758. *Systema Naturae*, 10th ed., 1:41.

This genus is represented in Israel by three species: *F. silvestris (lybica)*, *F. chaus* and *F. margarita*.

Dental formula is $\dfrac{3.1.3.1}{3.1.2.1} = 30$.

Felis silvestris Schreber, 1777

Figs 104–105; Plate XVII:3–4

Wild Cat

חתול בר

Felis (Catus) silvestris Schreber, 1777. *Die Saugethiere in Abbildungen nach der Natur.*, 3:397. Germany.

Synonymy in Harrison and Bates (1991).

Wild cats vary considerably in colour: brown-grey or light grey, with indistinct, dark grey spots and stripes. In the north they are darker than in the desert, where they are

Table 98
Body measurements (g and mm) of *Felis silvestris tristrami*

	Mass	Body Length	Tail	Ear	Foot
		Males (8)			
Mean	2600	528	330	59	129
SD	800	29	32	8	11
Range	1700–4250	480–560	275–360	45–70	110–150
		Females (8)			
n = 7					
Mean	2442	507	309	58.5	120
SD	742	72	23	9	10
Range	1425–3400	435–660	280–350	45–70	105–140

Table 99
Skull measurements (mm) of *Felis silvestris tristrami* (Figure 104)

	Males (10)			Females (8)		
	Mean	Range	SD	Mean	Range	SD
GTL	97.3	89.1–104.3	5.7	88.9	80.0–94.8	5.3
CBL	88.6	82.7–93.4	4.2	80.4	75.9–85.5	3.8
ZB	71.1	64.4–75.9	3.5	62.0	45.7–69.1	7.7
BB	44.6	42.8–47.8	1.6	42.6	37.3–44.8	2.4
IC	18.9	16.7–20.6	1.2	17.4	15.5–20.0	1.8
UT	31.2	29.0–34.1	1.6	28.8	26.8–30.3	1.1
LT	34.2	31.6–37.7	1.8	31.4	28.5–33.4	1.8
M	65.7	60.5–71.2	3.5	59.0	50.7–63.0	4.2
TB (F n=7)	21.9	20.7–24.3	1.2	20.5	19.0–22.2	1.2

Karyotype: 2N = 38, FN = 72, based on two males, one (identified as *F. silvestris*) from France and the other (identified as *F. lybica*) from Mauritania (Jotterand, 1971).

Figure 104: *Felis silvestris tristrami*. Dorsal, ventral and lateral views of the cranium, and a view of the mandible.

5 cm

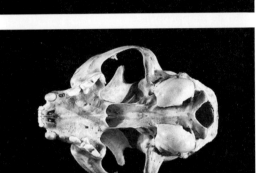

Figure 105: Distribution map of *Felis silvestris tristrami*.

● – museum records.

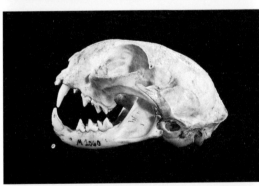

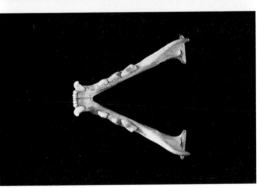

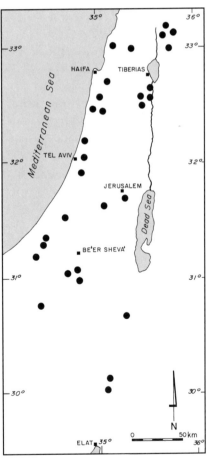

213

light grey or grey-yellow. The distal part of the tail has three black rings and the tip is black. The ear tips are dark, and occasionally have very short hair tufts.

This cat is very different from the European *silvestris*. It is smaller, more delicately built, and the tail is much thinner, because of the shorter hairs. Altogether, it appears to resemble an ordinary domestic cat (that developed from this species 5000 years ago in the Near East), but is smaller. The main difference is in colour. In grey domestic cats with a wild cat like pattern, this is pronouncedly black, and not indistinct, like in the wild cat. The animals seem to consider themselves as closely related, for hybridization between domestic and wild cats is much more common than in Europe. Therefore, the lumping of *Felis lybica* with *F. silvestris*, as suggested by Haltenorth and Diller (1980) seems to be inappropriate.

Felis silvestris tristrami Pocock, 1944 is the subspecies present in Israel (Harrison and Bates, 1991).

Distribution: This species is widely distributed in the Palaearctic region from western Europe to China and in most of Africa (Harrison and Bates, 1991).

In Israel it is found throughout the country, but is less common in the desert than in the Mediterranean region (Figure 105).

Fossil record: The earliest remains of this species found in Israel date to about 150,000 years BCE (Tchernov, 1988).

Habits: It is nocturnal and solitary. Individuals appear to have home ranges of about one km^2. It prefers rocky habitat and thickets, but also occurs in open habitat. It is an excellent climber.

Food: It feeds mainly on rodents, but also on young hares, birds, reptiles, including venomous snakes and occasionally on insects and other invertebrates.

Reproduction: Pregnancy lasts 63–66 days, and litter size is 1–5, commonly three. Females become oestrous every six weeks for 2–3 days each time. Young are born mainly during spring. In years with plentiful food (peak rodent years) females may give birth twice annually. Births take place in natural burrows or burrows dug by other animals, among rocks and in thick vegetation. The young open their eyes when one week old, and when 4–5 weeks old they start to feed on solid food and to leave their burrows. They start hunting when 10–12 weeks old. Weaning occurs when the young are three months old, and at five months the young become independent. They reach sexual maturity when 9–12 months old.

Parasites: This species is infected by fleas (*Ctenocephalides felis*), Diptera Pupipara (*Hippobosca longipenis*) and ticks (Ixodidae — *Rhipicephalus sanguineus, Haemaphysalis adleri, Ixodes redikorzevi*) (Theodor and Costa, 1967).

Relations with man: This species is the ancestor of the domestic cat, and in nature they interbreed freely. No pure stock of wild cats appears to exist at present in Israel, and the distinction between feral and wild cats is sometimes difficult. Feral domestic cats and hybrids, which live mainly on garbage and inhabit areas of high human population density live at much higher density than wild cats, exert therefore a much higher hunting pressure and have a harmful effect on wildlife, particularly reptiles, small mammals and ground nesting birds. The lizard *Lacerta trilineata israelica* was apparently exterminated from Mount Carmel and other areas due to predation by feral cats. The rufous bush robin (*Cercotrichas galactotes*), formerly a very common, low-nesting passerine of gardens and shrub-forest is now very rare, apparently also due to predation by this animal.

Felis chaus Guldenstaedt, 1776   Jungle Cat
Figs 106–107; Plate XVIII:1   חתול הביצה

Felis chaus Guldenstaedt, 1776. *Novi Comment. Acad. Sci. Imp. Petrop.*, 20:483. Terek River, north of the Caucasus.
Synonymy in Harrison and Bates (1991).

This species is a large, long-legged cat with a medium long tail. Males are almost equal in size to the caracal, but have much smaller paws, indicating that they prey mainly on relatively small prey. Males are much larger and heavier than females. The colour is very similar to that of *Felis lybica*, but the size, the long legs and the short tail make it easy to distinguish between the two species.
Felis chaus furax de Winton, 1898 is the subspecies present in Israel (Harrison and Bates, 1991).
Distribution: This species occurs from Egypt to Turkemenistan, India and Vietnam. In Israel it occurs in thickets, mainly near water and riverine habitats in the Mediterranean region, mostly along the Coastal Plain and the Jordan Valley, but also in oases along the shore of the Dead Sea as far south as Ne'ot HaKikar. The establishment of fish ponds and water reservoirs has enhanced their populations, and they are now also found in the mountains of the Galilee and the Northern Negev (Figure 107). The pupil of the eye contracts in light to a vertical slit, as in the small cats.
Fossil record: The earliest remains of this species found in Israel date to the Upper Palaeolithic period (Tchernov, 1988).
Habits: It is nocturnal, but occasionally active also during the day. When inactive it hides in thickets. It climbs and runs well, walks in shallow water and swims well and freely.
Unlike other cats, members of this species are often seen in pairs, or a pair with young. In captivity a male can be left in the same cage with a female and young without harming the young and will even defend them more intensely than the female does. Such observations indicate that this species is monogamous.
Food: It feeds on fish, rodents and birds, but also on snakes and amphibians. When hunting for fish it dives head first into the water, catching the fish with its mouth and not by striking it with a fore leg, as domestic cats do.
Reproduction: Gestation lasts 66 days and litter size ranges from 1–4, mainly 2–3. Young are born mainly during spring. At the Tel Aviv University Research Zoo females gave birth in March and September. Females may give birth twice annually. Births take place in dens located in thickets or burrows. The young open their eyes when 10–12 days old and are weaned when two months old. Sexual maturity is

215

Table 100
Body measurements (g and mm) of *Felis chaus furax*

| | Body | | | | |
	Mass	Length	Tail	Ear	Foot
	Males (18)				
	n = 10				
Mean	10000	710	281	72	172
SD	1400	88	20	10	7
Range	7400–12200	510–862	254–320	46–80	155–190
	Females (5)				
Mean	7643	712	265	71	158
SD	1585	80	34	3	6
Range	5865–9000	618–835	220–305	68–75	152–168

Table 101
Skull measurements (mm) of *Felis chaus furax* (Figure 106)

| | Males (10) | | | Females (6) | | |
	Mean	Range	SD	Mean	Range	SD
GTL	132.1	124.4–140.0	5.1	121.8	112.4–125.5	4.5
CBL	115.9	109.8–122.2	4.7	109.6	100.4–112.9	4.6
ZB	85.1	79.6–93.0	3.9	77.1	70.7–81.5	3.8
BB	52.6	49.4–54.7	1.7	49.3	48.3–50.3	0.8
IC	23.6	21.6–27.0	1.7	20.9	17.9–22.8	1.8
UT	42.3	40.4–44.6	1.5	39.6	38.0–42.2	1.6
LT	45.9	44.0–47.7	1.2	42.8	40.7–45.3	1.7
M	87.9	83.1–94.7	3.9	80.5	73.6–84.1	3.7
TB (F n = 4)	24.4	22.5–25.4	1.0	23.1	22.5–24.1	0.7

Karyotype: 2N = 38 (Zima and Kral, 1984).

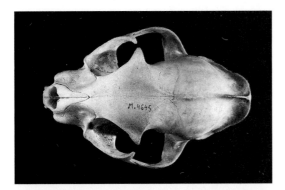

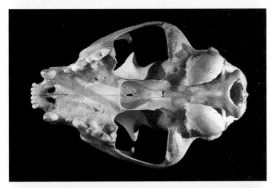

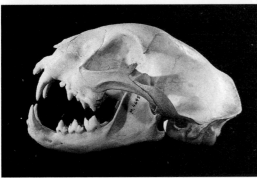

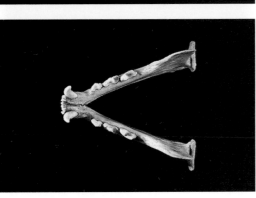

Figure 106: *Felis chaus furax*. Dorsal, ventral and lateral views of the cranium, and a view of the mandible.

5 cm

Figure 107: Distribution map of *Felis chaus furax*.

● – museum records.

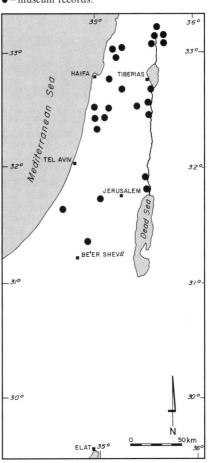

217

reached at the age of one year, but the male reaches full adult size only during the second year. In captivity they live up to 15 years.

Parasites: This species is infected by ticks (Ixodidae — *Haemaphysalis adleri* and *Ixodes kaiseri*) (Theodor and Costa, 1967).

Felis margarita Loche, 1858   Sand Cat
Figs 108–109; Plate XVIII:2   חתול החולות

Felis margarita Loche, 1858. *Revue Mag. Zool. Paris*, (2)10: 49, pl.1. Near Negonca, Algeria. Synonymy in Harrison and Bates (1991).

This is a small cat with soft, sandy-yellow fur with indistinct, dark spots and distinctive black stripes on the forelegs and less conspicuous ones on the hindlegs. The tail is grey with a black tip and several dark rings in front of it. The soles of the feet are covered with dense, stiff hairs, that prevent sinking in the sand.

Table 102
Body measurements (g and mm) of *Felis margarita harrisoni*

	Body		Tail	Ear	Foot
	Mass	Length			
Males (3)					
Mean	2330	449	260	63	113
Range	2139–2674	413–468	234–283	62–64	110–116
Females (1)					
Mean	1354	424	234	56	108

Table 103
Skull measurements (mm) of *Felis margarita harrisoni* (Figure 108)

	Males (3)		Females (1)
	Mean	Range	Mean
GTL	86.1	85.2–86.8	81.6
CBL	78.9	78.4–79.4	73.7
ZB	68.4	66.7–71.0	62.6
BB	42.7	42.4–43.2	42.6
IC	18.4	18.0–18.6	16.6
UT	27.2	26.7–27.6	27.0
LT	29.7	29.0–30.0	30.7
M	62.1	57.2–68.5	54.6
TB	23.7	23.3–24.1	22.5

Karyotype: $2N = 38$, $FN = 72$, based on a male from Baluchistan (Pakistan) (Jotterand, 1971; Schauenberg, 1974).

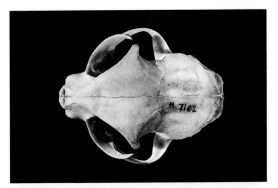

Figure 108: *Felis margarita harrisoni*. Dorsal, ventral and lateral views of the cranium, and a view of the mandible.

5 cm

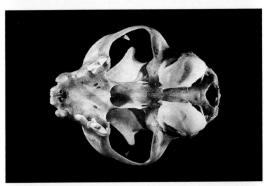

Figure 109: Distribution map of *Felis margarita harrisoni*.

● – museum records.

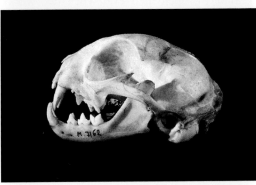

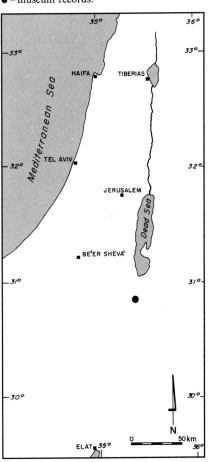

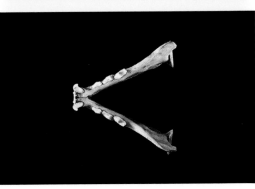

219

Felis margarita harrisoni Grubb and Groves, 1976 is the subspecies present in Israel (Harrison and Bates, 1991).

Distribution: This species occurs in north Africa from Morocco to Egypt, and in western Asia from Sinai and Arabia to Turkestan and Baluchistan (Corbet, 1978).

In Israel it was found on sands in the northern 'Arava Valley (Ḥazeva and 'En Yahav; Figure 109), but never in the sands of the western Negev, which is strange, as this species occurs in north-western as well as southern Sinai.

Habits: It is nocturnal. It has large, inflated tympanic bullae, which greatly improve its hearing ability. When inactive it lives in burrows dug by foxes or other animals in the sand. It likes to dig in the sand, but it is not known whether it digs its own burrows. In the 'Arava, an individual was observed to travel eight km during one night (Ilani, 1986). When frightened it crouches low on the ground, and may be approached and even picked up by hand.

Males have a distinctive barking call which may have territorial meaning. When angry it spits like other small cats.

Food: It feeds on rodents, birds, reptiles and insects. The gecko *Ceramodactylus doriae*, common on sands in its habitat in the 'Arava, may be one of its staple foods. Remnants of the large agamid lizard *Uromastix aegyptius* have been found near a burrow opening (M. Abadi, pers. comm.).

Reproduction: At the Tel Aviv University Research Zoo births occurred mainly during spring, but also during summer and autumn. Gestation lasts 63 days and litter size is 2–5. The young weigh 50–60 g at birth, open their eyes when two weeks old, start leaving their burrow when three weeks old and begin to eat solid food when five weeks old. They become independent when 3–4 months old and reach sexual maturity when one year old. In captivity they may live up to seven years.

Relations with man: As this species occurs only on sands in the 'Arava, its continued existence in Israel is doubtful due to the fact that its habitat is being destroyed and exploited for agriculture.

CARACAL Gray, 1843

Caracal Gray, 1843. *List Spec. Mamm. Brit. Mus.*, p. 46.

This genus is represented in Israel by a single species, *C. caracal*.

Caracal caracal (Schreber, 1776)   Caracal Lynx
Figs 110–111; Plate XVIII:3   קרקל

Felis caracal Schreber, 1776. *Die Saugethiere in Abbildungen nach der Natur.* ..., 3: pl. 106;
    text, 3:413, 587 (1777). Table Mountain, Cape Town, South Africa. (For discussion of
    type locality and author, see J.A. Allen, 1924, *Bulletin Amer. Mus. nat. Hist.* 47:279 and
    Pocock, 1939, *Fauna Br. India, Mamm.*, 1:306).
Synonymy in Harrison and Bates (1991).

This is a medium-sized (large males may reach 13 kg), long-legged, short-tailed cat. Its pupils are round rather than the elongated slits found in small cats and the jungle cat. The paws are relatively large, indicating that it is able to prey on relatively large animals. The soft fur is light brown, light yellowish-brown or light greyish-brown. In the area between Tel Aviv, Ramla and Gaza aberrant specimens are occasionally found, which are blackish-grey all over. Cubs of this morph are almost entirely black when born. The back of the ear is black with a light central spot. On the inner side of the front legs there are a few indistinct, dark spots. The tips of the ears feature well developed black tufts of hair, which are specially long in adult males. When excited the ears are twitched. A female leading cubs also twitches the ears so that the black back of the ears may have communication value. Males are significantly larger and heavier than females.

Caracal caracal schmitzi Matschie, 1912 is the subspecies present in Israel (Harrison and Bates, 1991).

Distribution: This species is widely distributed in much of Africa (but not in tropical rain forest) and in Asia from Sinai to India and north to Turkey and the Aral Sea (Corbet, 1978).

In Israel it was known to occur in the Negev Desert, but during the 1960s it was found also in the Mediterranean region of the country and today it appears to occur throughout most of the country (Figure 111). Its occurrence in the north coincided with the dramatic decline in carnivore populations due to a poisoning campaign carried out by the Ministry of Agriculture against jackals, but which also affected many other predators. After the decline of carnivores, hares, which are one of the main foods of caracals, increased considerably. The increased availability of their preferred food may have enabled the increase and expansion of the cararcal population. The lack of, or diminished competition appears to have enabled it to increase its range.

Habits: Its biology was studied in the northern 'Arava Valley by Weisbein (1989). It is crepuscular, being active mainly during the late afternoon and evening or early morning, but also during the night and in cool or cloudy weather it is also active during the day. Activity depends on ambient temperature: the higher the temperature, the later the onset of activity in the evening. Hence, activity starts at 3 pm and after 7 pm in December and July, respectively. Activity normally ends between 8–10 am, without a clear relationship to temperature or season. In the summer the caracal is active all night long. When resting it hides in thickets, caves and niches, as well as in burrows of porcupines and jackals.

In the 'Arava the size of its home range is inversely correlated with food availability and positively related to body mass of the owner. Mean home range of males (220 km^2) is significantly larger than that of females (57 km^2). There is a considerable overlap in individual home ranges — about 50% among males and 30% among females. Mean distance travelled in one night is 10 and 7 km for males and females, respectively. This distance normally does not change during the year, but females with young travel on average only 3 km per night. Caracals do not travel a fixed route. Sex ratio was 1:1.

221

Table 104
Body measurements (g and mm) of *Caracal caracal schmitzi*

	Body		Tail	Ear	Foot
	Mass	Length			
	Males (11)				
Mean	10650	809	274	87	187
SD	1600	98	21	10	10
Range	7400–12500	622–1050	228–300	73–110	168–205
	Females (9)				
	n = 8				
Mean	7256	711	269	77	176
SD	1050	77	20	6	10
Range	6000–8900	565–826	242–300	68–85	164–190

Mean body mass of six males and five females in the northern 'Arava was 9.77 and 6.15 kg, respectively (Weisbein, 1989).

Table 105
Skull measurements (mm) of *Caracal caracal schmitzi* (Figure 110)

	Males (10)			Females (5)		
	Mean	Range	SD	Mean	Range	SD
GTL	128.9	119.0–137.0	4.9	117.0	110.2–121.2	4.5
CBL (F n=4)	112.2	93.0–121.0	7.3	103.9	97.0–107.7	4.7
ZB	89.9	80.5–95.0	4.4	81.6	78.6–85.0	2.8
BB	55.9	54.0–57.4	1.1	52.0	50.1–53.7	1.4
IC	26.3	23.0–28.6	1.8	24.1	21.8–26.9	2.2
UT	41.8	39.2–42.9	1.6	40.1	37.8–42.3	1.9
LT	44.7	42.3–47.7	1.7	42.0	40.4–45.0	2.0
M	86.3	75.6–92.9	4.7	78.3	72.7–80.7	3.5

Dental formula is $\frac{3.1.2–3.1}{3.1.\ 2.\ 1} = 28$–30.

Karyotype: 2N = 38, FN = 72 (Jotterand, 1971).

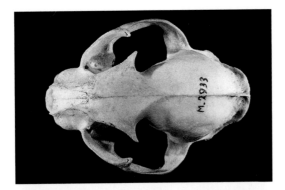

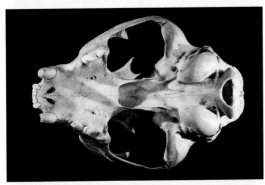

Figure 110: *Caracal caracal schmitzi.* Dorsal, ventral and lateral views of the cranium, and a view of the mandible.

5 cm

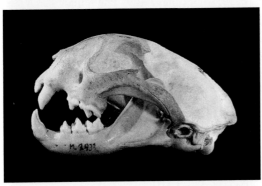

Figure 111: Distribution map of *Caracal caracal schmitzi.*

● – museum records.

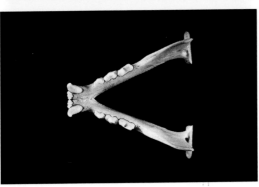

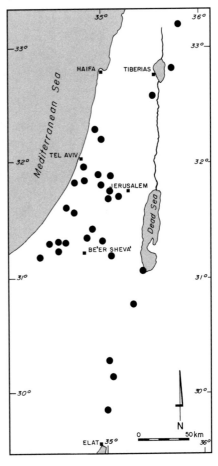

The caracal is very agile both on the ground and while climbing. One was observed to catch a partridge by jumping at the flying bird.

Food: It feeds on hares, rodents, hedgehogs, birds (mainly chuckar partridges), reptiles (Weisbein, 1989) and occasionally on carrion (Skinner, 1979). It also feeds on gazelles, even adult ones, and on domestic cats. In the northern 'Arava mammals and birds composed 62% and 24% of its diet, respectively, and in addition to the above items its food also included *Uromastix aegyptius* and *Herpestes ichneumon*.

Reproduction: The oestrous cycle lasts two weeks, and the oestrus itself 5–6 days. During one cycle, a female copulates with several males (in one case in the northern 'Arava, three males copulated with one female), and copulation order is determined by their size. Each series of copulation starts with the female and the male walking together and lasts 42–48 hours, during which there are many copulations. After each series of copulations both animals lick their sexual organs, and the female often rolls on its back from side to side, like other oestrous felids. After one male finishes, another series of copulations is begun with another male, which has waited without interfering about 0.5 km away.

On average, gestation lasts 78 days. Liter size ranges from 1–3 (mean two). In the northern 'Arava as well as in Tel Aviv University Research Zoo females gave birth all year round. The young weigh about 300 g at birth. They are reared in a thicket or a cave, generally one that has been deserted by a porcupine or other animal. The young open their eyes when 10 days old. When 3–4 weeks old they are transferred by their mother to another cave, sometimes daily. The female has a special bark-like call she uses to call the young. Weaning occurs when the young are 4–6 months old. When 9–10 months old they have all their permanent teeth, and disperse from the maternal home range.

Sexual maturity is reached when about 15 months old. Adult size is reached when 18 months (females) or 24 months (males) old. One male dispersed up to 60–90 km, but a female remained near the maternal home range. In captivity longevity is up to 17 years.

PANTHERA Oken, 1816

Panthera Oken, 1816. *Lehrbuch der Naturgeschichte Zool.*, 3 (2):1052.

This genus is represented in Israel by a single species, *P. pardus*.

Panthera pardus (Linnaeus, 1758)   Leopard
Figs 112–113; Plate XVIII:4; Plate XIX:1–2   נמר

Felis pardus Linnaeus, 1758. *Systema Naturae*, 10th ed., 1:41. Egypt.
Synonymy in Harrison and Bates (1991).

Two subspecies of leopards exist in Israel. *Panthera pardus tulliana* Valenciennes, 1856 is one of the largest subspecies of leopards, with a body mass apparently approaching 80 kg. Its fur is yellow-brown, and the centre of the spots are light in colour. We have no records of body measurements of this subspecies from Israel. However, skull measurements and preserved skins indicate it is a much larger subspecies than *Panthera pardus nimr* Hemprich and Ehrenberg, 1833.

P. p. nimr is one of the smallest subspecies of leopards. There is a marked sexual dimorphism, with the male being much larger than the female. The body mass of one male was 38 kg, while the maximum recorded mass of an obese female was 32 kg. The normal mass of males and females is 32 kg and 23–25 kg, respectively. Its colour is light yellow dorsally, with white flanks and belly. The colour fades during summer, the back becomes pale-yellow and the flanks and belly become yellowish, resembling *P. p. jarvisi* (the now extinct Sinai leopard) giving rise to the conclusion that these two subspecies are one (Harrison and Bates, 1991).

Table 106
Skull measurements (mm) of *Panthera pardus tulliana*

	Males (2)	
	Mean	Range
GTL	256.6	251.7–261.5
CBL	229.1	223.5–234.8
ZB	151.8	148.4–155.1
BB	86.4	85.2–87.6
IC	51.0	50.3–51.7
UT	78.4	77.1–79.6
LT	87.2	86.5–88.0
M (n = 1)	166.8	

Dental formula is $\frac{3.1.3.1}{3.1.2.1} = 30$.

Distribution: This species was formerly widely distributed in Asia apart from the Arctic tundra, and over most of Africa (Corbet, 1978).

Two subspecies are known from Israel: *P. p. tulliana* occurred in the Galilee, and *P. p. nimr* in the Negev and Judean Deserts. In addition, *P. p. jarvisi* was described from Sinai. There was apparently a well established population of *P. p. tulliana* in the Galilee, preying on wild boars, porcupines and hyrax. Occasionally livestock were taken and so the leopards were persecuted by the local human population. Many villages possessed a (badly preserved) leopard skin. By constant harassment, the population was eventually exterminated. The last specimen of *P. p. tulliana*, an old male, was killed by a shepherd in 1965 near Ḥanita in the north-western Galilee. The shepherd pushed his dagger into the mouth of the leopard, that had very worn teeth, and cut its throat from the inside. Reports of leopards exist from

Table 107
Body measurements (g and mm) of *Panthera pardus nimr*

| | Body | | Tail | Ear | Foot |
	Mass	Length			
Males (2)					
	n = 1	**n = 1**			
Mean	38000	1140	830	61	242.5
Range			780–880	46–76	235–250
Females (4)					
	n = 3				**n = 3**
Mean	25117	985.5	776	65	214
SD		51	48	6	
Range	16500–31850	912–1030	730–820	60–72	210–220

Table 108
Skull measurements (mm) of *Panthera pardus nimr* (Figure 112)

| | Males (2) | | Females (5) | | |
	Mean	Range	Mean	Range	SD
GTL (F n = 4)	211.3	210.5–212.0	190.6	172.5–235.0	29.8
CBL	192.5	191.8–193.1	172.4	161.0–195.0	14.8
ZB (F n = 4)	130.5	130.3–130.6	122.2	111.4–150.0	18.6
BB (F n = 4)	81.9	81.0–82.9	72.6	72.0–73.2	0.5
IC	40.8	40.5–41.1	38.8	35.8–44.4	3.5
UT (F n = 4)	65.5	65.3–65.7	60.6	55.0–69.6	6.4
LT (F n = 4)	74.7	73.7–75.7	67.1	61.2–79.5	8.3
M (F n = 4)	141.8	141.2–142.3	125.2	115.0–145.3	13.7

Karyotype: 2N = 38, FN = 68, for specimens of unknown origin (Pathak and Wurster-Hill, 1977).

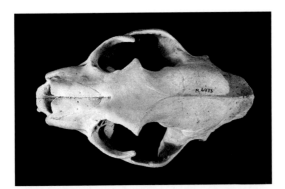

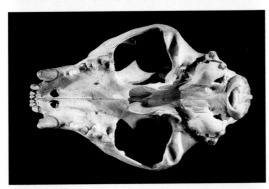

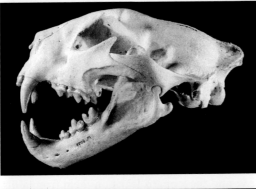

Figure 112: *Panthera pardus nimr*. Dorsal, ventral and lateral views of the cranium, and a view of the mandible.

5 cm

Figure 113: Distribution map of *Panthera pardus*. All records in the Galilee are of *Panthera pardus tulliana*, and all other records (from the Judean and Negev Deserts) are of *Panthera pardus nimr*.

● – museum records.

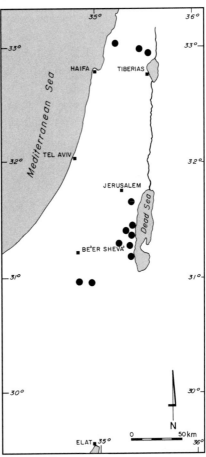

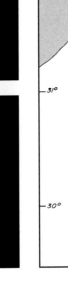

227

later dates from Mount Ḥermon, the southern Golan Heights and the Lower Galilee (near Shadmot Devora), but they have not been supported by concrete evidence, such as photos or footprints. In contrast, *P. p. nimr* still exists in the Negev and Judean Deserts. It is impossible to accurately estimate its population size, but between 15–20 still appear to occur there. Reichman and Shalmon (1992) estimated that in an area of 300 km^2 in the central Negev there are 5–10 leopards. Ayal (1990) considered that up to 25 leopards live in the Judean Desert and the areas adjacent to it, but this seems to be too high an estimate. A very old female died and was brought to Tel Aviv University in April 1995. There have been recent reports of this species from several places in the Negev (near 'En 'Avedat, in the Makhtesh Ramon, Biq'at 'Uvda, and near Elat) and the Judean Desert (Naḥal Og, 'En Gedi; Figure 113).

Fossil record: The earliest remains of this species found in Israel date to about 150,000 years BCE (Tchernov, 1988).

Habits: A small population of *P. p. nimr* was studied by Ilani (1986) in the Judean Desert, mainly at 'En Gedi. It is a solitary species. In the Judean Desert individuals guard a territory reaching up to 200 km^2 in size, but most activity is near wadis where water is available in and around oases in an area of 20–60 km^2. A few individuals maintain territories along the cliffs of the Dead Sea escarpment, whose length is about 30 km. There is no overlap in male territories, but a considerable overlap in those of the females. Territories are marked by urinating on trees, shrubs and rocks, or in a small depression dug in the ground with the hind feet. The odour of the urine is strong and during winter it can be smelled by humans even after 10 days. The excretion from the anal glands is also used for marking various objects. The leopards often scratch tree trunks with their claws, apparently marking them with excretion from glands in the soles of their paws.

The leopard is active mainly during night, but during cool days it can also be active during the day. It has several calls: roars, growls and a call reminiscent of a hoarse dog bark.

Food: It feeds mainly on hyrax and ibex, but faeces also contained remains of porcupines, gazelles, foxes, jackals, hares, wild boars (in the Galilee) and even rodents (*Acomys*), as well as birds (chuckar and sand partridges, rock pigeons and Tristram's grackle). Some of these food items might have been collected as carrion. Ibex and hyrax are ambushed from rocks or trees, and the leopard leaps on them from a distance of 3–4 m from a tree or the top of a rock. Over nine days one female hunted three ibex with a total mass of 136 kg, and she consumed about 25 kg of this prey. Prey is visited and consumed on consecutive days. At 'En Gedi there were two breeding females which chose to prey on domestic cats and dogs, in order to feed their cubs, causing alarm among members of the kibbutz, as a consequence of which these two females were taken into captivity. The removal of these two breeding females from the population caused the collapse of the leopard population in the 'En Gedi area. This population was remarkable, for they inhabited an area with three human settlements which are visited every week by thousands of visitors, without any negative human vs. leopard incidents occurring, other than the superstition of the kibbutz members. When the Nature Reserve Authority were asked to trap a male as well and to establish a captive breeding group of this endangered subspecies they replied: leopards are a protected species, they cannot be trapped.

Reproduction: During oestrus females mark their territories more often than at other times, and roll on the back often in the soil where they have urinated. In the Judean Desert oestrous females were observed mainly during winter, and also during spring and summer, but never during autumn. The length of the oestrus cycle is about one week, and occurs every 45 days. During the mating season a male and a female stay together for several days, copulating many times daily. In January 1982 a pair was observed to copulate more than 200 times over three days in Naḥal Ze'elim (Judean Desert). After each copulation the male roared, and during the entire period the animals did not eat.

Gestation lasts 90–105 days, and litter size in the Judean Desert was 1–2. Births occurred all year round, with a peak in spring. The cubs open their eyes when about a week old, and begin to eat meat when about one month old. When four months old they accompany their mother during hunting. The male seems to guard the female during the period after she gives birth. However, during the late 1980s several litters of the only fertile female remaining in the 'En Gedi area were killed by strange males, apparently as a mechanism to enhance her oestrus so that she could become impregnated by the killer male. However, this explanation was questioned by Ayal (1990), who summarised the observations in the Judean Desert for the Nature Reserve Authority. In one case an eight month old cub was killed. Cubs are also preyed on by hyaenas. A female may give birth every year. One of the females followed by Ilani (Ayal, 1990) gave birth to six cubs in five litters during the eight years of observations. Another female gave birth to nine cubs in seven litters. Sex ratio of the cubs was 1:1.

The female rears the cubs by herself. After giving birth she stays with the cubs continuously for 10 days. She then goes hunting for a day or more. Every few days she transfers the cubs to another hiding place. When they are small they are carried in her mouth and later they follow her to the new hiding place. Hiding places are caves or thickets. In the Judean Desert the young stay with the mother for a year, and female young remain apparently even after the mother has given birth to a new litter. Sexual maturity is reached at 2–3 years of age, and the youngest oestrous female observed in the Judean Desert was less than two years old. In captivity leopards reach 21 years of age.

Relations with man: In Israel there is no record of leopard attack on humans, and the damage they cause to herds is minimal. Nevertheless, hunting eliminated the Galilee population and reduced their number in the desert to the point of extinction. At present, the leopards in the Judean Desert are not shy, and there are many reports of them being observed by humans at a distance of several metres. In one case an old female leopard was shot by a soldier who claimed that he was afraid of attack. The leopard had blocked a narrow path on which a group of soldiers had to pass. As the leopard did not move and the soldiers were afraid that it might attack, they shot it. The post mortem showed that the leopard, an 18 year old female, was in any case moribund. She was blind, extremely emaciated, had only a few, very worn, teeth left and probably had not fed for weeks.

Leopards are mentioned many times in the Bible and the Talmud, in which they are a symbol of bravery. Several places in Israel carry the name of the leopard (namer in Hebrew, nimr (plural = nimrin) in Arabic: 'Ein Nimrin, Nimrin, Naḥal Namer, Ma'ale Namer).

HYRACOIDEA Hyraxes

PROCAVIIDAE שפניים

PROCAVIA Storr, 1780

Procavia Storr, 1780. *Prodromus Methodi Mamm.*, p. 40, pl. B.

This order is represented in Israel by a single species, *Procavia capensis*.

Procavia capensis (Pallas, 1766) Hyrax, Dassie, Coney
Figs 114–115; Plate XIX:3–4; Plate XX:1–3 שפן

Cavia capensis Pallas, 1766. *Miscellanea Zool.*, p. 30, pl. 3. Cape of Good Hope.
Synonymy in Harrison and Bates (1991).

A rabbit sized, robustly built mammal with a short neck, ears and legs and vestigial tail. Although the fur is generally brown-grey, individuals with lighter coloured fur, mainly on the head, can be found in the Negev. Third generation offspring of light coloured individuals which were bred in captivity in Tel Aviv have retained the light colour of their parents.

Procavia capensis syriaca Schreber, 1784 is the subspecies present in Israel (Harrison and Bates, 1991).

Distribution: This species is widely distributed in Africa from the Cape to the Mediterranean, penetrating into Asia in Arabia, Sinai, Jordan, Israel, southern Syria and Lebanon.

In Israel it occurs in rocky terrain throughout the country, in the Negev and Judean Deserts, the 'Arava, along the Jordan Valley and in the mountains of the Galilee, Samaria, Gilboa', Giv'at HaMore, Judea and Carmel (Figure 115). On Mount Ḥermon it occurs in summer up to 1300 m above sea level, and in Sinai up to 2000 m above sea level at Mount Katarina.

Fossil record: The earliest remains found in Israel date to about 150,000 years BP (Tchernov, 1988).

Habits: Hyrax are inhabitants of rocky terrain, in extreme desert and Mediterranean forest alike. They are excellent and agile climbers, often ascending narrow fissures by using all four legs spread to the sides, or by leaning on their backs and using their legs on the opposite side of the crag. They make use of small protuberances on the rocks while climbing, and when they fall they continue to run after hitting the

231

ground. They are able to jump vertically to about one and a half metres. They can run at the same speed as a human, but only for a very short distance. Hyrax always prefer to stay near rocks, to which they run for cover when threatened, but are sometimes found far from rocks, as can be seen when some specimens, mostly young-adult males dispersing from dense populations, have been found run over on roads in completely flat areas as far as 5 km from rocky terrain.

Their biology was studied by Meltzer (1967) in the Upper Galilee. The hyrax is diurnal. Their activity is limited to a narrow range of ambient temperatures, as they are not active in extreme hot or cold temperatures. When resting they hide in natural caves and crevasses. In summer they leave their caves at sunrise and start their activity immediately. In winter, after leaving the cave they bask in the sun for an hour to ninety minutes before foraging. In summer, there are two activity peaks, one in the morning and the other during late afternoon. At noon they rest in the shade of rocks or trees and in late afternoon they enter their caves, where they huddle together for warmth. They select caves in which the temperature rarely falls below 16°C, and huddling makes it even warmer. In captivity, they crowd together in the sleeping boxes in which the temperature can be 20°C higher than the ambient temperature. Because their oxygen consumption is relatively low, they do not suffer from the low oxygen pressure in their caves or sleeping boxes. At ambient temperatures between 25°C–26°C their own body temperature is 36.3°C, but it may rise or fall by 7.5°C when ambient temperatures rise or fall, respectively.

Hyrax are social animals, living in herds of up to 100 individuals comprising several families. Each family is composed of an adult male, several subordinate, younger males, females and cubs. Herds may reside in one locality or travel distances up to several kilometres in search of food or water. Such travelling may encompass a wadi, taking several days or weeks to complete.

Hyrax are generally not very vociferous animals, but they do have a large repertoire of calls (see Reproduction). When alarmed, hyrax emit a loud, shriek-like call. When fighting, they emit bark-like grunts. Suckling cubs produce calls similar to the contact calls of guinea pigs. After the mating season in July–August, territorial calling and fighting are discontinued in the family herds, because larger groups provide an advantage for thermoregulation towards winter when huddling together during low-ambient temperatures.

Their main natural enemies are the leopard (in the Judean and Negev Deserts), the bearded vulture, the stone marten and the eagle owl. Only the leopard is able to prey on adult hyrax, while other predators can take only young specimens. In cold winters they may die from exposure and from lack of food, as they are not active during cold, rainy days.

Food: The hyrax is generally herbivorous, feeding on a large range of plants, including those containing large quantities of secondary compounds such as Euphorbiacae, Ascelepiadacae, Solanaceae and Apocynaceae which are rarely if ever eaten fresh by other herbivores in Israel. Consumption of poisonous plants is made possible partly by the special structure of the hyrax intestine, which has more than one

Table 109
Body measurements (g and mm) of *Procavia capensis syriaca*

	Body		Ear	Foot
	Mass	Length		
Males (4)				
Mean	3049	500	35	76
SD	660	7	3	3
Range	2090–3555	495–510	33–39	72–80
Females (4)				
Mean	2034	475	34	70
SD	125	31	1.5	2
Range	1900–2150	450–514	32–35	67–72

Males may weigh up to 4 kg and females 2.5 kg.

Table 110
Skull measurements (mm) of *Procavia capensis syriaca* (Figure 114)

	Males (4)			Females (3)	
	Mean	Range	SD	Mean	Range
GTL	98.8	97.0–101.0	(n = 3)	87.9	84.7–90.1
CBL	94.9	91.2–98.0	3.3	84.9	82.7–86.5
ZB	59.3	55.4–63.2	3.2	51.1	49.7–52.0
BB	33.7	33–34.5	0.8	31.0	30.4–31.9
IC	24.5	22.3–27.7	2.4	19.5	18.6–20.6
UT	39.0	37.7–40.0	1.1	38.7	38.0–39.8
LT	39.6	36.4–41.7	2.5	38.5	38.1–39.0
M	73	72–74.7	(n = 3)	66.8	65.2–68.4
TB	7.3	7.1–7.5	(n = 3)	7.5	7.4–7.6

Dental formula is $\frac{1.0.4.3}{2.0.4.3} = 34$.

Karyotype: 2N = 54, based on zoo animals of unspecified origin (Hungerford and Snyder, 1969).

233

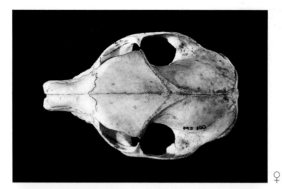

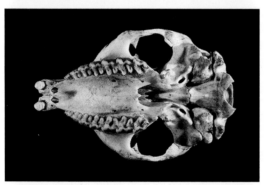

Figure 114: *Procavia capensis syriaca.* Dorsal, ventral and lateral views of the cranium, and a view of the mandible. Female.

5 cm

♀

♀

♀

♀

234

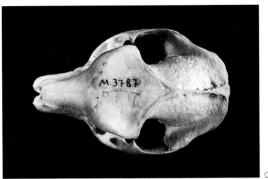

♂

Figure 114 (cont.): *Procavia capensis syriaca.* Dorsal, ventral and lateral views of the cranium, and a view of the mandible. Male.

|—————— 5 cm ——————|

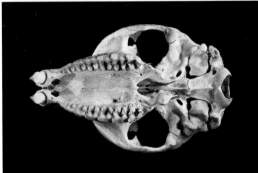

♂

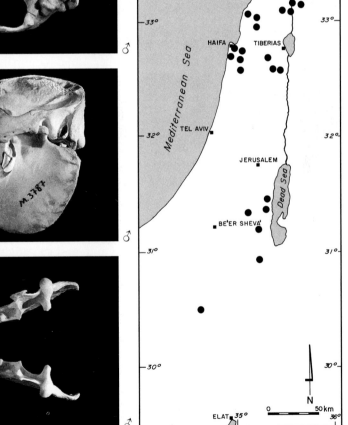

♂

♂

Figure 115: Distribution map of *Procavia capensis syriaca* .

● – museum records.

235

appendix. At times it may become partly carnivorous, feeding on insects and lizards. It consumes about 650 g a day. During winter and spring it eats mainly on the ground, consuming fresh grass and herbs, while in summer and autumn it eats leaves of bushes and trees which it obtains by climbing on even quite thin branches or by standing on its hind legs. It chews by moving its lower jaw up and to the sides.

Hyrax faeces are small and dry in summer, large and soft in winter. They defecate in permanent places, usually in caves and under rocks, where huge piles may accumulate through the years. Water is drunk when available, but most hyrax live where water is scarce and they use the moisture in their food. In the desert, hyrax are more numerous near water sources, but this may also be due to greater food availability.

Reproduction: Mating behaviour starts to appear in March–April. The dominant male often sits on a high rock, observing its herd. It threatens other males by moving its lower jaw sideways, sharpening its incisors in a movement reminiscent of chewing. This behaviour is the basis for the erroneous statement in the Bible that "The coney, though it chews the cud" (Leviticus 11:5). When sitting at their observation points, dominant males "sing" loudly, apparently declaring ownership of their herd and territory. The call is loud, indescribably weird, and far reaching.It consists of a series of rising and falling screams which can be heard from a large distance. Sometimes several males have adjacent territories and sing synchronously or alternately. During summer males start to fight before and after the foraging hours. Fights start with two males approaching one another, standing in parallel or anti-parallel positions and exposing their teeth by stretching the lips backwards. Concomitantly the body is stretched. Fighting commences with the combatants trying to bite each other on the head or back, pushing at one another with their backsides. While fighting they produce hoarse barking grunts. Only adult males participate in such fights, which may result in bleeding injuries. The participants strike at each other with wide opened mouth to reveal the sharp incisors. Because they push each other with their hind quarters most wounds are either on the head or the lower back. Defeated males aggregate in bachelor groups, in which the level of calling and fighting is low, although most members of the group still show the characteristic wounds, possibly received earlier when fighting for a territory. Fights continue until only one dominant male remains in each family, and most sub-adult ones, 2.5 years old, leave and form bachelor male herds. Copulation take place during July–August. Most young, 1.5 year old males, already sexually mature, are generally tolerated by the territorial dominant male and may serve estrous females, without the dominant male interfering. Generally, however, copulations in a family are performed by the dominant males, although sometimes a female may wander to another territory and copulate there.

Gestation lasts 7.5 months, and births occur during March–April. Litter size ranges from 1–6 (mean 3) pups. At the Tel Aviv University Research Zoo females gave birth (n = 77) between February–June and litter size was 1–5. In large litters, one or two young may die soon after birth. Newborn weigh about 200 gr, are precocious, and can walk and climb immediately after birth. When 4–7 days old they start eating herbs. They like climbing on their mother's back. Suckling lasts 10 minutes, during which the young produce a soft, low call. Weaning occurs 3 months after birth, but milk consumption is greatly reduced after the first month. Sexual maturity is reached when 16 months old, but in nature males do not generally mate before the age of 3.5 years. Adult size is reached when 3 years old. Females breed until 9 years of age. Longevity in captivity is up to 13 years.

Parasites: The hyrax is affected by internal parasites, mainly worms and spirochetes, and external ones such as ticks and fleas. They are mostly affected behind the ears and attempt to get rid of their parasites by scratching with teeth and toes and by wallowing on dry, loose soil. The following ectoparasites were found on this species in Israel: *Ctenocephalides arabicus* (Siphonaptera), *Dasyonyx diacanthus* (Malophaga), *Prolinognathus leptocephalus* (Anoplura), *Ornithodoros procaviae* (Argasidae) and *Rhipicephalus sanguineus* and *Hyalomma* sp. (Ixodidae) (Theodor and Costa, 1967).

Relations with man: The hyrax is mentioned in the Bible as a typical inhabitant of cliffs: "The high mountains belong to the ibex, the crags are a refuge for the coneys" (Psalms 104:18).

In the past, hyrax were hunted for food, but protection by law, elimination of most of their natural enemies (leopards, bearded vultures) and increased agriculture has caused a major explosion in their population size. When new areas in the Galilee and Golan Heights were prepared for agriculture by clearing them of rocks which were then piled in large piles, hyrax used these piles for retreats, causing considerable damage to agriculture and orchard trees. However, these rock piles provide little protection against cold and during cold winters many hyrax die in these new habitats. An electric fence is a good protection for crops.

Hyrax grow well in captivity and hand raised individuals become very tame. They have to be provided with well insulated boxes where they can huddle together on cold nights (Mendelssohn, 1965). Thought must also be given to their social structure, and it is important not to keep several adult males in one cage.

PERISSODACTYLA

This order is represented in Israel by one, introduced species, *Equus hemionus*.

EQUIDAE סוסיים

EQUUS Linnaeus, 1758

Equus hemionus Pallas, 1775   Onager Wild Ass
Figs 116; Plate XX:4   פרא

Equus hemionus Pallas, 1775. *Nova Comm. Imp. Acad. Sci. Petrop.*, 19:394.
Synonymy in Wilson and Reeder (1993).

The Mid-eastern onager *Equus hemionus hemippus* was the smallest of all wild horses. The measurements of two specimens kept in museums in Vienna and Paris are: body length 180–190 cm, tail 36–38 cm, leg 42–44 cm, ear 12.0–12.5 cm. The fur is light brownish-grey and is denser and darker in winter than in summer. The underparts, the inside of the legs and a narrow area along the back are white. There is a short dark brown mane on the neck, which continues as a dark line along the back.

Dental formula is $\frac{3.1.3-4.3}{3.1.3.3} = 40-42$.

Distribution: This species was widely distributed in central Asia, from the east of Turkey to Mongolia, but most populations have been exterminated. The last specimen from the Syrian Desert in Jordan was shot in the 1920s, in Syria and Iraq during the 1930s.

In Israel it was introduced by the Nature Reserve Authority into the Makhtesh Ramon and it can currently be seen in this area but also in other areas in the central Negev (Meishar, Naḥal Paran). The introduced population is a hybrid one of *E. h. onager* from Iran (three males and three females) and *E. h. kulan* from Turkmenistan (two males and three females). The latter were bought from a Dutch animal dealer. There were four releases between 1982–1987 in which 14 females and at least 14 males were released. In 1993 there were four female herds in the Negev: 20 females and nine foals ranging between territories of three males in the eastern part of the Makhtesh Ramon; a female and a foal ranging in a territory of a male in its western part; two females with two foals ranging in a territory of a male in the Meishar; and three females with one foal in Naḥal Paran. In addition to the above there were at least 30 non-territorial males. In 1995 there were four territorial males in the Makhtesh Ramon, two in the Meishar and one in Naḥal Paran with a total of 17 foals (David Saltz, pers. comm.; Figure 116).

Fossil record: The earliest remains of this species found in Israel date to the Upper Palaeolithic period (Tchernov, 1988).

Habits: The onager is a social animal. Females live in small groups composed of several females and their offspring, which wander through large territories occupied by single males. Non-territorial males live in bachelor herds. It is a typical desert animal, feeding during the cool hours of the morning and evening and traveling large distances in order to obtain food and water. Most animals range throughout the Makhtesh Ramon where annual rainfall is less than 70 mm, but come once every several days to a small spring in its eastern corner ('En Saharonim) in order to drink. This dependence on water enabled the Bedouin to hunt the original local subspecies to extinction. Bedouin would not eat donkey flesh, but ate onagers.

Food: The onager is herbivorous, preferring grasses (Gramineae), but also consuming various bushes such as *Atriplex halimus, Moricandia nitens, Lycium shawii, Centaurea aegyptiaca* and *Astragalus spinosus* (Y. Goldring, Report to the Nature Reserve Authority, 8 February 1990). The increasing population of the onager might have a negative effect on the vegetation of the Southern Negev, and this should be carefully monitored.

Reproduction: Mating takes place in spring, gestation lasts 11 months and a single foal is born. Females reach sexual maturity when two years old and may give birth at the age of three (Groves, 1974). Primiparous and old females produce primarily females and young nonprimiparous females produce mainly males (Saltz and Rubenstein, 1995). Reproductive rate is related to the amount of rainfall, and is higher in periods following more rainy years (D. Saltz, pers. comm.). Females survive up to 26 years in captivity.

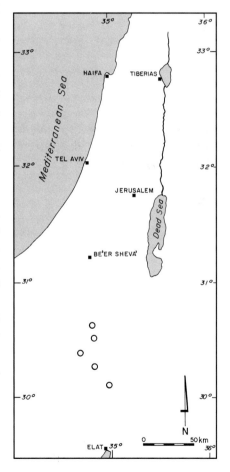

Figure 116: Distribution map of *Equus hemionus*.
O – verified field records.

Relations with man: This species was apparently never domesticated (the domestication in ancient Mesopotamia, 2500 BC is questionable), hence its Hebrew name "Pereh" (wild). In the Bible it is described as a wild desert animal ("A wild ass accustomed to the desert" Jeremiah 2:24; "Who let the wild ass go free? who untied his ropes? I gave him the wasteland as his home, the salt flats as his habitat" Job 39:5–6).

ARTIODACTYLA

This order is represented in Israel by two families, Suidae and Bovidae.

SUIDAE חזיריים

This family is represented in Israel by one species, *Sus scrofa*.

SUS Linnaeus, 1758

Sus Linnaeus, 1758. *Systema Naturae*, 10th ed., 1:49.

Sus scrofa Linnaeus, 1758   Wild Boar
Figs 117–118; Plate XXI:1–2   חזיר בר

Sus scrofa Linnaeus, 1758. *Systema Naturae*, 10th ed., 1:49. Germany.
Synonymy in Harrison and Bates (1991).

The wild boar is the largest wild mammal in Israel. It is heavily built, the neck and the legs are short, the muzzle is elongated and the eyes are small. Males are much larger than females. The feet have four well developed toes, but when standing on firm ground only the median pair touches the ground and the side hooves are used when walking on boggy ground.
Sus scrofa lybicus Gray, 1868 is the subspecies present in Israel (Harrison and Bates, 1991).
Distribution: The original distribution of this species was in the steppe and broad-leaved forests of the Palaearctic region in Europe, Asia and North Africa (Harrison and Bates, 1991). It was exterminated in several countries (Britain, Egypt), but introduced to North America. Domestic pigs which have frequently escaped or been released (Australia, many islands), have become feral and developed some similarity to the original, wild form.
In Israel the wild boar occurs in the Mediterranean region and in the swamps south of the Dead Sea, where a small isolated population of several dozen resides. During the 1930s and '40s wild boar populations were severely reduced by overhunting. They survived only in the Rift Valley from the Ḥula Swamp to the Dead Sea. After the establishment of the State of Israel (1948) hunting was greatly reduced and the remaining wild boar populations quickly increased and reoccupied their former habitats: Galilee, Yizre'el Valley, Mount Carmel and even the densely populated

Table 111
Body measurements (kg and mm) of *Sus scrofa lybicus*

	Body		Tail	Ear	Foot	Height
	Mass	Length				
			Males (9)			
	n = 18			n = 8		n = 3
Mean	100.9	1506	239	146	321	860
SD	20.6	95	30	16	18	
Range	73–150	1360–1640	200–300	125–170	295–340	780–920
			Females (12)			
	n = 7		n = 11	n = 10	n = 11	n = 2
Mean	61.4	1346	215.5	133	289	757.5
SD	11.1	57	20	12	43	
Range	53–82	1240–1450	185–240	120–160	210–320	755–760

Table 112
Skull measurements (mm) of *Sus scrofa lybicus* (Figure 117)

	Males (10)			Females (3)		
	Mean	Range	SD	Mean	Range	
GTL (M n=9)	428.1	414–451.5	12.5	371.0	368–374	(n = 2)
CBL (M n=9)	365.1	350–384	12.0	328.5	314–343	(n = 2)
ZB	160.5	148–178	9.7	144.0	140.9–146	
BB	90.4	81–100	6.7	85.0	80–88	
IC	87.8	82–98	6.3	78.9	75.5–84	
UT	166.1	157–173	5.7	148.1	145–154.5	
LT (M n=9)	179.8	173–188	4.7	160.9	157–164	
M (M n=9)	342.1	322–368	14.4	301.3	297–307	

Dental formula is $\dfrac{3.1.4.3}{3.1.4.3} = 44$.

The male canines are remarkable. Their roots remain open and they continue to grow throughout life. The upper ones turn upwards, whetting the lower ones with each chewing movement. The lower canines thus become pointed and triangular in cross section, with a sharp leading edge, and those of adult males are highly dangerous weapons. The female canines are simple, similar to those of carnivores.
Karyotype: 2N = 36–38, FN = 60, for specimens from Europe (Zima and Kral, 1984).

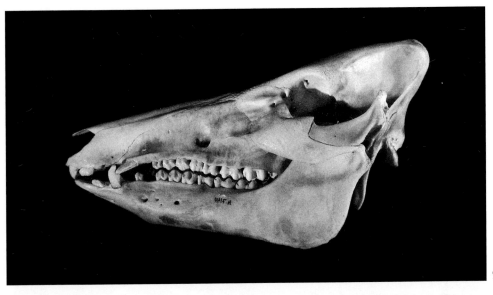

♀

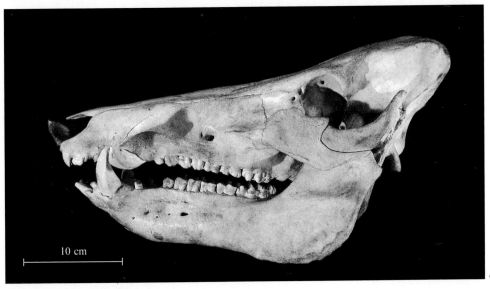

♂

Figure 117: *Sus scrofa lybicus.* Lateral views of the cranium.
Female (top), male (bottom).

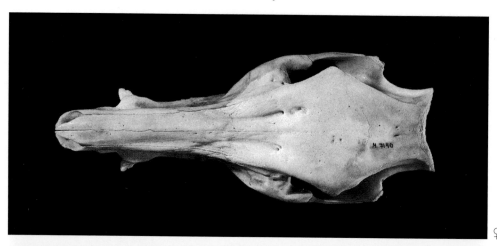

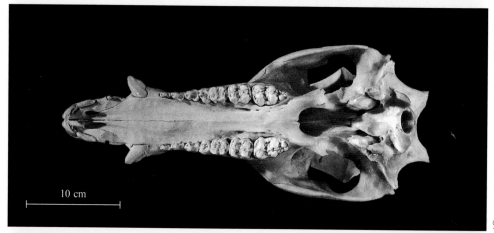

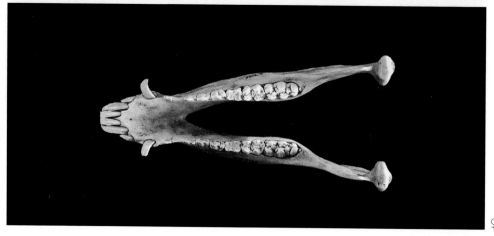

Figure 117 (cont.): *Sus scrofa lybicus*. Dorsal and ventral views of the cranium, and a view of the mandible. Female.

244

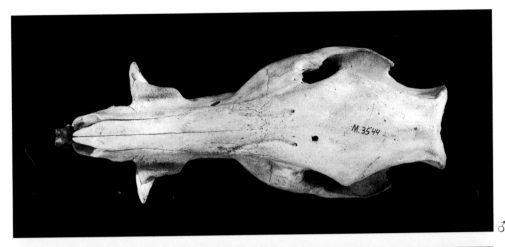

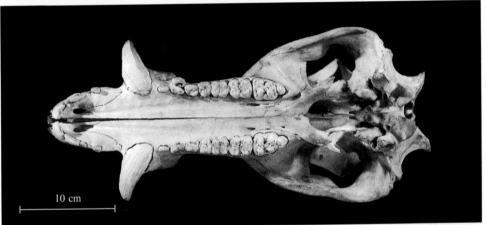

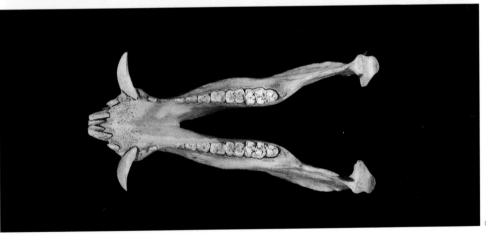

Figure 117 (cont.): *Sus scrofa lybicus*. Dorsal and ventral views of the cranium, and a view of the mandible. Male.

Coastal Plain. They have now been recorded from areas in which they had not previously been found in the present century — around Jerusalem and in the southern Coastal Plain as far south as Gedera. "Pigs" have also been spotted in Naḥal Besor, but those may have been introduced by man and are probably hybrids between wild and domestic "pigs" (Figure 118).

Fossil record: The earliest remains of this species found in Israel date to about 150,000 years BCE (Tchernov, 1988).

Habits: The wild boar prefers dense thickets, forest and riverine habitats. It is mainly nocturnal, but at low ambient temperatures it may also be active during the day. On Mount Ḥermon, at 1400–1600 m above sea level, it is common to see herds of up to 40 animals foraging together during the day. When not active they lay up in the shade, often in a shallow depression dug in the ground, which they may use repeatedly, sometimes for weeks. Wild boars like to rub their skin against rough tree trunks, especially conifers, apparently to get rid of external parasites, but this may also serve for marking. They defecate in fixed sites which function as "communication centers".

Wild boars live in family groups comprising a female with her offspring. Sometimes several such groups join together and form large herds. Within each family there is a hierarchy, with a female at the top. Herds of young males also occur, whereas adult males are solitary. Males usually do not attack females, as they have a strong innate inhibition from doing so, but fights between males occur, mainly during the breeding season (November-December). Males fight by striking at each other with strong, sideways blows of the heavy head. During these blows, the lower jaw is directed laterally towards their adversary, in order to employ the tusks. Little damage is caused, however, and any wounds are superficial, because of the layers of hard connective tissue, one to two cm thick, on the shoulders and sides of the body. These layers provide efficient shields, protecting the males from the tusks of other males. Females have no

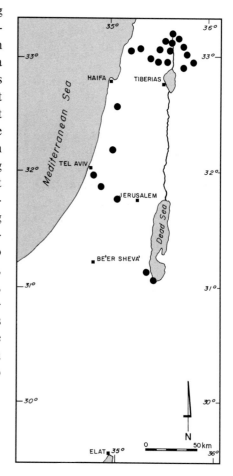

Figure 118: Distribution map of *Sus scrofa lybicus*.
● – museum records.

such shields and no protection, but males seldom attack them. Females also fight with sideways movement of the head, but they attempt to bite, using their incisors and their sharp canines. The same methods are used when fighting inter-specific enemies such as predators. Wild boars become extremely aggressive when aroused by the alarm shriek of young pigs. When attacking an unprotected enemy the pointed canines, penetrate easily and with a quick upwards movement of the head, cause large wounds. Females try to knock the opponent down and keep it down with the front feet, while biting. There are records of wild boars attacking humans, particularly when cornered or wounded and unable to flee.

Wild boars can cover large distances, but on Mount Meron they cover only about 1.5 km per night, and the diameter of the home range is about one km. Their population density there is about 3.2 individuals per km^2 (Cnaani, 1972). On the Carmel their population density was estimated to be 17.2 and 6.6 individual per km^2 in a Nature Reserve (Alona Forest) and a nearby unprotected (from hunting) forest (Rosenfeld, 1996).

They have several calls. The young scream when in danger, and such screams cause all members of the herd to come to their defense. Adults produce a snorting call. During fights they scream and when excited they bark. The alarm call of adults sounds like a loud snore. During copulation the male produces a low, rhythmic call, the "boar song".

Food: Wild boars are omnivorous, feeding on small vertebrates, insects and other invertebrates, carrion, berries, nuts, acorns (highly preferred food), seeds, rhizomes of various plants and greenery. They dig up much of the food with the snout, as well as overturning stones which may weigh dozens of kilograms and moving dead leaves from under trees; they chew young green grass and spit out the indigestible fibers in the form of an elongated lump.

Reproduction: During the mating season (November–December) males advertise by digging in the ground with their front legs and marking with urine mixed with a secretion from the prepucial glands. Copulation lasts 10–25 minutes. Gestation lasts about 4.5 months, and the young are born during March–April. Litter size ranges between 3–8, but one female from the Golan had 10 embryos. At the Tel Aviv University Research Zoo females gave birth between February–June, and litter size was 1–11 (n = 49). Females normally give birth for the first time when two years old, but under favourable conditions they may become estrous when eight months old and can produce the first litter at the age of one year. The litter size of first year mothers is three, occasionally four, while older females give birth to 6–8 young. In populations under intensive hunting pressure the proportion of one year olds giving birth is larger.

Before giving birth, the female uses her snout and front legs to dig a shallow depression in the ground, approximately one m by 1.5 m. The depression is then lined with vegetation and a nest about 50 cm high is built. This nest is used by the mother and young for several days after birth. The newborn weighs about 1 kg. Unlike most other mammals, the female does not lick the young. After delivery the young clean themselves, approach the mother from the front, and mother and young smell each other. The young can walk immediately after birth, have open eyes and are fully covered with hair. They are already able to play with each other 30 minutes after birth, sometimes even before the last of the litter has been born. They begin to nurse soon after they are born.

The female lies on her side to suckle. Normally each newborn suckles from a particular teat. During the first days, nursing takes place every 60 to 90 minutes, but the intervals between lactations increase with age. When the young are several weeks old, the mother stands during nursing. When young, the newborn huddle together to keep warm and to rest while in contact with their mother, as their thermoregulation is not yet developed. They start to consume solid food when one to two weeks old (they have 12 teeth at birth), and are weaned by three months old, at which age they have a full set of milk teeth. When they are two weeks old they follow their mother during her foraging. The young have a brown coat with yellow-brown side and back stripes, which disappear at four months of age. They stay with the mother until she gives birth to a new litter a year later, and the males then join herds of young boars. Sexual maturity is achieved at 1.5 years, but in nature males normally copulate later, as they cannot compete with older, stronger males. They have all permanent teeth when three years old. Maximum longevity in captivity is 20 years.

Parasites: On the Carmel, wild boars are infected by various internal parasites: *Metastrongylus* sp. (mainly in the lungs, 15.2% of the sample), *Ascaris suum* (2.5%), *Trichuris* sp. (3.8%), Spiruridae (2.5%) and Strongylidae (1.5%), all in the digestive system (Rosenfeld, 1996).

Relations with man: Wild boars can cause considerable damage to crops by eating or trampling them, by breaking branches of fruit trees and by digging in fields for rodents and insects. Due to their tasty flesh, wild boars are preferred game, and are hunted in northern and central Israel. It is estimated that despite the injunction against eating them by both the Jewish and Muslim faith, about 4000 wild boars are hunted in Israel annually.

During the 1960s and 1970s many wild boars were poisoned with alfa-chloralose (commercial name Tardemon) baited in dead chickens. Unfortunately, intestines which were left in the field caused secondary poisoning of carnivores and raptors.

BOVIDAE   Gazelles and Ibex   פריים

This family is represented in Israel by two genera, *Gazella* and *Capra*.

Dental formula is $\frac{0.0.3.3}{3.1.3.3} = 32$.

248

GAZELLA Blainville, 1816

Gazella Blainville, 1816. *Bulletin de la Société Philomathique, Paris*, p. 75.

These are small to medium-sized, graceful antelopes with large eyes and a short tail. The males are larger than the females and their horns are thicker and generally longer than those of the female. This genus is represented in Israel by two species: *G. gazella* and *G. dorcas*. In captivity, male *G. dorcas* have hybridized with female *G. gazella* . Most hybrids are males, and have under-developed testicles, or females which seldom reproduce. No hybrids were observed in nature.

Gazella gazella (Pallas, 1766)   Mountain Gazelle
Figs 119–120; Plate XXI:3–4; Plate XXII:1,3–4   צבי ארצישראלי

Antilope Gazella Pallas, 1766. *Miscellanea Zool.* p. 7. Syria.
Synonymy in Mendelssohn *et al.* (1995).

The following description is from Mendelssohn *et al.* (1995).
Compared with *G. dorcas*, *G. gazella* is larger on average; its ears are relatively shorter (10.2–13.5% of head and body length compared to at least 16% in *G. dorcas*). The horn cores are elliptical in cross-section (round in *G. dorcas*). The horns of adult males are of medium length (130–272 mm) compared with the longer horns (244–304 mm) of *G. dorcas*, always somewhat bowed outwards with the tips turned inward or forward; the widest span across the horns is 104–170 mm.
G. gazella is sexually dimorphic, with males larger than females and possessing longer and thicker horns. Desert subspecies are smaller: length of head and body of a male of the only recently described *G. g. acaciae* Mendelssohn, *et al.*, 1997 was 1000 mm, with a mass of only 16 kg; but length of tail was 155 mm and length of ear was 135 mm. The male horns have 14–20 prominent rings, are rather short (200–350 mm long) and thick basally. When viewed from the side they are more or less sigmoid, with the tips turned forward. When viewed from the front the horns are more or less divergent. Female horns are much smaller (generally 30% shorter then the male's horns in the same population), thinner and generally without rings; they are about 100–120 mm long and longer than 150 mm is rare. Female horns are irregular in shape, often bent, crooked or broken, but are still used for butting small predators threatening their offspring or for butting other females. In males, the horns form an angle of 60° to the basifacial axis; in females, about 30° (Mendelssohn *et al.*, 1995).
The largest gazelles in Israel, and those with the longest horns, are found in Upper Galilee, but large specimens also occur elsewhere. Mountain gazelles do not accumulate considerable fat reserves even under the most favourable conditions or in captivity. Desert subspecies are much lighter and longer-legged.
The pelage is short, sleek and glossy in summer, reflecting much of the sun radiation.

Table 113
Body measurements (kg and mm) of *Gazella gazella gazella*

	Body		Tail	Ear	Foot	Horns
	Mass	Length				
	Males (10)					
Mean	24,86	1053	105	119	336	248
SD	3,57	47	18	4	11	19
Range	17,1–29,5	1010–1150	80–130	112–123	320–354	220–291
	Females (10)					
	n = 9					**n = 8**
Mean	18,09	980	105	116	322	82
SD	2,64	36	15	5	5	22
Range	16,25–25	910–1010	90–130	110–125	320–340	58–115

Table 114
Skull measurements (mm) of *Gazella gazella gazella* (Figure 119)

	Males (10)			Females (10)		
	Mean	Range	SD	Mean	Range	SD
GTL (F n = 8)	192.2	186–196.1	3.6	179.8	172–186.2	4.5
CBL (F n = 8)	186.6	180–191.1	3.5	171.2	164–175	3.9
ZB (F n = 8)	82.6	77.5–85.9	2.6	73.9	69.8–77.4	2.9
BB (F n = 9)	57.1	55.3–57.9	0.9	53.4	50.5–55.9	1.9
IC (F n = 9)	58.2	55–60.4	1.9	51.9	46.4–55	2.5
UT (F n = 9)	58.5	56.6–60.5	1.5	57.9	54.2–59.6	2.4
LT	59.8	53.3–63.9	3.0	59.8	55.5–64	2.5
M	148.5	142.4–152.3	2.8	137.6	102.6–145.8	12.5
TB	27.9	24.4–30.5	2.3	23.9	22.3–27.9	1.7

Karyotype: In males, $2N = 35$; in females $2N = 34$ (Wahrman *et al.*, 1973).

250

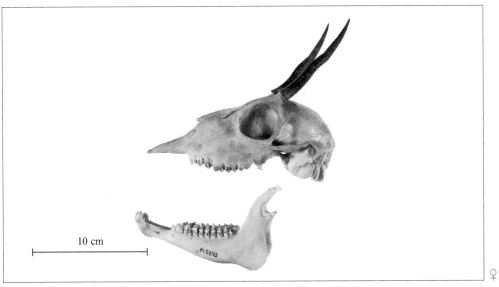

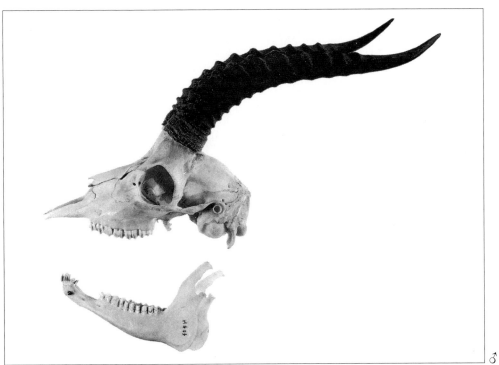

Figure 119: *Gazella gazella gazella.* Lateral views of the cranium.
Female (top), male (bottom).

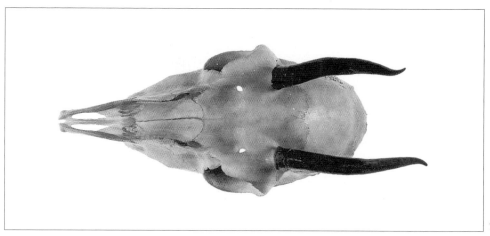

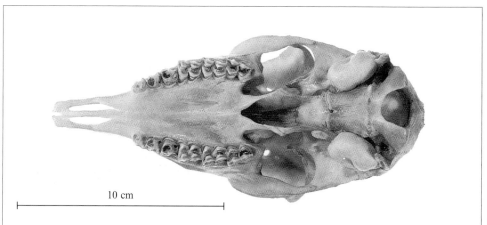

10 cm

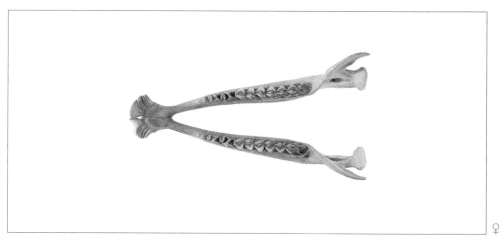

Figure 119 (cont.): *Gazella gazella gazella.* Dorsal and ventral views of the cranium, and a view of the mandible. Female.

252

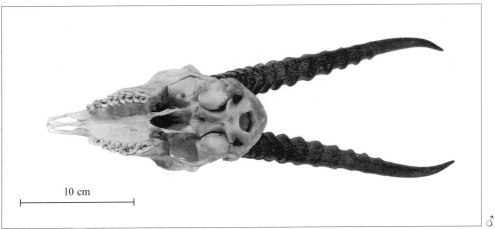

10 cm

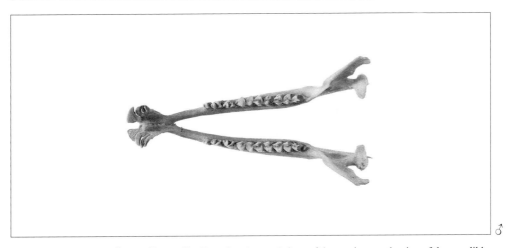

Figure 119 (cont.): *Gazella gazella gazella*. Dorsal and ventral views of the cranium, and a view of the mandible. Male.

253

In winter it is much longer, denser, and with thick underwool, rain proof and not glossy, enabling the gazelles to withstand the heavy winter rains (800–1000 mm) in Upper Galilee in their open habitats without need for protection. The fur in winter is especially long on the underside, forming a kind of blanket that insulates the body from the cold and wet soil, when the gazelle beds down. The colour of the dorsal pelage is earth-brown, but is lighter in drier regions. A dark brown side stripe commonly separates the brown upper side and the white underside, and a lighter brown stripe runs above this along the flanks. The legs are lighter than the body, and there is often a sharp transition between the dark body and light leg tones on the rump. The hairs on the rump are long and white, outlined by a dark brown line.

Distribution: The mountain gazelle is found in southern Lebanon, Syria, Israel, with west and south of the Arabian Peninsula, and isolated regions in Iran. Due to over-hunting there is now a hiatus in its distribution in the Northern Negev in Israel; it has probably extended into southern Lebanon.

Two populations of the mountain gazelle occur in Israel. *G.g. gazella* (Pallas, 1766) occurs north and west of the 150 mm isohyet, and a small population of *G.g. acaciae* is found in the 'Arava Rift Valley in the Negev, 50 km north of Elat (Mendelssohn *et al.*, 1997). In recent years *G. gazella* has appeared in the Judean Desert and has also penetrated into the Coastal Plain of the Dead Sea, following irrigated agriculture in these areas (Mendelssohn and Yom-Tov, 1987; Figure 120).

The estimated number of *G. g. gazella* in Israel in 1985 was about 10,000, a significant increase from the estimated less than 500 in 1948. The largest concentrations live in the southern Golan Heights (ca. 5000) and Ramot Yissakhar (ca. 3000). These populations, with free access to water and green food during the summer provided by modern agricultural practices, increase at an annual rate of 16% (Ayal and Baharav, 1985) to 25% (Mendelssohn

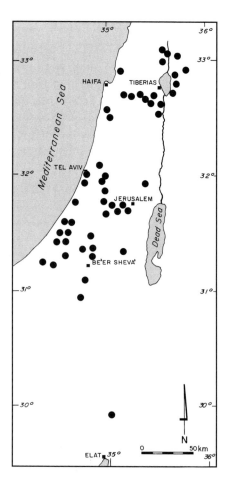

Figure 120: Distribution map of *Gazella gazella*. The record near Elat is of *G. g. acaciae*. All other records are of *G. g. gazella*.

● – museum records.

and Yom-Tov, 1987). Smaller populations exist in the Upper Galilee, mainly on the eastern slopes of Naftali Mountains, the Judean Hills, the Western Negev, the southern Coastal Plain and other places.

The isolated population of about 50 individuals, in 1987 (in 1996, it numbered only a dozen individuals) near Elat, southern Israel (of a subspecies), although having free access to water and a seemingly ample supply of high-quality food (*Acacia* trees and many bushes), has not increased in number since 1962 when the first annual count was conducted (Yom-Tov and Ilani, 1987). In autumn 1996 there were only about a dozen in this population (B. Shalmon, pers. comm.). Poaching in the past, and now predation mainly by foxes are apparently the causes of decline in this population. Fawns of *G. gazella* lay up for 4–6 weeks and are more exposed to predation by foxes than *G. dorcas*, that lay up for 1–3 days. A population of about 500 individuals of this subspecies lived near Ḥazeva in the northern 'Arava Valley until the 1950s, but was exterminated by military poaching between 1956–1963 (Mendelssohn, 1974).

Fossil record: *G. gazella* itself is only known from the Upper Pleistocene on. Taxa from the sites of Tabun, El Wad, Hayonim and Kebara, described by Bate (1940) under the names *Gazella decora, G. estraelonia, G. arista* and *Kobus cananites*, represent the extant species (Davis, 1980). Presumed early Holocene occurrences are from Kara'in Cave, Turkey; central Lebanon; Petra and Hesban, Jordan; east of Jericho, West Bank areas (where what has been described as G. *gazella* occurred with G. *dorcas* (Uerpmann, 1986). *G. gazella* was the only gazelle species documented in Israel until after the Pre-Pottery Neolithic B. periods (Tchernov *et al.*, 1986/87, Bar Oz, 1996). In the last glacial period its range expanded southward, reaching the southern Sinai Peninsula. The desert subspecies of *G. gazella* in the southern 'Arava Valley may be a glacial isolate.

Habits: Mountain gazelles can withstand severe climatic conditions. They live in the hot and dry Jordan Valley, the Negev Desert, where mid-day summer air temperatures often approach 45°C, and in the Upper Galilee where sub-zero temperatures are frequent on winter nights, and snow sometimes covers the ground for several days. They appear to be best adapted to an average annual temperature of 21°–23°C, average winter temperature of not lower than 14°C, and annual precipitation of 300–400 mm. During rainy periods they may seek refuge in places protected from wind, but they are usually exposed to the rain.

The mountain gazelle lives in many habitats, but not in dense forest or chaparral. It often dwells in very steep (up to 45°) terrain, but avoids rocky areas or walking on rocks. Gazelles prefer to walk on well-worn trails that follow the contour of slopes. When climbing a slope they walk on oblique trails, but when frightened they may run straight up the slope. Although they jump down from metre-high rock walls they jump upward only when they can see their landing place. The distribution of this species in the southern Negev coincides closely with that of dense *Acacia* stands. In Israel most natural predators of gazelles (the cheetah *Acinonyx jubatus*, the leopard *Panthera pardus* and the wolf *Canis lupus*) have either been exterminated (cheetah) or drastically reduced in number (wolf and leopard). Rarely is an adult gazelle in danger from predators, but caracals (*Caracal caracal*), wolves and packs of feral dogs are known to hunt adult gazelles, with the caracal appearing to be the most efficient. Fawns, especially while they are laid up, are taken by red foxes (*Vulpes vulpes*) which seem to be the most efficient fawn predators, jackals (*Canis aureus*), and hyaenas (*Hyaena hyaena*) and feral dogs. Although the first two of these three predators are common in areas inhabited by *G. g. gazella*, they do not significantly affect gazelle populations.

G. gazella have poorly developed preorbital glands that are apparently not used for marking in some subspecies. Glands at the carpal joints are associated with brushes of long, stiff hair. Well-developed interdigital and inguinal glands may leave odorous traces on the ground vegetation.

Population density varies between 30–50 individuals per km^2 in Ramot Yissakhar and the southern Golan Heights, respectively (Baharav, 1983) to about 10 per km^2 in the southern Carmel (Getraide and Pervolptzky, 1990) and 1.3–2.1 per km^2 in the southern Coastal Plain (Kooler *et al.*, 1995).

G. gazella are excellent runners and for several hundred metres may reach a speed of 80 km/hr. When frightened they may jump to a height of 2.4 m, but normally they cross fences by crawling under them; a space of 200 mm between fence and ground is sufficient. The high leaps (stotting) were noted as especially characteristic of this species by Vesey-Fitzgerald (1952). When excited or when fleeing, the white rump hair is erected, almost doubling the size of the white rump patch and making it, together with the dark frame and the wagging black tail, a conspicuous intraspecific communication mark. The dark side stripe, contrasting with the white underside, may also serve for visual communication because it emphasizes the white underside when the gazelles are standing and also somewhat when they rest in the normal ruminant position. A frightened gazelle crouches in a straight posture, head and neck stretched on the ground. In this posture no white is visible and the dark side and buttock stripe are close to the ground producing a perfect counter shading effect with the white hidden, likening the gazelle to a rock or a clod of earth (Mendelssohn and Yom-Tov, 1987). On hot days, gazelles rest while lying in the shade of bushes or trees (if available), whereas on cold and windy days they lie in shallow depressions which they dig with their fore legs, in places protected by rocks, or in the open if no protection is available.

The mountain gazelle posses acute vision, which seems to be its most important sense, used to identify enemies as well as conspecifics from a distance. They can see a large moving object from a distance of one km or more. They quickly learn to discriminate between a tractor (normally not a threat), and a jeep (sometimes with hunters). Their aural and olfactory senses are also excellent. The sense of smell is used mainly for food selection and intraspecific communication .

G. gazella has few calls. A sneeze-like voice serves as a warning call. While producing this call the nose is drawn down and forward. Females produce a low snoring voice to call the fawn. In extreme danger, for example while being taken by a predator, gazelles produce a strong bleating call. The bleating distress call of a fawn attracts the dam and releases aggressive behaviour.

G. *gazella* is gregarious and exhibits various social units: (a) Young male bachelor herds aged 0.5–2 years or more and numbering up to 40 individuals; (b) adult male bachelor herds, numbering up to 40 individuals; (c) female herds comprising up to 16 females of various ages and accompanied by their sons (up to 6 months old) and daughters; and (d) territorial males, aged 3 years or more. In the Golan Heights only 10% of the adult males (older than 3 years) hold territories at any one time (A. Lotan, pers. comm.). In northern Israel and the Golan Heights, territories vary in size between 0.2–0.5 km^2 (Baharav,, 1983; A. Lotan, pers. comm.). Female herds roam freely in and between the territories, and their home ranges vary between 0.2–2 km^2. Territories are marked by urinating and defecating in several spots along the perimeter of a territory; urination and defecation are performed in a ritualized sequence in the same spot, and urination always precedes defecation (Grau and Walther, 1976). Dung heaps in such marking stations may cover an area of 1 m or more in diameter. After they are deserted, a luxuriant nitrophilous vegetation develops around such marking stations (Danin and Yom-Tov, 1990). Marking is also done by rubbing the base of the horns against trees, possibly secreting odorous material from the frontal glands, but these glands do not swell during the mating season (Habibi, 1992b). It is possible that the branches from which the bark has been rubbed off confer visual as well as olfactory communication.

Territorial males may have encounters along the perimeter of their territories. They first perform a stiff-legged ceremonial walk side by side along the borders, sometimes standing in a parallel or anti-parallel position while displaying their horns. Fighting is carried out by pushing one another with horns interlocked. The rings on the horns ensure a strong hold. Occasional attempts to butt the unprotected flanks of the opponent are always met by the horns. A fight ends when one of the rival males retreats under the pressure of the other and flees, followed by the winner (Grau and Walther, 1976). Forelegs are never used in fight (Habibi, 1992b).

Mount Carmel provides a habitat for permanent groups of 3–6 females, which stay together all year long. Each group has a linear hierarchy. The mean area of the home range is 1.5 hectares, and there is a spatial, but not temporal, overlap between the home ranges of the groups, with the different groups defending their ranges against other groups (Geffen, 1995). On Mount Carmel, the proportion of time spent feeding increases from 40% in winter to 50% during autumn, probably due to lack of green food in the latter season. Most of the rest of the day is spent lying on the ground.

Food: The mountain gazelle is a grazer and browser (Geva, 1980). In northern Israel during winter gazelles feed mainly on grasses, and later in spring on dicotyledons. During summer when green herb food is scarce, they browse on the leaves and fruit of *Ziziphus lotus*, a common low spiny shrub; during autumn, before the rains, they feed on *Cynodon dactylon*, a common weed, and on acorns (*Quercus*) when available. Only a few plants are rejected altogether, and even plants that are known to be poisonous and not accepted by most herbivores, such as various Solanaceae, Oleander (*Nerium oleander*), *Urginia maritima* and leaves of the fig tree (*Ficus carica*), are eaten. Mountain gazelles were observed to live for weeks on end in potato fields with the leaves and flowers of potatoes apparently constituting their main (or only) food. However, they do not consume food whose tanin content is higher than 2.3% of dry mass, and when fed in captivity on tanin-rich food there was a significant decrease in food consumption (Geffen, 1995).

The small population of mountain gazelles in the southern 'Arava valley (*G. g. acaciae*) feeds mainly on leaves and pods of *Acacia* trees, commonly reached by standing on the hind legs and leaning on the branches with the front legs. Possibly as an adaptation to this browsing behaviour, *G. gazella acaciae* have relatively longer necks and legs than do *G. g. gazella* (Mendelssohn *et al.*, 1997). This population also feeds on leaves and young twigs of several shrubs (*Lycium arabicum, Nitraria retusa, Ochradaenus baccatus* and *Loranthus acaciae*) (Shalmon, 1988 and pers. comm.).

Feeding takes place mainly during the day but also during moonlit nights. When persecuted by humans, much of the gazelles' activity, feeding included, occurs at night. During the hot, dry summer much of the feeding (and other activities) is carried out in early morning and late afternoon and evening when temperatures are lower and the water content of their food is higher.

G. gazella prefer to drink water daily during summer, but do not always do so. Some populations exist in areas lacking in sources of surface water, consequently they improve their water balance by digging with their front legs for bulbs, corms and other succulent subterranean plant organs, or even by traveling to distant water sources. When dehydrated a gazelle may drink up to 2.5 litres of water within minutes. Gazelles are known not to feed on salt plants.

On Mount Carmel sex ratio was 1:1 (Getraide and Pervolotzky, 1990) and in the southern Coastal Plain it was two females per one male (Kooler *et al.*, 1995).

Reproduction: Under natural conditions mating takes place during autumn (October-November), but near agricultural areas, where green food and water are available throughout the year, mating takes place all year round. Such is the case in northern and central Israel, for example on Mount Carmel (Geffen, 1995), Ramot Yissakhar and on the southern Golan Heights, where young are born throughout the year, peaking in spring (April–June).

Males follow female herds passing or grazing in the male's territory. A male follows an oestrous female and performs the foreleg-kick. While following such a female, the male smells her genital area, and performs "Flehmen" when she urinates; he then mounts her repeatedly several times per minute until copulation is achieved. Copulation lasts only seconds, and is usually performed while walking. During copulation the male bends his front legs, and during ejaculation is in an almost upright position, pushing the female forward (Sambraus, 1973).

258

Oestrus occurs on average every 18 days (16–40 days) in captivity, and normally lasts 12–24 hours, occasionally 36 hours. During oestrus a female copulates several times, and in nature females were observed copulating with more than one male. If the female does not conceive she may become oestrous several times in succession until fertilization occurs. On Mount Carmel, oestrous females spent about 10% of their time in mating activities, thus reducing the time spent lying while feeding time was not shortened (Geffen 1995). Gestation lasts 180 days ($\pm$ 5). Before parturition the pregnant female leaves the herd and delivers in isolation. Mass of the newborn is about 1.75–2.5 kg, 11–12%, of its mother's mass. Neonate males are about 5% heavier than females. The newborn is well developed, having open eyes and fully developed pelage. Immediately after birth the female licks the young and eats the afterbirth tissues. The newborn tries to stand within the first hour after birth, and stands on the full length of the toes rather than on the tips. Normally the first fawn is born when the dam is two years old, but near agricultural areas many one-year old females give birth. In such areas, and in captivity, nursing females may become oestrous and pregnant, thus producing two fawns in one year. Only one instance of twins is known in Israel (Mendelssohn and Yom-Tov, 1987).

Nursing is carried out standing in an anti-parallel posture: the young suckles while the female licks the umbilical and genital region of the young. This enhances defecation and urination by the young, and the mother eats its faeces. Suckling lasts minutes (or less than a minute) and takes place several times a day. During the first weeks of its life the young spends most of the day lying curled up with eyes closed in a shallow depression in the soil, under bushes, or in high vegetation. The dam grazes or rests up to ca. 100 m from the hidden fawn, but guards it, attacking small potential predators (i.e. red fox), or trying to lead larger predators (e.g. jackal) away from it. A strong defense reaction is released by the alarm bleating of the fawn. After nursing, the dam observes the fawn seeking a hide and remembers its location (Mendelssohn, 1974). In the first few days after birth, fawns approach any adult, but the mother's extensive licking probably acts to establish the mother-young bond. At about 3–6 weeks of age the young gradually begins to accompany its mother and starts feeding on solid food. The suckling period lasts up to three months. At about that age the mother and young join a small female and young herd of 4–16 individuals. Females remain with their mother's herd possibly for life, but males leave the maternal herd when about six months old and join herds of young males (Mendelssohn and Yom-Tov, 1987). In Ramot Yissakhar 70% of the fawns survive to one year of age (Baharav, 1983), while in the Carmel only 34% do so, probably due to predation by feral dogs, jackals and foxes (Geffen, 1995).

At birth the fawn has three premolars and one molar on each side of both jaws. Second molars erupt by the age of two months, the third molars by six months and all are fully developed by 12 months. At 18 months, milk teeth start to be replaced, a process completed by 24 months. Dental attrition varies and depends greatly on diet (Mendelssohn and Yom-Tov, 1987).

Females reach adult mass at 18 months, males at three years. Under favourable food conditions (in captivity and in agricultural areas) females may conceive at 5–7 months and males may impregnate from 18 months on. Under natural conditions, however, females first conceive at 18 months and males are able to occupy territories when they are three years old. Life expectancy is 13 years in captivity, but only rarely more than eight years in nature. In several cases, captive females gave birth up to the age of 13 years, but most could not withstand the stress of pregnancy and lactation; their condition deteriorated after parturition and they died (Mendelssohn and Yom-Tov, 1987)

Parasites and disease: At normal population density, only a few parasites (*Linognathus* near *tibialis* and *Linognathus* near *pithoides*, Anoplura) were found on gazelles in Israel (Theodor and Costa, 1967). In a nature reserve in Israel, gazelles were found to harbour the fluke *Fasciola gigantica* (Nobel *et al.*, 1972). In the Tel Aviv zoo, cases of toxoplasmosis and of the esophageal nematode *Gongylonema pulchrum* have been recorded (Neumann and Nobel, 1978; Nobel *et al.*, 1969). At low population densities of less than 15 per km^2 in northern Israel gazelles are remarkably free of diseases or parasites. At densities of 30–40 individuals per km^2, however, infestation by ticks and worms (mainly lung worms) was noted (D. Barahav, pers. comm.).

An outbreak of foot and mouth disease, caused by virus type O, occurred in gazelle populations in northern Israel in the spring of 1985 (Shimshony *et al.*, 1986). It reached extensive proportions with an estimated mortality of 50% (1500 animals) in the Ramot Yissakhar Game Reserve; many other individuals suffered from various symptoms, including severe oral lesions, macroscopically visible muscular lesions in the heart, peeling of the hoofs and horns, typical buccal symptoms, and coronary and interdigital vesicles, but no diarrhea (Shimshony *et al.*, 1986). The outbreak subsided in May 1985. The unusually malignant form of the disease was attributed to a combination of the virulence of this particular virus strain and to an extremely susceptible host population, due to a high population density of about 35 individuals per km^2 (Shimsony *et al.*, 1986). Antibodies to bluetongue occur in wild *G. g. gazella*, but the virus causes no symptoms of the disease (Barzilai and Tadmor, 1972).

Relations with humans: Gazelles formed a large proportion of the animal remains found from the Neolithic period in Jericho (Clutton-Brock, 1979) and Mount Carmel (Bar Oz, 1996), and have been subject to hunting ever since. At the end of the British Mandate in Palestine, when overhunting and poaching were widespread, there were no more than 500 individuals left (Mendelssohn, 1974), but legal protection enabled the increase in numbers. At present, the Nature Reserve Authority of Israel conducts an annual count of gazelles in northern Israel, which is followed by culling these populations to a density of less than 30 individuals per km^2.

At high population densities gazelles may cause considerable damage to agriculture by eating the leaves of various crops and by rubbing the base of their horns (males), against fruit trees, thus killing young trees. Fencing and dogs provide effective protection against these damages.

In the Bible gazelles are symbols of beauty ("My lover ..... be like a gazelle on the rugged hills" (Song of Songs 2:17) and swiftness ("Asahel was as fleet-footed as a wild gazelle" (II Samuel 2:18).

Gazella dorcas (Linnaeus, 1758) Dorcas Gazelle

Figs 121–122; Plate XXII:2; Plate XXIII:1–2 צבי מדבר

Capra dorcas Linnaeus, 1758. *Systema Naturae*, 10th ed., 1:69. Lower Egypt.
Synonymy in Yom-Tov *et al.* (1995).

The following description is from Yom-Tov *et al.* (1995).

The Dorcas gazelle is one of the smallest living species of *Gazella*. Compared with its close relative, *G. gazella*, *G. dorcas* is smaller; its ears are longer (average ear length is 16% of head and body length, in comparison with a maximum of 13.5% in *G. gazella*).

G. dorcas is a light sandy-brown in colour. The lateral stripe is inconspicuous, but the facial pattern is well developed and the light supraorbital stripe is almost white. There is a black spot on the bridge of the nose in some individuals. The ears are long, about 140 mm, and are normally carried slanting laterally, but prick up when the gazelle is tense. The tail is long, about 120 mm, and bushier than in *G. gazella*, and the hooves are much narrower.

The horns are more compressed than in related species. The horns of the male, when viewed from the side, have a stronger curvature than do those of related species such as *G. gazella*, and when seen frontally are spread with the tips turning inward, showing the lyre-shape of gazelle horns more distinctly than those of most other species. They are about 244–304 mm long, and have 20–24 rings. The male horns of *G. dorcas* reach their maximum length at the age of two years, as in *G. gazella*. The male reaches sexual maturity at one and a half years; little length is added to the horns thereafter, recognizable by a few narrow rings at the base of the horn. The horns of the female are relatively strong, long and ringed, and are much longer and stronger than those of *G. gazella* females. They are straight or slightly lyre-shaped in frontal view, 170–190 mm long, and usually with 16–18 fairly conspicuous rings. Female horns continue to grow until about the third year, but horn growth slows after one and a half years of age and more so after the third year (Yom-Tov *et al.*, 1995).

Gazella dorcas littoralis Blaine, 1913 is the subspecies present in Israel (Harrison and Bates, 1991).

Distribution: *Gazella dorcas* ranges across northern Africa from Rio de Oro, Morocco east to Egypt, south to the Sudan, northern Ethiopia, Somalia, and Chad in Africa, and is found in Sinai and Israel.

Table 115
Body measurements (kg and mm) of *Gazella dorcas littoralis*

| | Body | | Tail | Ear | Foot | Horn length |
	Mass	Length				
	Males (6)					
	n = 5					**n = 10**
Mean	16.44	956	122.5	140	311	218
SD	11.59	21	11	5	22	18
Range	15–18.2	930–976	110–140	130–145	270–332	178.5–236
	Females (9)					
Mean	12.26	934	125	142	308	165
SD	1.86	69	14	5	14.5	22.5
Range	9.15–14.5	815–1010	104–145	135–150	290–337	123–199

Table 116
Skull measurements (mm) of *Gazella dorcas littoralis* (Figure 121)

| | Males (10) | | | Females (10) | | |
	Mean	Range	SD	Mean	Range	SD
(M n = 9); (F n = 5)						
GTL	179.1	170.1–190.0	6.2	173.2	168.4–177.7	3.9
(M n = 9); (F n = 5)						
CBL	175.4	166.7–185.3	5.8	168.0	162.5–170.9	3.6
ZB (M n = 9)	72.6	66.4–75.5	2.6	71.5	68.3–74.1	2.0
BB	53.6	50.0–55.8	1.8	52.3	50.4–55.1	1.9
IC (M n = 9)	48.7	45.0–52.5	2.3	46.5	42.8–51.1	2.5
UT	54.1	51.5–59.0	2.6	52.0	49.2–56.4	2.2
(M n = 7); (F n = 8)						
LT	57.2	54.4–60.0	1.9	55.8	52.4–59.2	2.2
(M n = 7); (F n = 4)						
M	145.8	140.5–151.3	4.7	144.0	142.8–146.8	1.9
TB	27.9	24.4–33.0	2.6	24.8	23.3–27.3	1.2

Karyotype: 2N = 31 for the male, and 30 for the female (Effron *et al.*, 1976).

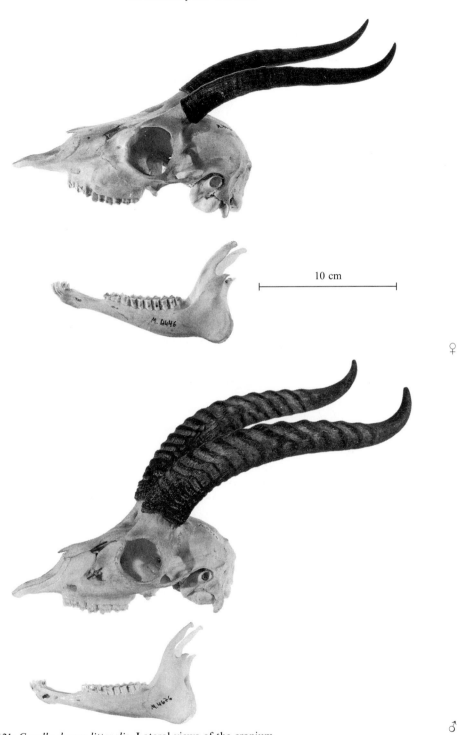

♀

♂

Figure 121: *Gazella dorcas littoralis*. Lateral views of the cranium.
Female (top), male (bottom).

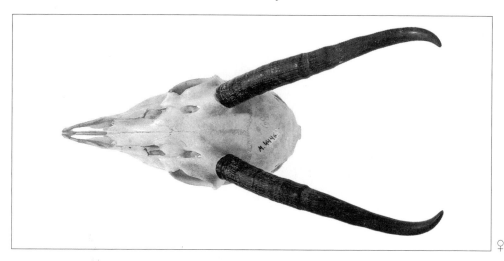

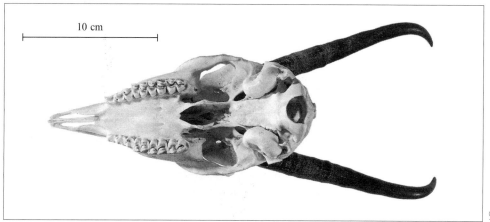

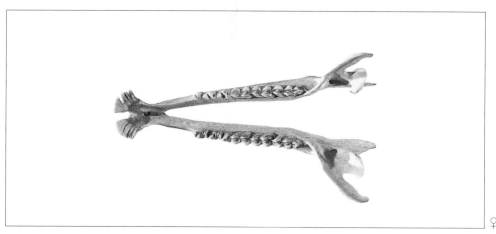

Figure 121 (cont.): *Gazella dorcas littoralis.* Dorsal and ventral views of the cranium, and a view of the mandible. Female.

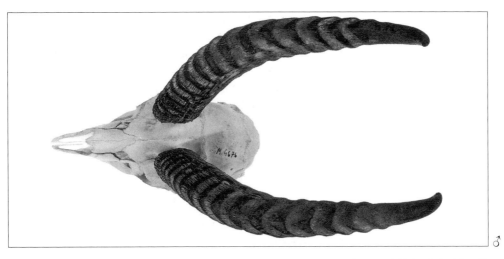

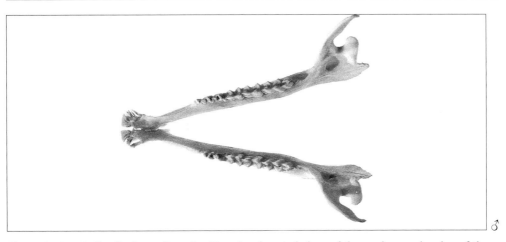

Figure 121 (cont.): *Gazella dorcas littoralis*. Dorsal and ventral views of the cranium, and a view of the mandible. Male.

265

In Israel it occurs south and east of the 150 mm isohyet in the Negev and Judean Deserts, as well as in the 'Arava (Figure 122). In parts of the western Negev *G. gazella* and *G. dorcas* may occur in the same area and may even form mixed herds. In Israel, Dorcas gazelle populations were almost exterminated by military poaching between 1956–1963 (Mendelssohn 1974), but subsequent effective protection from poaching by Bedouin and army personnel was followed by an average annual increase of 7% between 1964–1985 (Yom-Tov and Ilani 1987); this increase has slowed recently, probably due to caracal and wolf predation. The population in the 'Arava and southern Negev Desert was estimated to number 1500 in 1985 (Yom-Tov and Ilani, 1987).

Fossil record: Remains of *G. dorcas* have been found in late Palaeolithic sites near Kom Ombo, upper Egypt (Churcher 1972). There are subfossil occurrences of this species from Hesban, east of Jericho (where it occurred alongside *G. gazella*), and from sites in northern Africa (Uerpmann, 1986). Garrod and Bate (1937) identified what they considered to be two species of gazelles in the Mt. Carmel Kebara Caves, but Ducos (1968) raised doubts about this identification, and Davis (1977) identified only *G. gazella* from this cave. Hooijer (1961) identified *G. dorcas* in Kasr 'Akil, a Palaeolithic rock-shelter in Lebanon, but Davis (1974) rejected this identification on the basis of comparison with recent material. Clutton-Brock (1979) identified the Pre-Pottery Neolithic gazelle in Jericho as *G. gazella*, but these animals were later referred to as *G. dorcas* (Tchernov *et al.*, 1986/7). In the Levant and Israel, this species seems to be a post-glacial invader from northeastern Africa, displacing *G. gazella* by way of exclusion (Tchernov *et al.* 1986/7) in the more arid areas, and it may never have existed north of the Negev and Judean Deserts in Israel.

Habits: Gazelles have acute vision which seems to be their most important sense, allowing them to identify enemies and conspecifics from a distance — they can see an

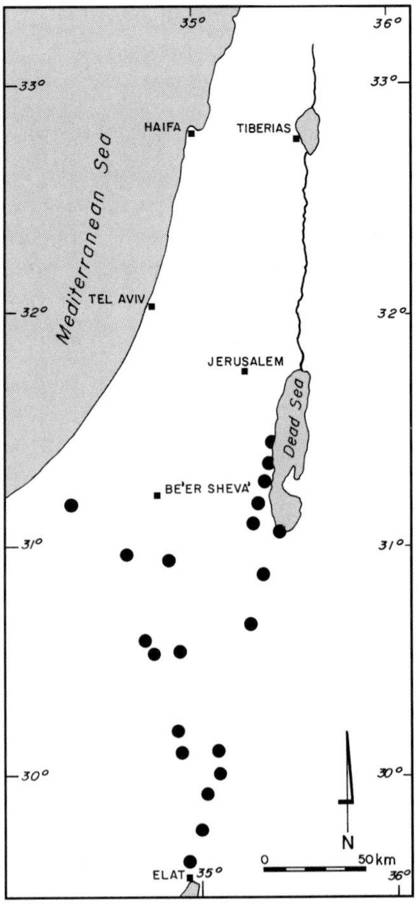

Figure 122: Distribution map of *Gazella dorcas littoralis*.

● – museum records.

arm waving from a distance of 1 km. Their hearing and sense of smell are apparently not used for detecting predators, and smelling is mainly for food selection and intraspecific communication (H. Mendelssohn, pers. obs.).

Dorcas gazelles are well adapted to the harsh climate of the desert. Captive *G. dorcas* fed on sorghum consumed an average of 420 g/day (= 1592 Kcal) during November, and 380 g/day (= 1440 KCal) in May, the hottest month in Khartoum. The average daily water consumption of these gazelles (from drinking, metabolic water and water in food) was 590 ml during November and 840 ml during May. A total of 1083 g of *Acacia* leaves provided all the necessary energy and more than the minimum water requirement in November (Carlisle and Ghobrial 1968). In Israel, the gross energy intake of captive *G. dorcas* was 600 KJ/Kg-1day-1 (Shkolnik, 1988). Animals maintained on dry sorghum had an average daily total water consumption of 4.1 L/100 kg body mass in winter and 5.6 L/100 kg body mass in summer. They withstood a lack of drinking water for 9–12 days in winter, with a body mass loss of 14%-17%, and for 3–4 days in summer, with body mass loss of 17%–20%. When deprived of drinking water they reduced their water loss by reducing urine output 3–4 fold and doubling its concentration. The thermal load of 15°– 20°C during summer caused a 30% reduction in the volume of urine produced with no significant effect on its concentration (Ghobrial, 1974). Gazelles must drink even in winter if fed on dry food only (Cloudsley-Thompson and Ghobrial, 1965; Ghobrial and Cloudsley-Thompson, 1966; Ghobrial, 1974, 1976a). The moisture absorbed by gazelles feeding on *Acacia* leaves is sufficient to cover their water loss during winter, but not during summer (Ghobrial, 1974). In Israel, however, most populations live in areas where drinking water is not freely available. In such areas gazelles appear to maintain their water balance by behavioural adaptations, and by feeding early in the morning when the water content of food plant tissues is higher than later in the day, and by feeding on green leaves and pods of *Acacia* and on various shrubs (Mendelssohn and Yom- Tov, 1987). Captive Dorcas gazelles drink an amount of water averaging 3.1% (1.0–4.9%) of their body mass daily (Ghobrial 1974, 1976b). In summer they drink daily the equivalent of 4.5% of their body mass. Preformed water and metabolic water supply on average 0.2% and 1.2% of their daily water requirement in winter, and 0.15% and 0.88% in summer, respectively. The daily output of urine averages 2.1% of body mass and the concentration of solubles in the urine averages 1.4 osmols/litre. The daily production of faeces averages 0.3% of body mass, with a water content of 52%. In summer, urine and faeces output decreases by about 30% and 50%, respectively, and the water content is reduced 3-fold. Gazelles begin to sweat at ambient temperatures of 25°C. Cutaneous water loss (from their body surface of 0.76 m^2) varies with ambient temperature and increases from 2.8–5.6 gh-1m^2 at 20°–22°C, to 81.8–87.4 gh-1m^2 at 26°–30°C. The respiratory rate increases with temperature from 45–55 breaths/min at 28°C, to 50–75 breaths/min at 29°C. Respiratory water loss is 0.15–0.35 mg/min. The total water loss during the day in summer reaches 300–400 ml/12h, dropping to 75–100 ml/12h at night (Ghobrial 1974, 1976b). When deprived of water, their body temperature fluctuates from 38.1°

to 40.2°C. Rectal temperature ranges from 38.6° to 39.2°C in winter, and from 38.8° to 39.8°C in summer. Maximal rectal temperature is 41°C (at ambient 47°C) and minimum is 37.7°C (at ambient 10°C; Ghobrial, 1974, 1976b).

Dorcas gazelles inhabit savannas, semi-deserts and deserts. They live in plains, broad wadis and occasionally in canyons and hilly country, but avoid steep terrain. They prefer hammada (stony desert) to sand, but do occur in sandy areas (although not on shifting dunes), and not in saline deserts. In all of these habitats they face extreme desert conditions where the ambient temperature may reach 45°C in Israel, and the average rainfall is 25 mm with large annual fluctuations (Yom-Tov and Ilani, 1987).

The daily activity of *G. dorcas* is determined mainly by climate. In summer they are active during the early morning (0500–0800) and the evening (1600–1800; Ghobrial and Cloudsley-Thompson, 1976), whereas in winter they may be active all day long provided the ambient temperature is not high. Where threatened, and on clear nights, they become nocturnal (Mendelssohn and Yom-Tov, pers. obs.). Dorcas gazelles are excellent runners and for several hundred metres may reach a speed of 80 km/hr (Yom-Tov and Mendelssohn, unpublished data). During midday in summer they rest while either standing or lying in the shade of *Acacia* trees or bushes. On cold and windy nights they lie in shallow depressions dug with their feet in places protected by rocks or bushes, but also in the open if no such places are available. They generally prefer to rest in places from which they are able to survey their surroundings well. The depressions in which they rest at night are generally found on elevated, flat hammadas. In Egypt, at high ambient temperatures they may seek shade near overhanging cliffs (Osborn and Helmy, 1980) as they also do in southern Sinai, where they dig their resting depressions in the shade of the cliffs (H. Mendelssohn, unpublished data).

Gazelles in Israel are hunted mainly by the caracal (*Felis caracal*) and the wolf (*Canis lupus*), but also by the hyaena (*Hyaena hyaena*). Fawns are also preyed on by smaller cats (*Felis* spp.), ratels (honey badger, *Mellivora capensis*), jackals (*Canis aureus*) and foxes (*Vulpes vulpes*). When agitated, *G. dorcas* react by tail twitching and later by erecting the tail (Aldos, 1986), skin shivering and making bouncing leaps with head high. The gazelles increase their alertness and flight distance when poaching begins to occur. The flight distance provides a fairly accurate reflection of poaching incidence.

Gazelles have few calls. The alarm call sounds like something between a short bark (Flower 1932) and a short, loud snort. The "annoyed" call is a long growling "rooo" (Haltenorth and Diller 1980). Females produce a low grunt to call their fawns. A loud bleating call is produced in extreme danger and pain (Mendelssohn and Yom-Tov, 1987).

In extreme desert conditions *G. dorcas* lives in pairs but where grazing is favourable they live in family herds consisting of an adult male and several females and young. Herds of 5–12 individuals are not uncommon in wide wadis with many *Acacia* trees and bushes in the southern Negev Desert (Baharav, 1980). Herd composition is determined largely by the distribution and abundance of food. After rains they tend to disperse on the wide plains and graze on annuals, whereas during the dry summers they concentrate in wide wadis where bushes and trees provide green food and shade.

Food: Dorcas gazelles feed on leaves, flowers and pods of various *Acacia* trees (*A. raddiana, A. tortilis*) and on leaves, young twigs and/or fruits of several species of bushes. They prefer *Astragalus vogelli, A. spinosus, Crotalaria aegyptia, Eragrostris bipinnata, Nitraria retusa, Ochradenus baccatus,* and *Zizyphus spina-christi,* but also eat *Argyrolobium saharae, Convolvulus tanatu, Farsettia ramosissima, Hippocrepis contricta,* and *Trichodesma africanus.* They rarely feed on salt plants such as *Anabasis articulata, Atriplex halimus* and *Suaeda* sp., and have been observed feeding on *Alhagi maurorum* (Baharav, 1980, 1982; Carlisle and Ghobrial, 1968; H. Mendelssohn, pers. obs.; Osborn and Helmy, 1980). They sometimes browse on trees while standing on their hind legs (Shalmon, 1987). After rains they graze on annuals and have been observed digging out bulbs of geophytes (Mendelssohn and Yom-Tov, 1987). Ward and Saltz (1994) found that in sand dunes in the Negev, this species feeds on the madonna lily *Pancratium sickenbergeri.* In summer, when all live plant material is subterranean, the gazelles dig holes in the sand to reach the stem and bulbs, while after the rains they eat the emerging tips of the leaves. The importance of *Acacia* trees for gazelles is also reflected in their population density: five individuals per km^2 where there are many *Acacia* trees, compared with 0.09 individuals per km^2 in areas without *Acacia* (Baharav, 1980).

In the southern Negev in spring, gazelles spend 2–8 hours a day browsing and grazing, and during this time may walk as far as 12 km. The time spent in a particular feeding area depends on food availability. During winter and spring, when green food (annuals and leaves of bushes) is abundant, they spend up to 40 min in one area, but much less during summer (Baharav, 1980). The time spent in each area is also a function of group size; the larger the group the less time it spends in each area. This is due to aggressive interactions between the females in the group (Baharav, 1982). In Israel, *G. dorcas* can live without drinking water provided they get moisture from the food. The high water content of *Acacia* leaves, bulbs and even dead annuals (at night and early morning) is sufficient, and *Acacia* leaves are the primary summer food (Baharav, 1980). In captivity, after four days of water deprivation and feeding on dry food, gazelles lost 14% of initial body mass, and afterwards replenished the loss of 2.2 litres within several minutes (Shkolnik *et al.*, 1979). They have a rather low water turnover (58 ml per kg body mass per day; Shkolnik *et al.*, 1979). Meinertzhagen's (1954) observation that gazelles drink sea water was contradicted by Ghobrial's (1976a) experimental results. However, when sea water was diluted by 50% or 25%, the animals maintained their original weight.

Reproduction: During the mating season an adult male may guard a small (several hundred m^2) territory, threatening or attacking approaching males. The territory is marked by concentrations of droppings and urination at permanent spots. Territorial males are found in areas with good grazing and may guard territories throughout the year, with maximum activity during the principal periods of sexual activity, September-November in the 'Arava, Israel (H. Mendelssohn, pers. obs.). Dung heaps are scattered throughout the territory rather than along its border (Essghaier and Johnson, 1981). They do not mark with secretion from their preorbital glands, which are quite small, but appear to use the interdigital and the frontal glands by digging in dung heaps and rubbing the base of the horns against trees and bushes, respectively. Bachelor herds consisting of 2–5 males were observed in the southern Negev and 'Arava, where *G. dorcas* spent the day grazing and browsing in the wide wadis, but at night remained in the foothills nearby or on the wide hammada (reg) plains. A harem consisting of 1–4 females and young accompanied by an adult male has a home range of 25 km^2 there (Baharav, 1980).

In most of its range and under natural conditions, mating takes place during September-November. Near agricultural areas, however, where green food and water are available throughout the year, mating occurs in other seasons (Mendelssohn and Yom-Tov, 1987). Females of a captive herd in Tel Aviv gave birth between March–June and September–November. In the southern 'Arava Valley in Israel, captive females occasionally mated two weeks after parturition, if they lost their fawn (Mendelssohn, pers. obs.). Pregnancy lasts six months (169–181 days; Dittrich, 1972; Slaughter, 1971), a few days more than *G. gazella*.

The dam gives birth to a single fawn with a body mass of 1.3–1.7 kg; twins are rare (Furley, 1986), occurring in Algerian populations, but have never been observed in Israel. The newborn is well developed, with eyes open and fully developed fur. Immediately after birth the female licks the young and eats the afterbirth tissues. The fawn tries to stand within the first hour, using the full length of the toes rather than just the tips. Unlike the mountain gazelle (*G. gazella*) the Dorcas fawn may nurse on its first day, while dam and fawn are lying down. The first fawn is usually born when the dam is two years old (Mendelssohn, 1974), but captive females have been observed to mate successfully at less than one year of age (Furley, 1986), although this has not been observed in Israeli *G. dorcas* either in the field or in captivity. In the Negev Desert about 90% of adult females become pregnant, but no pregnant yearlings have been observed (Baharav, 1983). Nursing is carried out with the mother and young standing in an anti-parallel posture. The young suckles while the female licks the umbilical and genital regions of the young. This enhances defecation and urination by the young, and the mother eats its faeces. Suckling lasts from less than a minute to about two minutes and takes places several times a day. During the first week of its life the young spends most of its time lying curled up with eyes closed in a shallow depression in the soil or under bushes (Baharav, 1983). The dam grazes and rests at some distance from the hidden fawn. The laying-up period of the *G. dorcas* fawn is shorter than that of the *G. gazella* fawn. At a few days to one week of age, the young gradually begins to follow its mother and to feed on solid food. The suckling period lasts for about three months and the fawn weighs about 7 kg at weaning (Baharav, 1983), at which time the mother and young join a small herd of up to 12 individuals.

In areas with much food (i.e., *Acacia* trees and bushes) average reproductive success was 0.24 fawns per mature female per year, compared with 0.16 fawns per female in relatively poor feeding areas (Baharav, 1983). The high mortality of fawns in 1964–85 resulted in a relatively low mean annual growth rate of 9% for *G. dorcas* populations in rich areas of the southern Negev and ʻArava in Israel, whereas in poor areas it was only 3% (Yom-Tov and Ilani, 1987). Longevity in captivity is up to 15 years, somewhat more than that of the mountain gazelle (Mendelssohn and Yom-Tov, 1987).

CAPRA Linnaeus, 1758

Capra Linnaeus, 1758. *Systema Naturae*, 10th ed., 1:68.

This genus is represented in Israel by one wild species, *C. ibex*.

Capra ibex Linnaeus, 1758   Nubian ibex
Figs 123–124; Plate XXIII:3–4; Plate XXIV:1–3   יעל

Capra ibex Linnaeus, 1758. *Systema Naturae*, 10th ed., 1:68. Valais, Switzerland.
Synonymy in Harrison and Bates (1991).

The male ibex is much larger than the female. It possesses longer and wider horns and is bearded. The beard starts growing in the second or third year, and becomes longer and darker with age.

Male horns may reach a length of 130 cm, being wider in front than at the back, and with bulges in front. A one year old male may have up to 10 small bulges on its horns, which erode with age and are not seen in the adult. However, when about one year old the first large bulge grows, and each year 1–3 bulges, sometimes more, are added. An old male may accumulate 30 or more, although the number of bulges is not an indicator of age, and may differ on the two horns of the same individual. The bulges serve during fights between males: when locked they prevent sliding of the horns. Age can be estimated by annual growth rings which are best seen on the back of male horns (less so in the front, and not on female horns). These rings are formed during the mating season, when the horn ceases to grow. Annual addition to the horn decreases with age, starting with up to 20 cm in the first year and less than one cm in old age. Female horns are much shorter and thinner than those of the males and are more pointed and elliptical in cross section. The longest female horn measured by us was 40 cm long. Although they grow only several mm annually, female horns are effective weapons against predators and conspecifics alike.

The fur is generally light yellowish-brown in the young, in females and in males in spring and summer. This coloration provides good protection in the pale desert habitat. The lower part of the legs has a conspicuous black and white pattern,

Table 117
Body measurements (kg and mm) of *Capra ibex nubiana*

	Body		Tail	Ear	Foot	Horn length
	Mass	Length				
	Males (4)					
	n = 1					**n = 5**
Mean	52	1252	116	124	299	485
SD		63	25	15	16	65
Range		1180–1270	95–150	105–140	280–320	425–580
	Females (3)					
Mean	25.3	1011	105	124	254	197
Range	21–28	993–1040	100–110	120–130	250–262	165–215

Males may weigh up to 60 kg and females up to 30 kg.

Table 118
Skull measurements (mm) of *Capra ibex nubiana* (Figure 123)

	Males (6)			Females (3)		
	Mean	Range	SD	Mean	Range	SD
GTL (M n = 5)	240.9	238.3–248.7	4.4	208.4	206.2–211.7	
CBL (M n = 5)	232.1	228–242.2	5.8	198.9	195.2–204	
ZB (M n = 4)	104.9	98.8–112.4	5.5	91.9	90.5–94.6	
BB	71.8	69–76.5	2.7	61.9	60.9–63.2	
IC	82.3	78–86	2.9	69.6	65.6–71.7	
UT	68.4	63.8–75.5	4.5	64.5	61.4–67.2	
LT	72.2	66.7–77.9	4.9	69.6	67.6–71.9	
M (M n = 4)	192.2	186–200	6.3	176.5	176.4–176.5	

Karyotype: $2N = 60$, $FN = 58$, for specimens of unknown origin (from zoos. Zima and Kral, 1984).

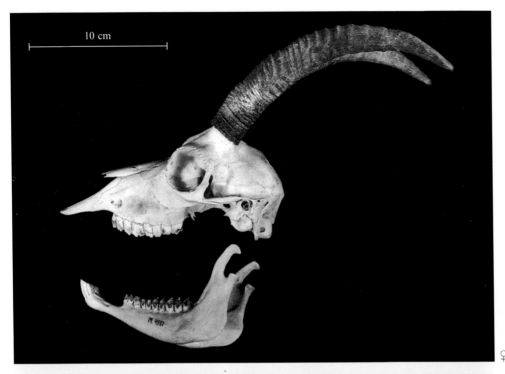

Figure 123a: *Capra ibex nubiana*. Lateral views of the cranium.
Female (top), male (bottom).

273

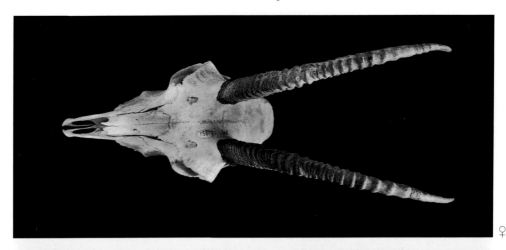

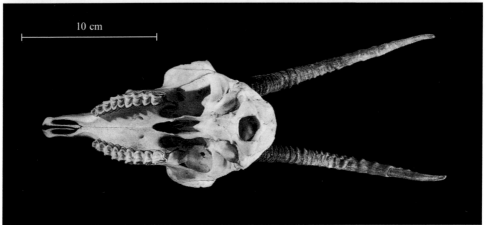

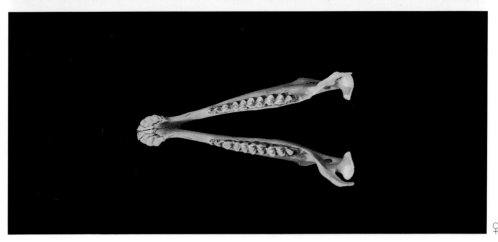

Figure 123a (cont.): *Capra ibex nubiana*. Dorsal and ventral views of the cranium, and a view of the mandible. Female.

that may have a signal value, while the animal is standing, walking or running. When the ibex lies down, the conspicuous legs are hidden and the animal blends with its surroundings. The rump is slightly lighter than the rest of the fur and the short tail is darker, but this pattern is not as conspicuous as in gazelles.

Towards the breeding season in autumn males develop a "rutting fur". The upper part of the legs, the breast and the sides of the body become dark brown, almost black. Together with the black beard and the large, dark horns the adult males are now very conspicuous in the light desert. This coloration begins to appear at the age of three or four years, but only late in the season, in October. The older the male, the earlier the rutting fur appears, and with fully adult males of six years or more it already appears in August, or even in late July. For reasons as yet unknown some fully adult males do not adopt the rutting fur and remain with the light protective coloration throughout the year. Whether such males never change to the rutting fur or whether this phenomenon occurs only once in their life time, is not known.

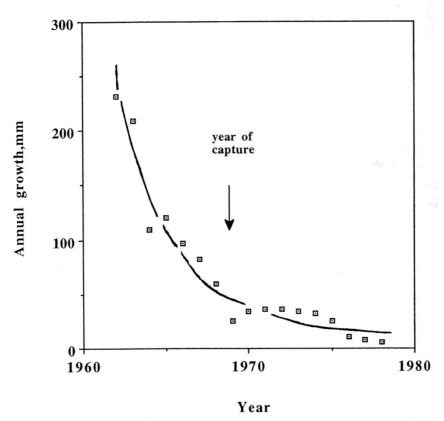

Figure 123b: Annual growth of the horns of a male *Capra ibex nubiana* between 1962–1978. The animal was caught in 'En Gedi on 24 December 1968. Note the reduced growth rate 1969, the year following its capture and transfer to a zoo.

Capra ibex nubiana F. Cuvier, 1825 is the subspecies present in Israel (Harrison and Bates, 1991).

Distribution: This species occurs in the mountain ranges of central Europe, Arabia, Ethiopia, Egypt, Sinai, Turkestan, Caucasus, central Siberia, Afghanistan, Pakistan and north western India (Corbet, 1978).

In Israel it occurs in the Judean and Negev Deserts, mainly east and south of the 100 isohyet. In the Judean Desert they presently occur as far north as Naḥal Prat near Jericho, and in winter 1996 a male ibex was observed near Ma'ale Efraim, about 30 km north of Jericho (D. Nissim, pers. comm.). Six ibex were introduced by the Nature Reserve Authority to the southern Golan Heights in 1970 and their number there currently stands at more than about one hundred (D. Pevzner, pers. comm.; Figure 124). This introduced population developped well not withstanding the very different climate, demonstrating the great adaptability of this species. The number of ibex in Israel has risen rather dramatically since they became legally protected from hunting. For example, in the oasis of 'En Gedi there were about 80 during summer 1970, about 160 in 1980 (Hakahm, 1982), about 200 in 1985 and currently (1995) at least 350. It is estimated that today there are about 1500 ibexes in Israel, about half of which live in the Judean Desert, 400 are near the springs of Ramat 'Avedat in the central Negev, 150 near Elat and the rest in various localities in the Negev. The present number would appear to be the carrying capacity of their habitat in Israel.

Fossil record: The earliest remains of this species found in Israel date to about 150,000 years BCE (Tchernov, 1988).

Habits: The behaviour of this species was studied in 'En Gedi by Aronson (1982). Ibexes are inhabitants of steep cliffs, climbing and jumping with great agility. They can jump several metres from one side of a wadi to the other, over great depths. When walking on cliffs they usually form a single file and prefer to walk on horizontal, well

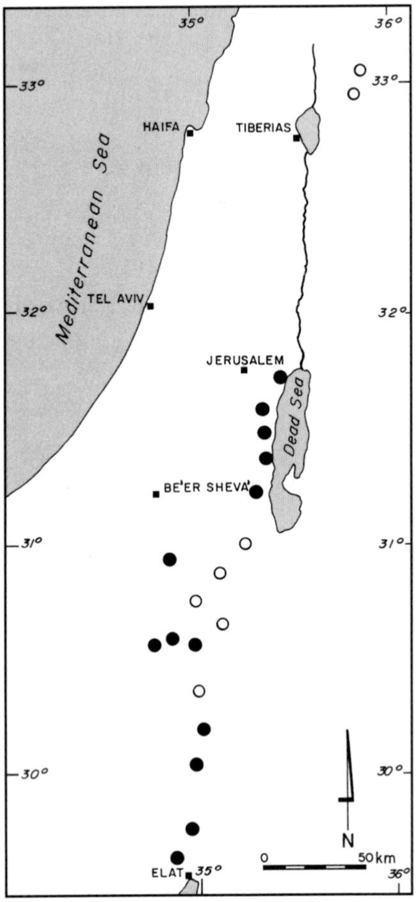

Figure 124: Distribution map of *Capra ibex nubiana*.
● – museum records, ○ – verified field records.

276

used trails, which have probably been used for generations. They may jump vertically up to three m height. Ibex, particularly females, climb on trees in order to
browse on fresh leaves. There are recorded cases of the horns of large males becoming lodged in the branches of trees, with the males being unable to free themselves
and having died in this position.

Baharav and Meiboom (1982) noted that in Sinai they follow the sun's path during
the day, starting on eastward facing slopes in the morning and by the evening being
found on southwesterly slopes, apparently in order to gain maximum solar radiation
during cold winter days. When resting, they bed down on the upper part of slopes,
which provide a good survey of the surroundings. A group of adult males was
observed in such a place at midday on a hot August day (air temperature 34°C), resting exposed to the full glare of the sun. Their short, glossy summer fur would appear
to reflect much of the solar radiation. Despite their thick winter fur being rain proof,
they are shy of rain, at the first sign of which they run to find shelter in caves or
under overhanging rocks.

Ibex have several home ranges during the year. In winter and early spring the herds
are usually spread out over desert plateaus where there are fresh annuals, while during summer they concentrate in areas near water, particularly oases and deep ravines
where they can find shade and browse. In 'En Gedi, for example, there are about
three times as many ibex in the summer than in winter.

In 'En Gedi ibex spend about 50% of their time feeding, 24% resting and 26% walking and standing (Hahakm, 1982). They sleep on steep slopes, on which each individual excavates with its front legs a shallow depression in the ground, clearing it
of debris and stones before bedding down. They usually sleep in herds, sometimes
numbering 90 individuals.

The ibex is a social animal, normally living in either female herds with or without
young (up to 2 years of age) or male herds, numbering 2–20, but at times even 50
individuals. Usually only very old males are seen alone. Female herds appear to
be composed of related individuals: a mother, her daughters and grand-daughters.
Greenberg-Cohen *et al.* (1994), working in 'En 'Avedat in the Negev have shown
that a female herd features a linear dominance hierarchy. Social rank correlated
with aggression and age, horn size and other morphological traits. Rank was established by various antagonistic interactions, which took place mainly during the
morning and late afternoon, when the animals were actively moving and feeding.
There is also a clear hierarchy in male herds. A fight starts with each individual displaying its horns, while turning the head to the side. It proceeds with frontal horn
clashes, followed by pushing. Often the opponents stand on their hind legs and
fall upon one another, thus increasing the impact of the horn clash, which is
heard from a great distance. Females also try to stab each other with the tips of
their horns. At times a horn gets broken in a fight, and occasionally the horns
become entangled and the males die with their horns locked together. Fights deter-
mine the hierarchy of the herd. Males seem to be at their peak when 6–10 years old,
after which they cannot compete with younger males.

When alarmed, ibex utter a shrill whistle. They are preyed upon by leopards, and possibly also by caracals and wolves when they are far from their cliffs. When attacked, the kids produce a bleating call.

Food: The ibex is herbivorous, during winter feeding mainly on grass and herbs, and when these dry switching to bushes (Hakham, 1982). In the 'En Gedi area they prefer grass, but when grass is largely unavailable during summer they browse on leaves and branches of bushes and trees, preferring (in diminishing order of preference): *Acacia* spp., *Zizyphus spina-christi, Moringa peregrina, Salvadora persica, Atriplex halimus, Ochradenus baccatus* and *Salsola* spp. The following plants are generally rejected (in diminishing order of rejection): *Solanum incanum, Calotropis procera, Aerva persica, Withania somnifera* and *Tamarix* spp. (Hakahm, 1982). Species which are rejected when green may be consumed when dry, apparently because their secondary compounds no longer function. Males eat larger and coarser food items than females.

Baharav and Meiboom (1982) analysed ibex faeces from the mountains of southern Sinai during winter, and found that the main species consumed were *Globularia arabica, Helianthemum lipii, Ephedra foliata, Zilla spinosa, Thymus decussatus, Gymnocarpus decander* and *Echinops glaberrimus*, while *Juncus* sp. and *Pyrethrum santolinoides* were consumed at a lower rate.

During summer ibex must drink at least once every few days. However, in the Negev there are herds which dwell far from any water source (for example in Naḥal Zihor), and which must walk every few days to seek water or dig for it in wadi beds. Such water sources, known by the Bedouin as "tmile", are generally dug below dry water falls where gravel covers a water holding layer of soil. The dependence of ibexes on water is exploited by the Bedouin, who build ambush sites of stones above such sources and shoot ibex that come to drink.

Reproduction: Mating season occurs during October–November, and most females are served by a dominant male (Hakham, 1982). During oestrus the males produce a strong odorous secretion from glands in the anal region and the underside of the tail which is easily smelt by humans. Courting takes hours on end: the male walks after the female, smells her anal region sometimes with his upper lip curled backward and up (Flehmen) and with his head and neck thrust forward so that the horn tips touch his back, and with his tail raised. From time to time the male protrudes its tongue, quivers it rapidly and then withdraws it . Copulation lasts seconds, and is rarely observed. Although males reach sexual maturity when two years old, they are able to mate only when at least five years old, when they can compete with other adult males.

Litter size is 1–2, and gestation lasts five months. In 'En Gedi births are quite synchronous and all take place within 3–4 weeks (Hakham, 1983). In nature the young are born in spring, but in a captive herd at the Tel Aviv University Research Zoo births occurred between March–July. Before delivery, the female leaves her herd and seeks a hiding place. This behaviour was noted in the Bible: "Do you know when the ibex give birth?" (Job 39:1). Delivery takes about an hour, while the female lies on the ground. The newborn stands within 15 minutes, and suckles within two hours after being born. Suckling lasts less than a minute. For their first day or two the young tend to hide among rocks or below cliffs while the mother stands in the vicinity. The young are able to follow their mother from almost immediately after birth, and several days after delivery they return with her to the mother's herd. When about a week old the kid starts to nibble green food, but they are only weaned when four months old. The young spend much time playing among themselves jumping, fighting and pushing each other.

In 'En Gedi, females tend to rear their young (up to one month old) in crèches numbering several kids. In 1981 there were up to 40 crèches (Hakham 1983). An interesting phenomenon was observed for several years between 1987 and 1991 at 'En 'Avedat in the central Negev, where more than 20 females brought their kids to a certain place on a steep slope of difficult access and formed a "nursery band" (Levi and Bernadsky, 1991). The nursery occupied about ten hectares and was about 50 m above the wadi bed, on a north facing slope. The kids were brought there when they were between 2–15 days old, and stayed for about 12 days on average. The nursery was active for about a month. Females arrived there several times a day for suckling, and left to feed at the wadi afterwards, but there were always at least three females at any one time at the nursery. The maximal number of kids at the nursery was 48. The function of the nursery appears to have been to enable the females to feed while their kids were protected from predation. The nursery area was used for night rest also by young males and non-breeding females. However, Müller *et al.* (1995) claim that the site is not a "crèche" as suggested by Levi and Bernasky, 1991), but a natural trap in which the kids accidentally enter, but are not capable of leaving until they are several weeks old. Müller *et al.* (1995) base their judgment on several facts: dams attended only their own kids when in the nursery and were often aggressive towards other kids and females approaching the nursery and three kids fell to their death from the site. These authors also suggest that kids and their mothers enter the trap due to human disturbance.

In nature females become pregnant when 18–30 months old, but in captivity they may do so when six months old. Adult size is reached when six years old in males and 3–4 years in females. Under good environmental conditions females give birth annually, and there is a high proportion of twins. In 'En Gedi, reproductive success depends on rainfall: during poor rainfall years about 20% of the females are accompanied by kids, while during a good year this proportion rises to 40%. Although sex ratio at birth is apparently 1:1, more females than males are normally observed in 'En Gedi (Hakham, 1982). Longevity in the wild is up to 12 years, but in captivity it is up to 18 years.

Parasites: Ibex are known to be infested by several external parasites, such as the ibex fly (*Lipoptema chalcomelanea*), fleas and ticks (*Hyalommina rhipicephaloides*) (Theodor and Costa, 1967).

Relations with man: The main enemy of the ibex is man. In Israel they were overhunted and their population reduced to a very low number (probably less than

400) during the British Mandate period, until protection by law enabled them to recover. A herd of about 50, which was observed in 1934 near 'En Fashkha at the north-west end of the Dead Sea, was almost completely exterminated by 1948. This occurred despite their flight distance of 800 m and, according to the game ordinance of the Mandatory Government, their being completely protected. When protected they become very accustomed to man, and in nature reserves such as 'En Gedi and 'En 'Avedat they allow people to approach them to a distance of less than 5 m, although females are shyer than males. In such areas they can cause damage to agriculture. Large males are very assertive and may occasionally adopt threat behaviour towards approaching humans. However, where hunting takes place they keep a distance of several hundred metres.

Bedouin sometimes raise ibex kids in their goat herds (Murray, 1912 and Y. Yom-Tov, personal observations). Ibex can breed with goats, and the offspring are fertile. An experimental herd of such hybrids is kept at Kibbutz Lahav in the Northern Negev.

The ibex is well known in ancient history in the Middle East. They are depicted on walls in tombs of ancient Egypt and are a frequent motif in rock paintings in the Negev and Sinai deserts. In Naḥal Mishmar, Judean Desert, a Chalcolithic treasury contained bronze artifacts of ibex heads. Ibexes are mentioned several times in the Bible as typical inhabitants of cliffs: "The high mountains belong to the ibex" (Psalms 104:18). Females are a symbol of beauty: "A loving female ibex" (Proverbs 5:19).

LAGOMORPHA Hares

LEPORIDAE ארנביים

This order is represented in Israel by the genus *Lepus*.

LEPUS Linnaeus, 1758

Lepus Linnaeus, 1758. *Systema Naturae*, 10th ed., 1:57.

This genus is represented in Israel by one, very variable, species — *Lepus capensis*.

Lepus capensis Linneus, 1758   Cape Hare
Figs 2–3, 125–127; Plate XXIV:4; Plate XXV:1–2   ארנבת

Lepus capensis Linneus, 1758. *Systema Naturae*, 10th ed., 1:58. Cape of Good Hope.
Synonymy in Harrison and Bates (1991).

A medium size mammal with long ears, hind legs longer than fore legs, and a short tail. This species is highly polymorphic in Israel. Its colour varies from brown to yellow, and specimens from northern and central Israel tend to have black or grey hair on the centre of the back. Fur colour varies in relation to latitude, aridity and soil. In the Mediterranean region of Israel and in areas of dark soil (normally corresponding to the Mediterranean region and on basaltic soils), the fur tends to be darker than in desert areas or areas of light soil (in the Negev or on sand dunes). In areas where light and dark soils occur side by side, as in the central Coastal Plain, both dark and light coloured hares are found.

Body parameters vary greatly: mean body mass of hares from the 'Arava valley is 1650 g and mean body length is about 50 cm, while in Bet She'an valley in northern Israel they are about 3000 g and 60 cm, respectively. The decrease in body dimensions are gradual, conforming with Bergmann's rule (Yom-Tov, 1967; Figures 2 and 125). Body proportions of Israeli hares also conform with Allen's rule, and body extremities (ears, legs and tail) of northern hares are relatively smaller than those of southern ones (Figure 3). The relative length of the tympanic bullae is also related to ambient temperature, being about 16% and 13.5% of the greatest length of the skull in the 'Arava and northern Israel, respectively. The great variety in body colour, dimensions and relative size of the extremities is undoubtedly the cause for the many species and subspecies of hares described from Israel and the surrounding countries (*capensis, europaeus, arabicus, judeae, syriacus and sinaiticus*).

281

Table 119
Body measurements (g and mm) of *Lepus capensis*

a) Specimens from central Israel between 31°30′N and 32°30′N.

	Body				
	Mass	Length	Tail	Ear	Foot
Males (10)					
Mean	2640	518	84	123	131
SD	320	44	9	9	7
Range	2400–3000	440–570	80–90	112–140	120–140
Females (10)					
Mean	2625	518	89	120	131
SD	342	23	10	6	5
Range	2250–3200	480–555	85–110	113–130	125–140

b) Specimens from the Negev and the 'Arava valley in southern Israel.

	Body				
	Mass	Length	Tail	Ear	Foot
Males (10)					
Mean	1490	434	78	121	110
SD	154	26	9	7	4
Range	1250–1700	410–475	70–110	110–130	110–115
Females (10)					
Mean	1817	457	77	125	111
SD	293	35	7	8	7
Range	1450–2300	380–510	65–90	110–135	100–120

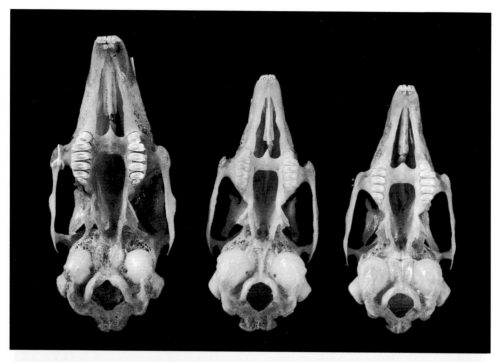

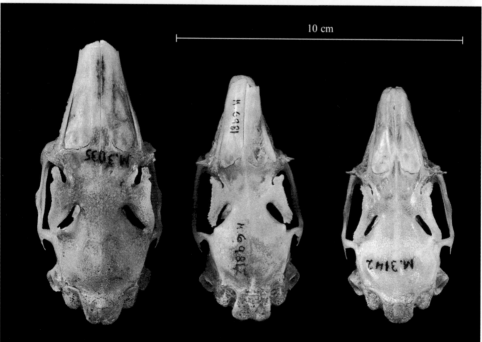

Figure 125: Skulls of hares (*Lepus capensis*) from the Mitle Pass, western Sinai (smallest), Yotvata, ʿArava Valley and Palmaḥim, Coastal Plain (largest).

Table 120
Skull measurements (mm) of *Lepus capensis* (Figure 126)

a) Specimens from central Israel between 31°30′N and 32°30′N.

	Males (10)			Females (10)		
	Mean	Range	SD	Mean	Range	SD
GTL	92.2	88.8–94.9	2.3	91.2	87.5–96.4	3.5
CBL	81.2	77.8–83.5	2.2	80.1	77.2–86.5	2.1
ZB	43.5	42.2–45.0	1.0	43.5	42.1–46.0	1.4
BB	31.2	29.2–32.4	1.1	30.8	28.5–32.4	1.1
IC	21.9	19.1–24.6	1.6	21.0	16.5–26.4	3.1
UT	15.9	15.1–16.9	0.6	15.9	14.8–17.4	0.7
LT	17.0	16.5–18.2	0.5	17.1	16.3–18.8	1.0
M	70.7	67.4–73.8	2.3	70.1	67.4–73.4	2.3
TB	12.8	11.3–14.2	0.9	12.3	11.6–13.0	0.4

b) Specimens from the Negev and the ʿArava Valley in southern Israel.

	Males (10)			Females (10)		
	Mean	Range	SD	Mean	Range	SD
GTL	81.0	75.1–85.0	2.9	84.3	79.4–93.3	3.9
CBL	71.5	67.1–76.0	2.6	74.4	70.8–82.6	3.4
ZB	38.7	37.1–39.8	1.0	39.4	36.5–43.3	1.8
BB	28.7	26.9–29.5	0.8	28.2	26.3–29.3	1.1
IC	20.0	17.1–23.6	2.5	17.8	15.8–21.6	1.6
UT	14.1	12.9–15.5	0.7	14.5	13.6–17.0	1.0
LT	14.7	13.4–15.9	0.7	15.2	13.9–17.0	1.0
M	62.0	58.2–65.8	2.5	62.4	60.1–63.6	1.2
TB	13.3	12.7–14.0	0.4	12.9	12.2–13.4	0.4

In a sample of 25 males and 21 females from the ʿArava Valley, males and females did not differ in skull size.

Dental formula is $\frac{2.0\ 3.3}{1.0.2.3} = 28$.

Karyotype: 2N = 48 for specimens from Israel (Stavy *et al.*, 1982).

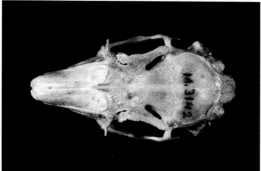

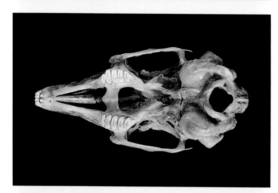

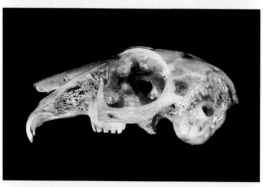

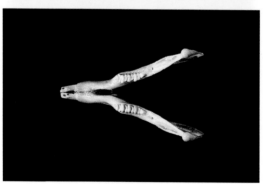

Figure 126: *Lepus capensis*. Dorsal, ventral and lateral views of the cranium, and a view of the mandible.

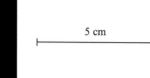

5 cm

Figure 127: Distribution map of *Lepus capensis*.

● – museum records.

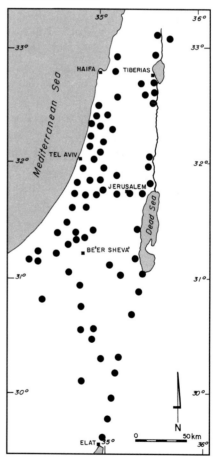

285

Stavy *et al*. (1982) reported on hybridization experiments between specimens of the northern and southern populations. The three out of four desert females and two out of four northern females which were inseminated with sperm from the opposite population gave birth to hybrids, which were intermediate in colour. We believe that all hares in Israel belong to one species, *Lepus capensis*.

Distribution: This species, as understood here (following Harrison and Bates, 1991) is widely distributed in Africa from the Cape north to Egypt and west to Senegal. It is also found throughout much of Eurasia and was introduced to North and South America and Australia.

It is common in open habitats throughout Israel, and is more common in plains than in mountainous areas (Figure 127).

Fossil record: The earliest remains of this species found in Israel date to the Acheulo-Yabrudian period, about 150,000 years BCE (Tchernov, 1988).

Habits: The Cape hare is nocturnal, hiding during the day under bushes and within dense vegetation cover, often laying up in a shallow form which it digs in the ground. In desert areas hares often hide in holes, under rocks or shallow burrows of up to 50 cm depth. These burrows often fit the animal so well, that it is possible that they are excavated by the occupant. Such digging behaviour has, however, not yet been observed. When resting in cold weather they lie on the ground, the body contracted and covering all four legs, with the ears flat on the back. In warm weather they spread their fore legs in front of the body, hind legs to the sides, and ears erect. These behaviours serve to thermoregulate body temperature by changing the body surface area according to ambient temperatures. When curious they sit on the hind legs with the front legs drooping, or stand on all fours with front legs and ears erect.

Hares escape their predators by running swiftly (up to 70 km/hr) with twists and turns. At times they jump to heights of 1 m or more. When running, the long hind legs touch the ground in front of the forelegs, as is well seen in the tracks. When moving slowly, for instance while feeding and moving from one plant to the next, hares advance by small hops or by walking, moving the legs alternately. The normal resting posture is either to remain seated on the folded hind legs with the fore legs held straight, or alternately, with the fore legs bent and extended on the ground.

Hares are normally solitary, but particularly during the breeding season males and females are often observed close to one another, and it is therefore possible that during this season they live in pairs. Home ranges of individual hares overlap to a large extent. They fight by striking each other with their front legs. Females also defend their young in this way, and on such occasions may also bite the attacker. When caught or hit they sometimes scream loudly. Young hares also emit similar alarm calls which serve to call their mothers. When attacked, hares also make hissing and gargling noises. When cornered in captivity or when alarmed they use their hind legs to drum on the ground.

Dense populations occur near agricultural settlements, even in desert areas. For example, in the 'Arava valley where the normal density is one individual per km^2, population density in agricultural fields is up to ten per km^2, and in a 25 acre alfalfa field in central Israel more than 100 hares were counted during one night.

In Israel hares seem to be the staple prey food of the caracal (*Felis caracal*) and of the golden eagle (*Aquila chrysaetos*).

Food: Hares are herbivorous, feeding on a great variety of plants, including some (i. e. *Retama raetam* and *Pulicaria* sp.) which are rarely eaten by other animals.

Reproduction: In central and northern Israel hares reproduce all year round, with a marked peak between February and July, but in the desert they normally reproduce only during winter and spring. A female may give birth several times (probably 2–4) annually, and in good conditions with plenty of food it can become oestrous on day 38 of its gravidity and become pregnant with a new litter (superfoetation). Pregnancy lasts 42 days, sometimes several days longer. Litter size is 1–5 (usually 2–3 in central and northern Israel, 1–3 in the desert). The young are born in a shallow form in the ground which the female digs prior to delivery, or in a hidden place in dense vegetation. They are born fully covered with hair, with a white spot on the forehead and with eyes and ears open. The nursing period lasts about a month, when the young become independent, but they already start feeding on grasses when 5 days old. The mother does not remain near the young, and suckles them once or twice nightly. The young lies on its back to nurse attached to the mother in a parallel position, and finishes nursing within a minute or so.

Parasites: Hares in Israel were found to be infected by several ectoparasites such as lice (*Haemodipsus* sp.), fleas (*Synosternus pallidus* and *Xenopsylla conformis*) and ticks (*Rhinicephalus sanguineus* and *Hyalomma excavatum*) (Theodor and Costa, 1967).

Relations with man: In Israel hare populations increased tremendously after 1964, following a dramatic decrease in the population size of their predators as a result of a poisoning campaign by the Ministry of Agriculture against jackals and other carnivores. Damage by hares to agricultural crops increased until most of the carnivore populations had recovered by the end of the 1960s. During the period of dense populations hares were heavily infested with internal parasites, particularly nematodes and *Taenia piciformis* (Cestoda) which infect the abdominal cavity.

The hare's habit of making chewing movements even when it is not eating convinced the author of Leviticus (11, 4) that the hare was chewing the cud, which it does not do.

Raising hares in captivity is not easy, since hares, particularly young ones, are extremely sensitive to contamination and often die from diarrhea. In small cages the female is dominant over the male and may kill him. Leverets can be raised on fresh cow's milk or liquid baby food. It is advisable to keep hares on a suitable wire-netting floor, so that their food does not become soiled by excrement.

RODENTIA Rodents מכרסמים

This order is represented in Israel by seven native families: Sciuridae, Cricetidae (including Cricetinae, Microtinae, and Gerbillinae), Muridae, Spalacidae, Gliridae, Dipodidae and Hystricidae; and one introduced family: Capromyidae.

SCIURIDAE Squirrels סנאיים

This family is represented in the Middle East by one species, *Sciurus anomalus*.

SCIURUS Linnaeus, 1758

Sciurus Linnaeus, 1758. *Systema Naturae*, 10th ed., 1:63.

Sciurus anomalus Gmelin, 1778   Persian Squirrel
Figs 128; Plate XXV:3   סנאי

Sciurus anomalus Gmelin, 1778. *Systema Naturae*, 13th ed, 1:148. Sabeka, 25 km south-west
   of Kutais, Georgia, Caucasus.
Synonymy in Harrison and Bates (1991).

This is a small arboreal squirrel with a bushy, flattened tail and prominent ears. The colour is yellowish-buff, the tail has a reddish tinge and the underparts are buff to white. In winter, it grows a tuft of elongated hairs on the tip of the ear, that is, however, much less developed than in *S. vulgaris*. No specimens from Israel are available in museums, but some data from Lebanon and Jordan are provided by Harrison (1972). Body measurements are: length 348 and 352 mm (n=2), tail 140 mm (n=2), hind foot 46-60 mm (n=7), ear 29-31 mm (n=2). Skull measurements are: GTL 47.1-50.9 mm (n=2), CBL 29.3-45.2 mm (n=2), ZB 26.2-30.0 mm (n=3), BB 22.2-23.8 mm (n=3), IC 14.6-17.1 mm (n=30, UT 9.0-9.3 (n=3), LT 9.1-9.6 mm (n=3) and M 30.8-34.2 mm (n=3).
Sciurus anomalus syriacus Gray, 1867 is the subspecies present in the study area (Harrison and Bates, 1991).

Dental formula is $\dfrac{1.\ 0.1-2\ .3}{1.0.\ 1.\ 3} = 20\text{-}22$

Karyotype: 2N=40, FN=76, for a specimen from Iran (Nadler and Hoffmann, 1970).

Distribution: This species occurs in Asia Minor, Iraq, Transcaucasia, western Iran and south to Syria, Lebanon and northern Jordan (Harrison and Bates, 1991). It also occurs on the southern slopes of Mount Ḥermon (Gavish, 1995; Figure 128). There is no record that this species occurred in recent years (at least during this century) in the area of Israel.

Fossil record: The earliest remains of this species found in Israel are from the Late Acheulian period (about 150,000 years BCE. Tchernov, 1988).

In the Carmel it existed at least until about 12,500 years ago (Bar Oz, 1996).

Habits: The Persian squirrel is a diurnal, arboreal species. According to recent observations on Mount Ḥermon (L. Gavish, pers. comm.) it spends much time foraging on the ground. It builds nests in hollow trees and crevices in rocks.

Food: On Mount Ḥermon it feeds mainly on acorns (*Quercus* ssp.) (Gavish, 1995), but also on buds, mushrooms and other vegetable matter, on insects (Lewis *et al.*, 1967) and occasionally on avian eggs and chicks (Ognev, 1940). Captive specimens preferred acorns to walnuts and hazel-nuts.

Reproduction: On Mount Ḥermon it seems to breed several times annually (L. Gavish pers. comm.). In Lebanon young squirrels were observed from late April (Harrison and Bates, 1991).

Relations with man: In Jordan squirrels are eaten and live specimens are sold in the markets (L. Gavish, personal communication).

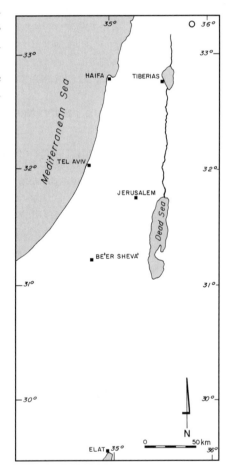

Figure 128: Distribution map of *Sciurus anomalus syriacus*.

○ – verified field records.

CRICETIDAE   Hamsters, Gerbils and Voles

This large family is represented in Israel by three subfamilies: Cricetinae, Gerbillinae and Microtinae.

Dental formula of all Cricetidae is $\dfrac{1.0\ 0.3}{1.0.0.3} = 16$.

Key to the Cricetidae in Israel
(After Harrison and Bates, 1991)

1. Cheekteeth complex and prismatic; m^3 subequal or larger than m^2.
   Subfamily Microtinae
- Cheekteeth composed of simple transverse laminae in adults; m^3 clearly shorter than m^2. **Subfamily Gerbillinae**
- Cheek-pouches present. Tail very short, less than one-third of the head and body length and without a terminal tuft. Tail at least one and a half times as long as hind foot.
   Cricetulus migratorius

CRICETINAE   Hamsters   אוגריים

This sub-family is represented in Israel by one species, *Cricetulus migratorius*.

CRICETULUS Milne-Edwards, 1867

Cricetulus Milne-Edwards,1867. *Annales Sci. Nat. Paris,* (5), 7:375.

Cricetulus migratorius (Pallas, 1773)   Grey Hamster
Figs 129–130; Plate XXV:4   בר אוגר

Mus migratorius Pallas, 1773. *Reise Prov. Russ. Reichs.*, 2:703. Lower Ural, Western Siberia. Synonymy in Harrison and Bates (1991).

The grey hamster is much smaller and less robust than the golden hamster *Mesocricetus auratus*. Its tail is short, even shorter than that of a vole, but its ears are relatively large. Its colour is ash-grey, with white underparts.
Our sample showed no significant difference in body and skull size between females and males or between individuals from Mount Ḥermon and from the Northern Negev, at the southernmost part of the distribution range of this species in Israel. However, the lack of geographical variation might be due to the small sample size. *Cricetulus migratorius cinerascens* Wagner, 1848 is the subspecies present in Israel (Harrison and Bates, 1991).

Table 121
Body measurements (g and mm) of *Cricetulus migratorius cinerascens*

	Mass	Body Length	Tail	Ear	Foot
		Males (13)			
Mean	24	96	21	17	16
SD	7	11	3	1.5	3
Range	18–30	81–120	18–26	14–19	12–18
		Females (9)			
Mean	27	95	22	17	15
SD	4	10	4	2	0.5
Range	22–33	80–110	19–30	12–19	14–16

Table 122
Skull measurements (mm) of *Cricetulus migratorius cinerascens* (Figure 129)

	Males (9)			Females (5)		
	Mean	Range	SD	Mean	Range	SD
GTL	25.9	24.6–26.8	0.8	26.4	25.2–27.7	1.1
CBL	24.2	22.5–25.4	1.0	24.8	23.3–26.0	1.2
ZB	13.6	12.9–13.9	0.6	14.0	13.1–14.8	0.7
BB	11.3	10.7–12.0	0.4	11.3	11.0–11.7	0.3
IC	4.2	4.0–4.5	0.2	4.0	3.9–4.1	0.1
UT	3.8	3.5–3.9	0.1	0.2	3.6–4.1	0.2
LT	3.7	3.5–4.0	0.1	3.8	3.7–4.0	0.1
M	16.3	15.0–17.0	0.8	17.0	15.8–18.1	1.1

Karyotype: 2N = 22, FN = 40, for a specimen of unknown origin (Lavappa, 1977).

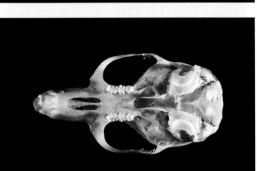

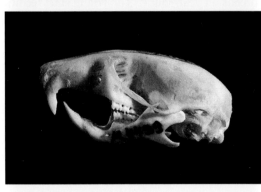

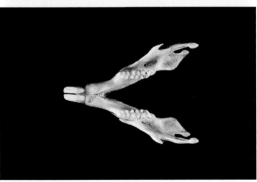

Figure 129: *Cricetulus migratorius cinerascens*. Dorsal, ventral and lateral views of the cranium, and a view of the mandible.

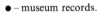

1 cm

Figure 130: Distribution map of *Cricetulus migratorius cinerascens*.

● – museum records.

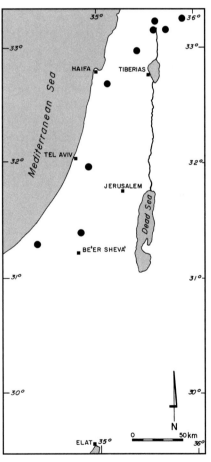

293

Distribution: The grey hamster occurs in the Balkan and east to Pakistan and Turkestan, and south to Israel and Jordan.

In Israel it is found in the mountains and hills of the Mediterranean region, on Mount Carmel, in Judea, Galilee and on the Golan and Mount Ḥermon. It prefers the well drained soils of the Upper Eocene and Cenonian periods, and in the Golan it also occurs on basaltic soils. It is more common on Mount Ḥermon than in the other areas of its distribution (Figure 130).

Fossil record: The earliest remains of this species found in Israel are from the Lower Musterian period (Tchernov, 1988).

Habits: It is a nocturnal rodent, but during low ambient daytime temperatures it has been found active during the day. It is slow and easy to catch by hand during day and night. It lives in burrows, but prefers natural ones, such as holes and fissures in the soil; if necessary, however, it digs its own burrows.

Food: It feeds mainly on seeds, which it stuffs into its cheek pouches which sag to the shoulders during foraging and which it then empties in its burrow by pressing them with its paws. However, it also feeds on other vegetable matter, on insects and even small vertebrates (Lewis *et al.*, 1967, Lay, 1967). It stores the food in underground chambers in its burrow.

Reproduction: Gestation lasts 16–19 days. Litter size ranges from 3–5 (mean of seven litters at Tel Aviv University Zoo was five; occasionally up to 11; Harrison and Bates, 1991), and females may give birth 2–3 times annually. In nature the young are born in spring, but in captivity it breeds all year round. The young start feeding on solid food when two weeks old, and reach sexual maturity at three months of age. In captivity longevity is up to five years.

Parasites: This species is infected by fleas (*Nosopsyllus sincerus*) and mites (*Laelaps ekstremi*) (Theodor and Costa, 1967).

MICROTINAE Voles נברניים

This subfamily is represented in Israel by two genera: *Microtus* and *Arvicola*. *Arvicola* is now perhaps extinct.

Key to the Species of Microtinae in Israel
(After Harrison and Bates, 1991)

1.	Robust form, greatest length of skull in excess of 38 mm.	**Arvicola terrestris**
–	Smaller forms, greatest length of skull less than 31 mm.	2
2.	Tail longer, subequal to half the length of the head and body.	**Microtus nivalis**
–	Tail longer, clearly less than half the head and body length.	**Microtus socialis**

MICROTUS Schrank, 1798

Microtus Schrank, 1798. *Fauna Boica*, 1:72.

This genus is represented in Israel by two species, *M. socialis* and *M. (Chionomys) nivalis*.

Microtus is characterized by a cylindrical body with short legs, short ears (hence its scientific name "micro" — small, "otus" — ear in Greek) and tail, small eyes and rootless molar teeth which grow throughout life. This is an adaptation to the high rate of attrition of the molars, for voles feed on green, succulent food, grasses and herbs that have a high cellulose, lignin and silica content. Most voles have high reproductive rates.

Microtus socialis (Pallas, 1773)   Levant Vole
Figs 131–132; Plate XXVI:1–2   נברן השדה

Mus socialis Pallas, 1773. *Reise Prov. Russ. Reichs.*, 2:705. "Grassy regions of desert by Ural River".
Synonymy in Harrison and Bates (1991).

This species has a typical vole shape: a cylindrical body with short legs, short tail (less than 25% of body length) and small ears. Its colour is earth-brown, with lighter underparts.

Microtus socialis guentheri Danford and Alston, 1880 is the subspecies present in Israel (Harrison and Bates, 1991).

Distribution: The Levant vole occurs in the Balkans, parts of North Africa, Asia Minor, Syria, Lebanon and Israel and east to Iran and Turkestan. This species is the most southerly in its distribution within the genus in the Old World.

In Israel it occurs in the Mediterranean region north of latitude 31°20′, where it is common in heavy, semi-heavy and other poorly drained soils in the Coastal Plain

Table 123
Body measurements (g and mm) of *Microtus socialis guentheri*

	Body		Tail	Ear	Foot
	Mass	Length			
Males (10)					
Mean	50	121	25	13	18.5
SD	14.7	12.0	3.2	1.8	0.8
Range	29–76	102–140	20–30	10–15	17–20
Females (10)					
Mean	37	109	23	13	18
SD	7.1	12	4	1	1
Range	29–49	92–120	17–25	11–14	17–20

Table 124
Skull measurements (mm) of *Microtus socialis guentheri* (Figure 131)

	Males (10)			Females (10)		
	Mean	Range	SD	Mean	Range	SD
GTL	28.0	26.9–29.6	1.0	27.6	25.9–29.4	1.2
CBL	27.4	26.1–29.3	1.1	26.8	25.0–28.4	1.1
ZB	16.0	14.1–17.6	1.0	15.4	14.5–16.3	0.6
BB	11.5	11.0–12.4	0.5	11.5	10.9–12.0	0.3
IC	3.8	3.5–3.9	0.1	3.8	3.6–4.0	0.2
UT	6.5	6.1–6.9	0.2	6.4	6.1–7.0	0.3
LT	6.3	6.0–6.7	0.2	6.3	5.9–6.7	0.3
M	19.0	17.7–20.2	0.9	18.8	17.5–20.7	1.1

In the field, body mass is heaviest in winter and spring (December–March), reaching a median value of about 60 g for both sexes, while in summer and autumn it is about 40 g (Cohen-Shlagman *et al.*, 1984a).

Karyotype: $2N = 54$, $FN = 52$, for specimens from Bulgaria (Belcheva *et al.*, 1980).

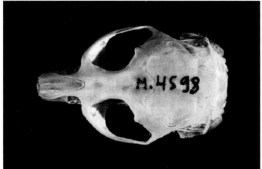

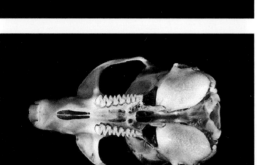

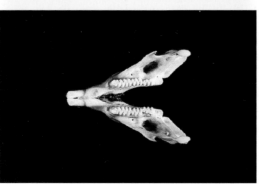

Figure 131: *Microtus socialis guentheri.* Dorsal, ventral and lateral views of the cranium, and a view of the mandible.

1 cm

Figure 132: Distribution map of *Microtus socialis guentheri.*

● – museum records.

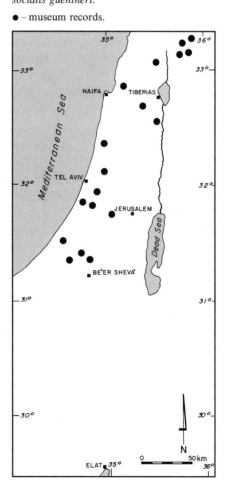

297

and northern valleys (but not in the central and southern Jordan Valley) and in mountain valleys in Judea, Samaria, on Mount Carmel and the Galilee. On Mount Hermon it inhabits dolinas up to 2000 m above sea level, whereas *Microtus nivalis* lives outside the dolinas in rocky habitats. Sometimes the burrows of both species are separated by as little as two metres (Figure 132).

Fossil record: The earliest remains of this species found in Israel are from the Late Acheulian period (about 150,000 years BCE. Tchernov, 1988).

Habits: The vole is nocturnal, but during winter it is also active during the day, particularly on cloudy days and in areas with high vegetation cover or high population density. While active they commonly squeak, and this call is easily heard in such densely populated areas.

It is a very social rodent (hence the species name), with a social hierarchy pertaining among the members of the burrow. Each family lives in a series of horizontal underground tunnels (c. 5 cm diameter, depth 20–25 cm during winter, 60 cm during summer). The shallow depth of the tunnels is apparently due to the fact that the vole prefers heavy, poorly drained soils. Each burrow system has up to 10 openings, several chambers for storing food (mainly seeds, at times up to 1.5 kg) and a nesting chamber. During summer voles can be found in deep fissures in heavy soil. Between the burrow openings above ground they use fixed trails on which they cut down the vegetation. The burrows are easily located by the bare soil around the openings, where the voles have eaten all the grass and herbs. Sex ratio in the field does not differ from parity (Bodenheimer, 1953; Cohen-Shlagman *et al.*, 1984a, b).

Their home range is very small. In the Coastal Plain about 75% of the permanent residents were caught within a diameter of 15 m. The population density ranges widely: in summer there may be only one individual per dunam (1000 m^2), rising in winter to several dozen, and in years of a mass increase several hundreds can be found in the same small area. In 1985 there were areas in the central Golan Heights where population density was estimated to be up to 1000 per dunam (Z. Zook-Ramon, pers. comm.) and the entire vegetation, other than a few non-palatable species, was completely eaten. During peak population years voles become thin, and by the end of the mass increase weigh only about 20–30 g.

Voles have many natural enemies. They are preyed on by various mammals, such as wolves, jackals, foxes, marbled polecats, mongooses, wild cats, etc., as well as by nocturnal and diurnal birds of prey. Remains of voles are the most common in pellets of barn owls (*Tyto alba*) in the Mediterranean region of Israel. In 1985, in the Golan, voles were apparently the sole source of food for jackals and foxes, whose scats contained exclusively vole hairs.

Food: The vole is herbivorous, feeding mainly on grasses and herbs, but also on seeds after the green vegetation has withered. Seeds are then stored in the burrows. It also eats bulbs, roots and even tree bark.

Reproduction: Gestation lasts 21 days, and post-partum oestrus may occur, especially at the beginning of a mass increase in population. In captivity reproduction occurs all year round, but in nature mainly between November–April. The female prepares a nest of dry grass and other plant material in one of the chambers of its burrow. Litter size ranges between 1–10 (mean 6; up to 13 in Lebanon, Lewis *et al.*, 1967), but in years of mass reproduction litters of 14 can occur. However, because there are only eight nipples, there is considerable pup mortality among the larger litters. The eyes open when the pups are a week old, and they are suckled for 2–3 weeks. In one season a female may give birth up to seven times and produce 60 young (Cohen-Shlagman *et al.*, 1984a, b).

Pups may reach sexual maturity at the age of one month when weighing about 20 g, but generally at two months, and in captivity they may reproduce until they are three years old. In a population studied in the southern Coastal Plain of Israel most individuals survived for only a few months, several lived more than one year, but there were a few individuals that lived 2.5 years. In captivity they may live up to four years (Cohen-Shlagman *et al.*, 1984a, b).

Parasites: This species is affected by fleas (*Ctenophthalmus congener, Nosopsyllus sincerus, Stenoponia tripectinata*), ticks (Ixodidae — *Ixodes redikorzevi*), mites (*Eulaelpas stabularis, Androlaelaps glasgowi, Laelaps pachypus, Hirstionnyssus arcuatus*) and Protozoa (*Grahamella* sp.) (Theodor and Costa, 1967).

Relations with man: Voles may occasionally become agricultural pests. In Israel populations used to peak every decade, with a smaller peak in the middle of the decade. During 1930 severe damage was caused to the wheat crop in the Yizre'el Valley. Voles stored large stocks of wheat grain in their burrows, and after the vegetation dried some farmers dug up these stocks for their own consumption. During 1985 most of the annual vegetation in the central Golan was eaten by voles, causing great damage to cattle farmers who had to feed their range cattle. During this event voles also ate food which they normally avoid, such as the bark of eucalyptus trees. They established dense populations in the drained areas of the Ḥula Valley, where they caused considerable damage to alfalfa fields. The drained peat soil, that is easy to dig and retains a high humidity, was apparently optimal for voles. Alfalfa is a highly preferred food and voles living in alfalfa fields ignore all poison bait.

Many raptor species wintering in Israel were dependent on voles that bred in winter and provided a food source that renewed itself throughout winter and spring. The very numerous wintering raptors prevented mass increase of voles. During the 1950s the Plant Protection Department tried to eradicate voles completely by repeated poisoning campaigns in the Mediterranean zone of Israel using grain soaked with thallium sulfate spread in the fields. This persistent poison caused paralysis. The poisoned, slow moving voles were easy prey for carnivores and raptors, which were affected by secondary poisoning (Mendelssohn, 1972). The near extermination of their natural predators enabled voles to increase their numbers in the following years, as their reproductive rate is much higher than that of their predators. This, in turn, led to the phenomenon of vole populations after the poisoning campaign becoming much higher than prior to it. Altogether, as voles prefer green food, controlling them with poisoned grain bait is generally not successful. The most effective way to reduce vole populations is the modern deep ploughing of fields, that destroys the burrow systems, thus exposing the voles to predators and adverse weather conditions. Such ploughing is more effective during summer, as voles are susceptible to heat and their population density is relatively low. Recently several kibbutzim tried succesfully to control voles by increasing the density of one of their natural enemies (barn owls) by providing nest boxes for the birds.

Microtus (Chionomys) nivalis (Martins, 1842)   Snow Vole
Figs 133–134; Plate XXVI:3   נברן השלג

Arvicola nivalis Martins, 1842. *Revue zool., Paris*, p. 331. Faulhorn, Bernise Oberland, Switzerland.
Synonymy in Harrison and Bates (1991).

This vole has a tail length of about half its body length. Its colour is light brownish-grey, much greyer than the Levant vole.

Table 125
Body measurements (g and mm) of *Microtus nivalis hermonis*

	Body		Tail	Ear	Foot
	Mass	Length			
Males (9)					
Mean	39	112	57	15	20
SD	8	9	6	1	1
Range	27–44	105–125	45–63	14–16	18–22
Females (8)					
Mean	37	113	57	15	19
SD	13	7	8	1.5	1
Range	22–56	100–121	42–68	12–17	18–20

Table 126
Skull measurements (mm) of *Microtus nivalis hermonis* (Figure 133)

	Males (10)			Females (10)		
	Mean	Range	SD	Mean	Range	SD
GTL	29.2	26.4–30.9	1.5	29.1	27.4–30.8	1.2
CBL	28.4	25.4–30.5	1.5	28.1	26.5–30.3	1.4
ZB	16.1	14.3–16.7	0.7	16.1	14.6–17.0	0.9
BB	12.7	11.8–13.5	0.6	12.6	11.5–13.8	0.7
IC	4.3	4.1–4.6	0.2	4.3	4.1–4.6	0.1
UT	6.7	6.2–7.2	0.3	6.6	6.2–7.2	0.3
LT	6.5	5.9–6.9	0.3	6.6	6.0–7.9	0.5
M (male n=9)	20.1	17.9–21.2	1.1	20.1	18.8–21.4	1.1
TB (male n=9)	8.3	7.4–8.9	0.6	7.8 (n=8)	7.4–8.6	0.4

Karyotype: 2N = 54, FN = 52, for specimens from Europe (Burgos *et al.*, 1989).

300

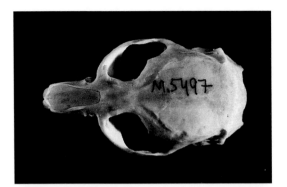

Figure 133: *Microtus nivalis hermonis.* Dorsal, ventral and lateral views of the cranium, and a view of the mandible.

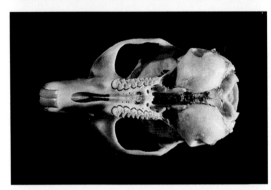

1 cm

Figure 134: Distribution map of *Microtus nivalis hermonis.*

● – museum records.

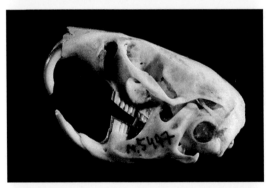

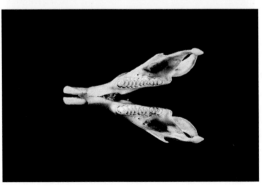

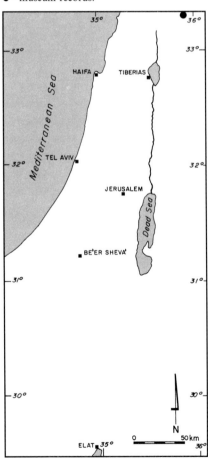

301

Microtus nivalis hermonis Miller, 1908 is the subspecies present in Israel (Harrison and Bates, 1991).

Distribution: This species has a wide, but disjunctive distribution on mountain ranges in the west Palaearctic, in Spain, the Alps, Apennines, Balkans, Turkey, Lebanon, Syria, Caucasus and Iran.

In Israel it occurs only on Mount Ḥermon, at heights above 1500 m above sea level, mostly in rocky areas. In Lebanon it occurs above 1150 m above sea level (Atallah, 1977), probably due to the cooler climate there. Mount Ḥermon is the southernmost area of its distribution (Figure 134).

Habits: On Mount Ḥermon this vole inhabits areas of eroded rocks, while the Levant vole there lives in deep soil in valleys. They may dwell side by side, but each species is restricted to its characteristic microhabitat. It is mostly nocturnal, but at times, if ambient temperatures are not too high, it is also active during the day. It lives in rock crevices, but also digs burrows in the soil, but in snow covered areas it has a tunnel system above the ground and under the snow ("nivalis" — snowy in Latin), which is exposed when the snow melts.

Food: It feeds on grass, herbs and seeds, but also eats insects; altogether it is much more omnivorous than the Levant vole.

Reproduction: The gestation period is 21 days and the 2–7 young are born in spring (May) (Atallah, 1977). The pups are relatively big, weighing about 4 g at birth.

ARVICOLA Lacépède, 1799

Arvicola Lacépède, 1799. *Tableau des divisions, sous divisions, ordres et genres des mammifères*, p. 10.

This genus is represented in Israel by one species, *Arvicola terrestris*.

Arvicola terrestris (Linneus, 1758)   Water Vole
Figs 135–136   נברן המים

Mus terrestris Linneus, 1758. *Systema Naturae*, 10th ed., 1:61. Uppsala, Sweden.
Synonymy in Harrison and Bates (1991).

This is a big vole. No complete specimens from Israel are available, only several skulls that were found in regurgitated barn owl pellets in the Ḥula Valley. However, skull measurements are very similar to those given by Harrison and Bates (1991) for a sample from Iraq. In Europe there exist two forms of this species, that are morphologically identical, but one is aquatic while the other is more terrestrial. It is not known to which form the Israeli population belonged, but it was probably the aquatic one, for digging in the dry, hard soil would have been difficult and the terrestrial form would have faced competition from the well adapted mole rat (*Spalax leucodon*).

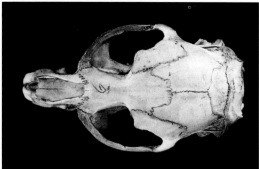

Figure 135: *Arvicola terrestris hintoni*. Dorsal, ventral and lateral views of the cranium, and a view of the mandible.

1 cm

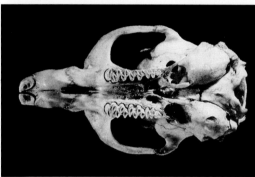

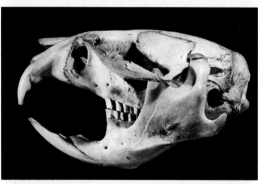

Figure 136: Distribution map of *Arvicola terrestris hintoni*.

● – museum records.

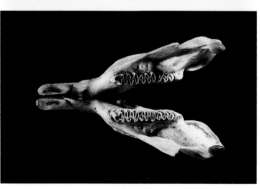

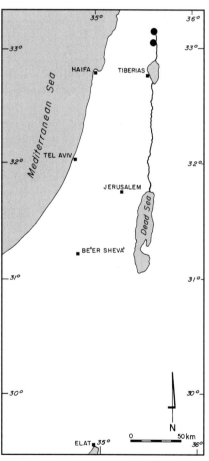

Arvicola terrestris hintoni Aharoni, 1932 is the subspecies present in Israel (Harrison and Bates, 1991).

Table 127

Skull measurements (mm) of unsexed *Arvicola terrestris hintoni* found in owls' pellets in the Hula Valley (Figure 135). All these skulls are kept at the Mammal Collection of The Hebrew University of Jerusalem

	Mean	Range	SD	N
GTL	40.3	38.5–41.8	1.2	10
CBL	40.1	37.7–42.2	1.5	10
ZB	23.1	21.0–25.0	1.1	19
BB	14.4	13.1–15.2	0.6	16
IC	5.0	4.3–5.4	0.3	21
UT	9.9	9.2–10.6	0.4	19
LT	9.7	8.9–10.3	0.5	12
M	28.4	24.2–30.3	2.3	11
TB	9.6	9.3–10.2	0.3	8

Karyotype: $2N = 36$, $FN = 60–68$ for specimens from Europe (Zima and Kral, 1984).

Distribution: The water vole is widely distributed in the Palaearctic, where it occurs in most of Europe and east to Siberia, south to Lake Baikal and the Tien Shan mountains.

In Israel it was found only in barn owl pellets on the banks of the Hula Lake before this was drained during the early 1950s (Dor, 1947; Figure 136). About 30 skulls were found, but the water vole has never been caught alive in Israel. Although it appears to have been exterminated when the Hula Swamp was drained, there is still a possibility that small populations may survive in pockets of suitable habitat.

Fossil record: The earliest remains of this species found in Israel are from the Kebaran period (Tchernov, 1988).

Habits, Food, Reproduction: No data are available from Israel.

Rodentia

GERBILLINAE   Gerbils   גרביליים

This subfamily is represented in Israel by four genera: *Gerbillus, Sekeetamys, Meriones* and *Psammomys*. Several species of this sub-family co-exist on sandy soils, where morphological character displacement has been demonstrated (Yom-Tov, 1992).

Key to the Species of Gerbillinae in Israel
(After Harrison and Bates, 1991)

1.   Upper cheekteeth without trace of cusps in all stages of wear; laminae always joined.
     2
–   Upper cheekteeth with clear traces of cusps in juveniles; laminae of molars not always joined.   7
2.   Tail relatively short, not exceeding 80% of head and body length. Upper incisors without longitudinal grooves (in adults).   **Psammomys obesus**
–   Tail longer, exceeding 80% of head and body length. Upper incisors with grooves.   3
3.   Tail bushy for more than half its length.   **Sekeetamys calurus**
–   Tail with only a terminal tuft, not bushy for more than half its length.   4
4.   Body relatively large. Greatest length of skull more than 33.0 mm   5
–   Body relatively small. Greatest length of skull rarely exceeds 33.0 mm   7
5.   Tympanic bullae smaller, mastoid portions do not project behind supraoccipital. Soles of hind feet without dense tuft of brown hairs.   **Meriones tristrami**
–   Tympanic bullae larger, mastoid portions project behind supraoccipital.   6
6.   Tail with a scanty terminal black tuft.   **Meriones crassus**
–   Tail with well developed terminal black tuft and median black line along dorsal aspect.
     Meriones sacramenti
7.   Zygomatic plates do not project far forwards. Size small,. Tail not bicoloured. Soles of feet predominantly hairy.   8
–   Soles of feet predominantly naked.   10
8.   Tympanic bullae smaller, mastoid portions do not project behind the supraoccipital. Larger species, greatest length of skull more than 29.0 mm.   **Gerbillus pyramidum**
–   Smaller species, greatest length of skull less than 29.0 mm.   9
9.   Incisive foramina smaller, subequal in length with maxillary cheekteeth.
     Gerbillus gerbillus
–   Incisive foramina larger, clearly exceeding length of maxillary cheekteeth.
     Gerbillus andersoni
10. Pygmy species, greatest length of skull less than 23.0 mm and maxillary cheekteeth less than 3.1 mm in total length.   **Gerbillus henleyi**
–   Larger species, greatest length of skull more than 23.0 mm and maxillary cheekteeth more than 3.0 mm.   12
12. Greatest length of skull never exceeds 29.6 mm. Antero-superior rim of each bony

auditory meatus not inflated. Ossicles concealed in the meatus by bony downgrowth of tympanic annulus. The dark colour of the upper side extends to below the eyes.

Gerbillus dasyurus

- Antero-superior rim of each bony auditory meatus is inflated. Ossicles not concealed in the meatus. Tympanic bullae larger, mastoid portions project beyond supraoccipital. The dark colour of the upper side does not extend to below the eye.   **Gerbillus nanus**

GERBILLUS Desmarset, 1804

Gerbillus Desmarset, 1804. *Nouveau Dictionnaire d'Hist. nat. Paris*, 1:24, tabl. Meth., p. 22.

This genus is represented in Israel by six species belonging to two sub-genera, *Gerbillus* and *Dipodillus* (= *Hendecapleura*): *G. pyramidum, G. andersoni, G. gerbillus, D. henleyi, D. nanus* and *D. dasyurus*. Members of this genus have relatively large hindfeet, and delicate, soft fur which is yellow or yellowish-grey in colour, lighter in desert species. All three species of *Gerbillus* inhabit sandy soils and have long, stiff hairs on the toes of their hind legs that increase the foot area and prevent sinking into the loose sand. Most species inhabit desert areas and only *G. dasyurus, G. pyramidum* and *G. andersoni* occur in the Mediterranean region, the former in rocky habitats and the latter on sand along the Coastal Plain.

Gerbillus (Gerbillus) pyramidum (I. Geoffroy, 1825)   Greater Egyptian Gerbil
Figs 137–138; Plate XXVI:4; Plate XXVII:1   גרביל החולות

Gerbillus pyramidum 1825. I. Geoffroy, *Dictionnaire Classique d'Hist. Nat.*, 7:321. Giza Province, Egypt.
Synonymy in Harrison and Bates (1991).

The fur is sand-yellow and the sides of the head and underparts are white. The toes of the hindfeet possess long, stiff, white hairs, that increase the foot area and prevent sinking into the loose sand on which these gerbils live. The tuft of hairs on the tip of the tail is poorly developed and greyish-white, and probably has a less communicative importance than in other gerbils, which have a well developed, dark or black tail tuft. The large, dark eyes are a most conspicuous feature of this species.
Gerbillus pyramidum floweri Thomas, 1919 is the subspecies present in Israel (Harrison and Bates, 1991).
Distribution: This gerbil occurs in much of the Sahara and east to Sinai and Israel. In Israel it occurs on sandy soil in the western Negev where the population has 64–66 chromosomes, and along the Coastal Plain originally as far north as the Yarqon River where the population has only 50–52 chromosomes. However, during the 1980s a population was discovered on sandy soil north of 'Akko (Acre) (Figure 138). This population was apparently introduced from the western Negev, as its members have 66 chromosomes.

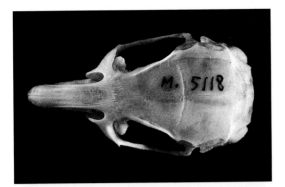

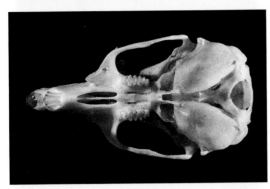

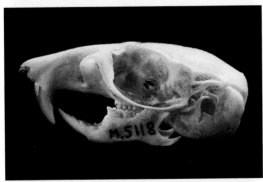

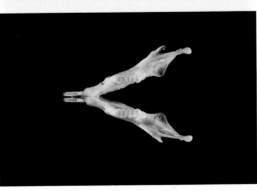

Figure 137: *Gerbillus pyramidum floweri.* Dorsal, ventral and lateral views of the cranium, and a view of the mandible.

1 cm

Figure 138: Distribution map of *Gerbillus pyramidum floweri.*

● – museum records.

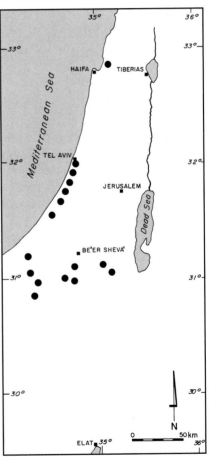

307

Table 128
Body measurements (g and mm) of *Gerbillus pyramidum floweri*

	Body		Tail	Ear	Foot
	Mass	Length			
Males (33)					
Mean	40	108	139	15	33
SD	6.5	7	13	1.5	2.5
Range	23–58	92–125	105–158	11–17	25–37
Females (33)					
Mean	35	101	134	14	31
SD	7	9	10	1.5	3
Range	23–57	108–117	110–151	11–17	26–38

Table 129
Skull measurements (mm) of *Gerbillus pyramidum floweri* (Figure 137)

	Males (8)			Females (8)		
	Mean	Range	SD	Mean	Range	SD
GTL	33.2	32.7–34.0	0.5	31.1	29.3–32.4	1.3
CBL (F n = 7)	29.1	28.0–29.7	0.6	26.8	25.0–28.4	1.3
ZB	18.2	17.8–18.6	0.3	17.3	15.9–18.2	0.8
(M n = 5; F n = 7)						
BB	15.3	15.1–15.7	0.2	14.8	14.1–15.3	0.4
IC	6.7	6.2–7.1	0.3	6.6	6.2–7.0	0.3
UT	4.6	4.3–4.9	0.3	4.4	4.1–4.6	0.2
LT	4.0	4.4–4.8	0.2	4.4	4.1–4.7	0.2
M	19.5	18.9–20.3	0.4	18.2	17.1–19.3	0.3
TB	11.7	11.3–12.3	0.3	11.1	10.4–11.9	0.5

Brand and Abramsky (1987) observed that mean body mass of southern (western Negev) populations is smaller by as much as 10% than that in the southern Coastal Plain. These authors stated that the above differences in body mass probably reflect a response to environmental conditions, and have little or no genetic basis. Yom-Tov (1991) found no such difference between the Israeli populations of this species, thus confirming Brand and Abramsky's (1987) conclusions. Brand and Abramsky (1987) also found seasonal fluctuations in body mass in this species: body mass in spring was about 10% larger than in autumn. However, such seasonal differences were not found by Wahrman and Gurevitz (1973).

Karyotype: Two chromosomal forms are known in Israel. The one in the western Negev has $2N = 64–66$; the other, along the Coastal Plain has $2N = 50–52$, $FN = 76$, (Wahrman *et al.*, 1988). The hybrid zone between the two chromosomal races, which is about 50 km wide, consists of a cline of karyotypes with the proportion of acrocentric chromosomes increasing southwards, while the number of metacentric chromosomes decreases (Wahrman and Gurevitz, 1973). Its preferred habitat is dunes and dune slopes with only sparse vegetation, rather than the valleys between dunes which have denser vegetation.

Fossil record: The earliest remains of this species found in Israel are from the Natufian period (Tchernov, 1988).

Habits: This gerbil is nocturnal. During the day it hides in burrows dug in the sand, plugging the openings by pushing sand with its head. This serves to maintain high air humidity and a relatively stable temperature in the burrow, as well as protection against predators, mainly snakes.

This psammophilic species prefers to inhabit dunes and exposed sandy areas with shifting sand (but not the poorest sands inhabited by *G. gerbillus*), and not the more vegetated areas in the depressions between the dunes, which are the typical habitat of *G. andersoni*. In places where it is sympatric with *G. gerbillus* and *G. andersoni* it is in an intermediate position between the two other species (Abramsky *et al.*, 1985). Where it is sympatric with *G. andersoni*, *G. pyramidum* is dominant and is active during the first half of the night while *G. andersoni* is active in the second half (Ziv *et al.*, 1993).

Food : It feeds on seeds which it sieves from the sand with its front feet and on leaves, flowers and insects. In a study in the western Negev (Bar *et al.*, 1984) seeds, leaves and insects comprised 63%, 30% and 7% of the annual diet, respectively. The proportion of seeds in the diet increased to 85% during spring. The preferred food species were Gramineae, *Erodium* and *Plantago*. Like many other gerbillids it is able to survive for long periods only on seeds, provided that ambient temperature and humidity are not too extreme. For reproduction, however, green, succulent food is obligatory.

Reproduction: Young are born between February–June, the gestation period is 24 days and litter size ranges from 3–6. The eyes of the pups open at 19–20 days old and ears at 8–16 days (Happold, 1968). The young are weaned when about one month old. Sexual maturity occurs early: a captive male fertilized a female when it was three months old. A captive male lived for six years.

Parasites : This species is affected by fleas (*Synosternus cleopatrae, Xenopsylla ramesis, X. nubica, Stenoponia tripectinata*), ticks (Ixodidae — *Rhipicephalus sanguineus*), mites (*Androlaelaps centrocarpus, Androlaelaps insculptus, A. marshalli, Hirstionyssus craticulatu*), Protozoa (*Nuttallia* sp., *Grahamella* sp.) and spirochaetes (*crocidurae* group) (Theodor and Costa, 1967).

Gerbillus (Gerbillus) andersoni (de Winton, 1902)   Anderson's Gerbil
Figs 139–140; Plate XXVII:2   גרביל החוף

Gerbillus andersoni de Winton, 1902. *Annals Mag. nat. Hist.*, (7), 9:45. Mandara, Egypt.
Synonymy in Harrison and Bates (1991).

This gerbil is yellowish-brown with white underparts. The colour of specimens in the more southern areas is lighter. The tuft of the tail is dark brown, and the back of the ear is characteristically dark. This dark-coloured ear is a good diagnostic feature to distinguish between light-coloured specimens of this species and other gerbils.

Table 130
Body measurements (g and mm) of *Gerbillus andersoni allenbyi*

	Body		Tail	Ear	Foot
	Mass	Length			
Males (22)					
Mean	25	89	119	15	27
SD	4	6	11	1	1.5
Range	17–34	80–110	85–135	14–16	25–30
Females (13)					
Mean	26	84	122	15	27
SD	5	9	9	1	2
Range	19–38	65–100	100–132	12–16	24–31

Table 131
Skull measurements (mm) of *Gerbillus andersoni allenbyi* (Figure 139)

	Males (10)			Females (10)		
	Mean	Range	SD	Mean	Range	SD
GTL	28.2	27–29.7	0.7	28.8	28–29.8	0.7
CBL	24.3	23.1–25.7	0.7	24.9	23.7–25.9	0.8
ZB	15.3	14.6–16.5	0.7	15.7	15–17.2	0.8
BB	13.2	12.7–13.5	0.3	13.2	12.8–13.8	0.3
IC	5.5	5.0–6.0	0.3	5.5	5.1–6.2	0.3
UT	4.0	3.8–4.1	0.1	3.9	3.8–4.1	0.1
LT	4.0	3.8–4.3	0.2	4.0	3.8–4.1	0.1
M	16.2	15.6–16.7	0.4	16.7	16.3–17.3	0.4
TB	9.9	9.4–10.4	0.3	10	8.9–10.5	0.5

Karyotype: 2n = 40, FN = 76, for specimens from Israel (Wahrman *et al.*, 1988).

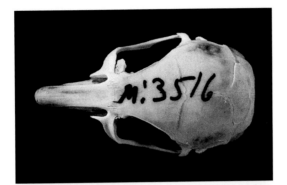

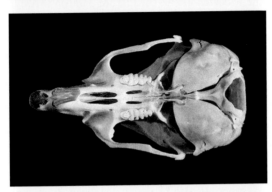

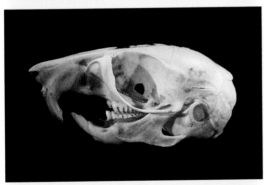

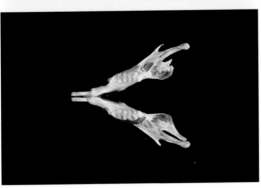

Figure 139: *Gerbillus andersoni allenbyi.* Dorsal, ventral and lateral views of the cranium, and a view of the mandible.

1 cm

Figure 140: Distribution map of *Gerbillus andersoni allenbyi.*

● – museum records.

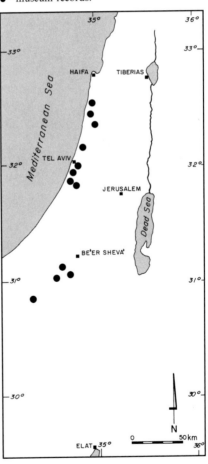

311

Brand and Abramsky (1987) noted that the mean body mass of southern (western Negev) populations is smaller by as much as 15% than that in the Sharon, but Yom-Tov (1991) found no such difference between the Israeli populations of this species. Brand and Abramsky (1987) also found seasonal fluctuations in body mass in this species: body mass in spring was about 15% larger than in autumn, as well as much larger fluctuations in body mass (32%) between winter-spring and summer-autumn of 1979 in a study area south of Mount Carmel (Abramsky, 1984).

Gerbillus andersoni allenbyi Thomas, 1918 is the subspecies present in Israel (Harrison and Bates, 1991).

Distribution: This gerbil occurs along the Mediterranean coast from Tunisia through Libya and Egypt to Sinai and Israel.

In Israel it occurs in sands that are covered with vegetation, as well as on fossilized sandstone hills along the Coastal Plain from the Egyptian border and north to Mount Carmel, and in the western Negev (Figure 140). It prefers vegetation rich valleys between dunes or dune slopes with vegetation.

Fossil record: The earliest remains of this species found in Israel are from the Natufian period (Tchernov, 1988).

Habits: This species is found mainly in depressions between dunes rather than on dunes and exposed sandy areas, which is the typical habitat of *G. pyramidum* and even more so of *G. gerbillus*. However, in areas where it is not sympatric with *G. pyramidum* it is found also on exposed dunes. It also inhabits stabilized dunes and fossilized sand-stone hills. Where it is sympatric with *G. pyramidum*, *G. andersoni* is inferior and is active during the second half of the night while *G. pyramidum* is active in the first half (Ziv *et al.*, 1993). The population density south of Mount Carmel was 4–12 individuals per 1000 m^2 (Abramsky and Sellah 1982, Abramsky, 1984). Mean diameter of home range in a study area south of Mount Carmel was 34.7 and 32.5 m for males and females, respectively (Abramsky, 1984).

Food : This gerbil feeds on seeds which are sieved from the sand as well as on other parts of plants and on insects. In a study in the western Negev (Bar *et al.*, 1984) these foods comprised 46%, 46% and 7% of the annual diet, respectively, but during spring the proportion of seeds in the diet increased to nearly 90%, while in winter green food formed 82% of the diet. (Abramsky, 1980). The preferred foods were species of Gramineae, *Erodium* and *Plantago*.

Reproduction: In the Coastal Plain, young are born between February–June (Abramsky, 1984). The gestation period is about three weeks, and litter size ranges from 3–5.

Parasites: This species is affected by fleas (*Synosternus cleopatrae, Stenoponia tripectinata*), Anoplura (*Polyplax gerbilli*), ticks (Ixodidae — *Rhipicephalus sanguineus*), mites (*Androlaelaps centrocarpus, A. hirsti, A. insculptus, A. marshalli*) and Protozoa (*Grahamella* sp.) (Theodor and Costa, 1967).

Gerbillus (Gerbillus) gerbillus (Olivier, 1801)   Egyptian Gerbil
Figs 141–142; Plate XXVII:3   גרביל דרומי

Dipus gerbillus Olivier, 1801. *Bulletin Sci. Soc. philom. Paris,* 2:121. Giza Province, Egypt.
Synonymy in Harrison and Bates (1991).

The coloration of this species is very similar to that of *Gerbillus pyramidum*, but the lower part of the sides of the body is white. The poorly developed tail tuft is almost white.

Brand and Abramsky (1987) showed mean body mass of a population in the southern 'Arava Valley (Samar) to be about 14% larger than that of two populations in the western Negev. This difference, which is the opposite of the usual north-south pattern, was attributed to ecological character release: in the western Negev *G. gerbillus* is sympatric with two other gerbils (*G. pyramidum* and *G. andersoni*), while in Samar it has no competitors. However, Yom-Tov (1991) found no morphological differences between the Israeli populations of this species. Brand and Abramsky (1987) also found seasonal fluctuations in body mass in this species: body mass in spring was about 10% greater than in autumn.

Gerbillus gerbillus asyutensis Setzer, 1960 is the subspecies present in Israel (Harrison and Bates, 1991).

Distribution: This gerbil occurs in the Sahara from Algeria and Nigeria to Sinai and Israel.

In Israel it occurs in sandy desert areas, in the western Negev and in the 'Arava Valley (Figure 142). In the 'Arava it can sometimes be found also on flats of solidified sand which are inhabited by *G. nanus* (Zahavi and Wahrman, 1957). In Sinai it occurs at altitudes below 1100 m, in correlation with bushes of *Hammada salicornica* which are absent above this height (Haim and Tchernov, 1974).

Habits : This species occurs in a variety of soils: dunes with sparse vegetation, including small, isolated dunes on stone plains (hamadas), alluvial fans and arkose (Haim and Borut, 1974). In areas where it is sympatric with *G. pyramidum* its population density is highest in the poorest habitat, apparently because it is excluded from better areas by its larger competitor (Abramsky *et al.*, 1985)

Food : It feeds on seeds which are sieved from the sand, on green food and insects. In a study in the western Negev (Bar *et al.*, 1984) these foods comprised 69%, 26% and 5% of the annual diet, respectively, but during spring it ate mainly seeds (82%) and insects (16%). The preferred food species were Gramineae, *Erodium* and *Plantago*.

Reproduction: Litter size ranges from 2–4. In captivity it breeds throughout the year.

Parasites: This species is infected by fleas (*Synosternus cleopatrae, Stenoponia tripectinata, Nosopsyllus (Nosinius) sinaiensis, Xenopsylla conformis* and *Leptopsylla algira*), Anoplura (*Polyplax gerbilli*), ticks (Ixodidae — *Rhipicephalus sanguineus*), mites (*Androlaelaps centrocarpus*), Protozoa (*Nuttallia* sp., *Grahamella* sp., *Eperythrozoon* sp.) and Spirochaetes (Theodor and Costa, 1967, and Krasnov *et al.*, 1997).

313

Table 132
Body measurements (g and mm) of *Gerbillus gerbillus asyutensis*

	Body		Tail	Ear	Foot
	Mass	Length			
Males (37)					
Mean	23	82	118	12	27
SD	4	7.5	10	1.5	2.5
Range	15–35	66–99	95–149	10–16	21–35
Females (23)					
Mean	23	84	117	12	27
SD	4	7	7	1.5	2.5
Range	16–33	75–105	100–126	10–15	22–32

Table 133
Skull measurements (mm) of *Gerbillus gerbillus asyutensis* (Figure 141)

	Males (10)			Females (10)		
	Mean	Range	SD	Mean	Range	SD
GTL	27.9	27–28.7	0.6	27.2	25.7–29.0	1.0
CBL	23.9	23.1–24.9	0.7	23.0	21.7–24.3	1.0
ZB	15.0	14.1–16.0	0.7	14.4	13.8–15.0	0.4
BB	13.3	12.7–13.8	0.4	13.2	12.7–13.6	0.3
IC	5.4	4.8–6.0	0.4	5.6	5.1–6.0	0.3
UT	3.6	3.2–3.9	0.3	3.4	3.1–3.7	0.2
LT	3.6	3.2–4.1	0.3	3.4	3.1–3.7	0.2
M	15.8	15.0–17.0	0.7	15.4	14.3–16.2	0.6
TB	10.1	9.1–10.5	0.5	10.1	9.3–20.8	0.5

Karyotype: Males: 2N = 43; Females 2N = 42, FN = 76–78, for specimens from Israel (Wahrman *et al.*, 1988).

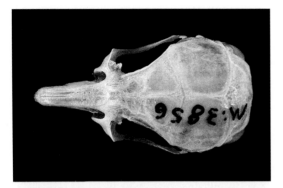

Figure 141: *Gerbillus gerbillus asyutensis.* Dorsal, ventral and lateral views of the cranium, and a view of the mandible.

1 cm

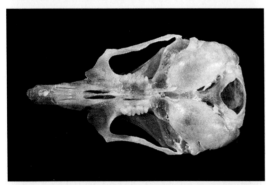

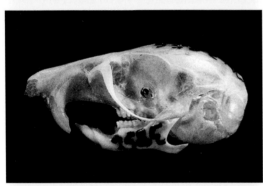

Figure 142: Distribution map of *Gerbillus gerbillus asyutensis.*

● – museum records.

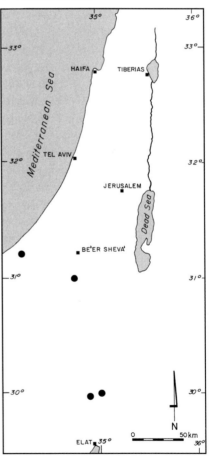

315

Gerbillus (Dipodillus = Hendecapleura) henleyi de Winton, 1903   Pygmy Gerbil
Figs 143–144; Plate XXVII:4   גרביל זעיר

Dipodillus henleyi de Winton, 1903. *Novitates zool.*, 10:284. Zaghig, Wadi Natron, Egypt.
Synonymy in Harrison and Bates (1991).

This is the smallest gerbil and the smallest rodent in Israel. Its colour is generally
light brown, with white underparts. The sides of the head are also whitish with
some brown hair. The tuft of the tail tip is brown, and well developed. Like other
species of the sub-genus *Dipodillus* it has naked soles.
Gerbillus henleyi mariae Bonhote, 1909 is the subspecies present in Israel (Harrison
and Bates, 1991).
Distribution: The pygmy gerbil occurs throughout the Sahara from Senegal and
Algeria to Egypt, south to Chad and east to Sinai and Arabia.
In Israel it occurs in most of the Negev in various habitats, ranging from the sand
dunes in the north, gravel plains, narrow and wide wadis in the Makhtesh Ramon,
and wide wadis in the southern Negev where it digs its burrows below bushes around
which wind-blown sand has accumulated (Figure 144).
Habits: This gerbil was studied by Shenbrot *et al.* (1994) in the Makhtesh Ramon in
the Negev. The average density was one individual per hectare in sand dunes, 1.8 on
gravel plains, 2.3 in narrow wadis and 0.9 in wide wadis. In a peak year (1995) summer densities were 4.7 individuals per hectare in sand dunes, 4.0 on gravel and 2.4 in
wadis (G. Shenbrot and B. Krasnov, pers. comm.). Densities also fluctuated between
seasons: during winter some individuals relocate from gravel plains to wadis and an
opposite relocation was observed during spring. This was the most abundant of the
gerbil species on gravel plains of the Ramon. There was a very high level of individual turnover: on average 81% of animals were newly marked at each of three
months trapping sessions. There were permanent shifts of home ranges of individuals even in preferred habitats. Sex ratio did not differ from parity. The proportion
of young varied during the year between 7–26%, except for September (76%). Molt-
ing takes place during September–October.
Food: This species is a typical seed eater, more so than other species of the same
genus in the Negev.
Reproduction: Reproduction of this gerbil was studied by Shenbrot *et al.* (1994) in
the Negev. There are two distinct periods of breeding — spring (February–May)
and summer (August–September). During spring 80% of adult females breed, but
only 21% in summer. Pregnancy lasts 19 days. At birth, newborn weigh about
one gram, and their body length is 23.5 mm. Litter size is 3–6 (Mendelssohn and
Yom-Tov, 1987). Maximum recorded longevity in the field was 241 days.
Parasites: In the field, about 15% of the population are slightly infected by fleas
(*Xenopsylla conformis, X. dipodilli* and *Nosopsyllus theodori*) (Shenbrot *et al.*, 1994).

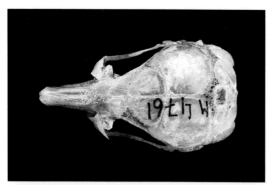

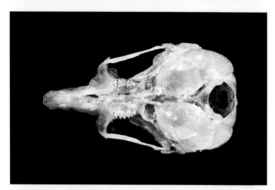

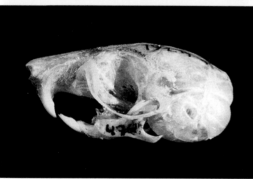

Figure 143: *Gerbillus henleyi mariae*. Dorsal, ventral and lateral views of the cranium, and a view of the mandible.

1 cm

Figure 144: Distribution map of *Gerbillus henleyi mariae*.

● – museum records.

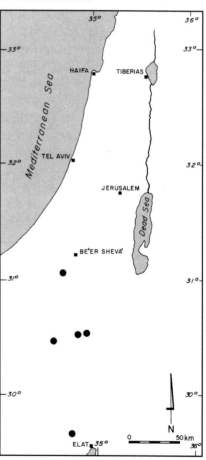

Table 134
Body measurements (g and mm) of *Gerbillus henleyi mariae*

| | Body | | Tail | Ear | Foot |
	Mass	Length			
		Males (4)			
				n = 3	
Mean	7.9	59.5	90	8	17
SD	1.2	4.2	11.5	1.0	2.2
Range	7.0–9.5	55–65	78–105	7–9	15–20
		Females (4)			
	n = 3				
Mean	10.8	63	92	8	18
SD	2.8	6.2	5.5	0.8	0.5
Range	7.5–12.5	55–68	84–95	7–9	17–18

Very similar results were presented for a much larger sample by Shenbrot *et al.* (1994), who also showed that body mass was significantly lower in winter and spring (9.2 g) than in summer and autumn (10.5 g).

Table 135
Skull measurements (mm) of *Gerbillus henleyi mariae* (Figure 143)

| | Males (4) | | | Females (4) | | |
	Mean	Range	SD	Mean	Range	SD
GTL	22.0	21.7–22.4	0.4	22.0	21.3–22.5	0.5
CBL	19.2	18.5–19.2	0.6	19.4	18.9–19.6	0.3
ZB	12.2	12.0–12.5	0.2	12.4	12.3–12.4	0.1
BB	11.3	11.0–11.9	0.4	11.3	11.3–11.4	0.1
IC	4.4	3.9–4.8	0.4	4.2	3.9–4.7	0.3
UT	2.8	2.6–3.2	0.3	2.8	2.7–3.1	0.2
LT	2.9	2.8–3.0	0.1	2.8	2.7–2.9	0.1
M	12.2	11.9–12.4	0.2	12.3	12.1–12.4	0.1
TB	8.9	8.6–9.1	0.3	8.9	8.8–9.1	0.1

Karyotype: 2N = 52 for specimens from Israel (Wahrman *et al.*, 1988).

Rodentia: Gerbillus

Gerbillus (Dipodillus = Hendecapleura) nanus Blanford, 1875   Baluchistan Gerbil
Figs 145–146; Plate XXVIII:1   גרביל דרומי

Gerbillus nanus Blanford, 1903. *Annals Mag. nat. Hist.,* 16:312. Gedrosia, west of Gwadar, Baluchistan.
Synonymy in Harrison and Bates (1991).

Its colour is mainly light brown, with the white colour of the underparts extending to the lower part of the sides of the body and to the sides of the head, to above the eyes. The white colour above the eyes distinguishes this species from *G. dasyurus.* The tuft of the tail is well developed, bushy and dark brown, almost black. The soles of the feet are naked.
Gerbillus nanus arabium Thomas, 1918 is the subspecies present in Israel (Harrison and Bates, 1991).

Table 136
Body measurements (g and mm) of *Gerbillus nanus arabium*

	Body		Tail	Ear	Foot
	Mass	Length	Tail	Ear	Foot
Males (35)					
Mean	26	90	127	13	24
SD	5	6.5	13	1.5	2
Range	10–36	77–105	88–150	8.5–19	20–27
Females (26)					
Mean	24	90	119	13	24
SD	5	7	13	2	1.5
Range	16–36	72–103	,90–145	6–18	20–26

Table 137
Skull measurements (mm) of *Gerbillus nanus arabium* (Figure 145)

	Males (10)			Females (10)		
	Mean	Range	SD	Mean	Range	SD
GTL	28.3	27.1–29.3	0.7	28.1	26.5–29.9	1.2
CBL	24.4	23.2–29.6	0.7	24.9	23.0–28.8	1.6
ZB	15.2	14.2–15.6	0.5	15.1	14.9–15.4	0.2
BB	12.8	12.5–13.1	0.2	12.7	12.3–13.4	0.4
IC	4.8	4.4–5.2	0.3	4.8	4.4–5.1	0.3
UT	3.7	3.5–3.8	0.1	3.6	3.4–4.0	0.2
LT	3.6	3.4–3.7	0.1	3.6	3.3–3.8	0.2
M	15.8	14.6–16.2	0.5	15.7	15–16.6	0.6
TB	11.1	10.5–11.7	0.4	11.0	10.5–11.7	0.4

Karyotype: 2N = 52, FN = 60, for specimens from Israel (Wahrman *et al.*, 1988).

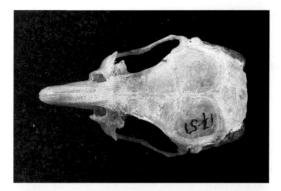

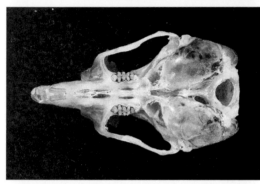

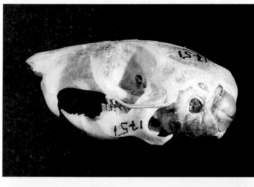

Figure 145: *Gerbillus nanus arabium*. Dorsal, ventral and lateral views of the cranium, and a view of the mandible.

1 cm

Figure 146: Distribution map of *Gerbillus nanus arabium*.

● – museum records.

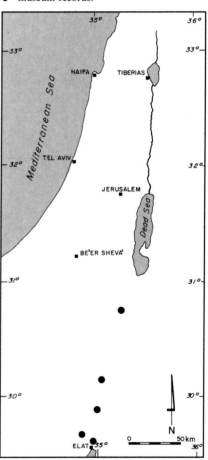

Distribution: It has a wide distribution in deserts from North Africa to Pakistan. In Israel it occurs in the 'Arava valley (Figure 146).

Habits: This gerbil occurs on solid sand in desert areas, but not on loose sand, on which *G. gerbillus* is found. It digs its burrows beneath *Nitraria retusa* and other bushes (Zahavi and Wahrman, 1957).

Reproduction: Litter size ranges from 2–5. In captivity it breeds during spring and summer.

Parasites: This species is affected by fleas (*Synosternus cleopatrae, Nosopsyllus pumilionis, N. theodori, Xenopsylla conformis*), Anoplura (*Polyplax kaiseri*), ticks (Ixodidae — *Hyalomma dromedarii*) and Protozoa (*Grahamella* sp.) (Theodor and Costa, 1967, and Krasnov *et al.*, 1997).

Gerbillus (Dipodillus = Hendecapleura) dasyurus (Wagner, 1842)   Wagner's Gerbil
Figs 147–148; Plate XXVIII:2–3   גרביל הסלעים

Meriones dasyurus Wagner, 1842. *Archiv Natrugesch.*, 8(1):20. Arabian West Coast.
Synonymy in Harrison and Bates (1991).

The coloration of this gerbil is very different in the different populations. In the desert they are light greyish-yellow, similar to *Dipodillus nanus* and the bushy tuft of the tail tip is brown. In the Galilee and the Golan they are dark reddish-brown, similar to the sympatric *Apodemus sylvaticus,* and the tuft of the tail is black. Specimens from intermediate areas are also intermediate in coloration. The dark colour of the upperside always envelops the eye, in contrast to the similar *G. nanus*, in which the white colour of the underside extends to above the eye. There are no significant size differences between the desert and Mediterranean populations, but the tympanic bullae are larger in the desert population.

Gerbillus dasyurus dasyurus is the subspecies present in Israel (Harrison and Bates, 1991).

Distribution: It occurs in Arabia and north to Syria, Iraq and north eastern Egypt. In Israel it is widespread in rocky areas (but normally not on cliffs) from the Galilee to the southern Negev (Figure 148). In many areas it is sympatric with *Acomys cahirinus.*

Fossil record: The earliest remains of this species found in Israel are from the Late Acheulian period (about 250,000 years BCE (Tchernov, 1988).

Habits: Wagner's gerbil lives on rocky slopes and digs its burrows under rocks, as well as in natural crevices and in fissures in rocks. It climbs well, sometimes using all four legs spread sideways to climb up a vertical crevice. In the Mediterranean zone of Israel it lives in areas with sparse vegetation. In the Judean hills it is active mainly for two hours after sunset, and again two hours before sunrise. The mean diameter of home ranges of adult males and females in this area were 10.7 and 8.9 m, respectively, and between 6–7 m for young animals. Population peaked in autumn and was lowest in spring (Ritte, 1964). The biology of this species was

Table 138
Body measurements (g and mm) of *Gerbillus dasyurus dasyurus*

		Body			
	Mass	Length	Tail	Ear	Foot
		Males (64)			
Mean	24	89	116	13	24.5
SD	4.5	11	11	2	3
Range	15–34	73–110	83–145	11–16	22–29
		Females (51)			
Mean	23	84	117	13	24
SD	4	7	11	1	2
Range	15–32	70–100	85–145	10–14	21–29

Table 139
Skull measurements (mm) of *Gerbillus dasyurus dasyurus* (Figure 147)

	Males (13)			Females (13)		
	Mean	Range	SD	Mean	Range	SD
GTL	25.8	26.6–28.9	0.6	27.5	26.6–28.6	0.6
CBL	23.8	23.3–25.0	0.6	23.9	23.3–25.0	0.5
ZB	14.3	13.8–14.6	0.4	14.3	13.6–14.8	0.5
BB	13.0	12.4–13.6	0.4	12.9	12.6–13.4	0.2
IC	4.9	4.6–5.3	0.2	5.1	4.7–5.3	1.2
UT	3.8	3.5–4.2	0.2	3.8	3.5–3.9	0.2
LT	3.8	3.5–4.2	0.2	3.7	3.4–4.0	0.2
M	15.7	15.0–16.7	0.5	15.7	15.3–16.2	0.3
TB	9.5	9.1–9.9	0.4	9.6	9.0–9.8	0.2

Karyotype: $2N = 60$ for specimens from Israel (Wahrman *et al.*, 1988).

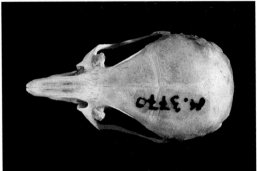

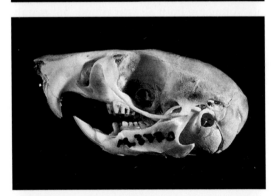

Figure 147: *Gerbillus dasyurus dasyurus.* Dorsal, ventral and lateral views of the cranium, and a view of the mandible.

1 cm

Figure 148: Distribution map of *Gerbillus dasyurus dasyurus.*

● – museum records.

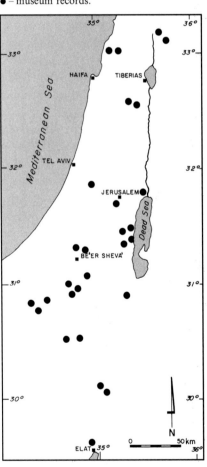

323

studied by Shenbrot *et al.* (in press) in Makhtesh Ramon, central Negev. Population density was related to annual precipitation and fluctuated between 0.2–5.0 individuals per dunam. The main factor affecting the distribution of this species was soil structure variables. It avoided patches with sandy soils and dense vegetation in wadies, and occurred mainly on loess hills with rocks and least on gravel plains. Two types of burrows were recorded: simple ones with 1–3 openings and without a nest were more abundant than complex burrows with 4–5 openings and a chamber with a nest. The burrow's diameter was 25–30 mm. Sex ratio was skewed in favour of males, which formed about 60% of the population.

Food: It feeds on seeds and leaves, but also on invertebrates, particularly snails, which are hoarded in piles near the burrow entrance. However, as this gerbil is often sympatric with *Acomys cahirinus*, it is possible that it inhabits rock crevices, formerly used by *Acomys*, that collected the snails. *Dipodillus dasyurus* generally eats fewer snails than *Acomys*, for it feeds mainly on seeds.

Reproduction: In Makhtesh Ramon, males in reproductive condition were found throughout the year, apart of December. Reproductive females were captured between February and October, with a peak in May. Pregnancy lasted 18–22 days, litter size ranged between 3–6 and the weight of newborn ranged between 1.3–1.9 g. The young started to eat solid food when 17–23 days old. Both parents warmed the pups in the burrow. In Saudi Arabia, gestation lasts 24–26 days (Al-Khalili and Delany, 1986). Litter size ranges from 2–6 (mean of 7 litters at Tel Aviv University Zoo was 4). In captivity young are born all year round, but mainly in spring (April–May) and autumn (September–November). In the Judean hills young are first caught in traps when they already weigh 15 g, indicating a long development period in the nest (Ritte, 1964). Longevity in captivity was more than 4 years and up to 2.5 years in the field (Shenbrot *et al.*, 1997).

Parasites: This species is infested by fleas (*Xenopsylla dipodilli*, *X. conformis*, *X. ramesis*, *Coptopsylla africana*, *Rhadinopsylla masculana*, *Nosopsyllus theodori*, *Parapulex chephrenis*, *Stenoponia tripectinata*), Anoplura (*Polyplax kaiseri*), ticks (Ixodidae — *Rhipicephalus sanguineus*, *Hyalomma* sp.), mites (*Androlaelaps androgynus*, *A. longipes*, *Hirstionyssus arcuatus*, Protozoa (*Nuttallia* sp., *Grahamella* sp.) and Spirochaetes (*crocidurae* group) (Theodor and Costa, 1967; and Krasnov *et al.*, 1997).

SEKEETAMYS Ellerman, 1947

Sekeetamys Ellerman, 1947. *Proceedings zool. Soc. Lond.*, p. 271.

This genus is represented in Israel by one species, *S. calurus.*

Sekeetamys calurus (Thomas, 1892)   Bushy-tailed Jird
Figs 149–150; Plate XXVIII:4   יפה זנב

Gerbillus calurus Thomas, 1892. *Annals Mag. nat. Hist.*, 9:76. Near Tor, Sinai.

This species is distinguished from other gerbilline species by its very bushy tail, which is longer than the body. The species scientific (Greek) and Hebrew names

Table 140
Body measurements (g and mm) of *Sekeetamys calurus calurus*

	Mass	Length	Tail	Ear	Foot
		Body			
				Males (11)	
				n = 9	
Mean	54	116	150	20	31
SD	13	7	8	1	2
Range	42–82	110–130	140–160	18.5–21	29–34
			Females (8)		
	n = 5		**n = 7**		
Mean	50	111	137	19	30.5
SD	7	5	13	1	1
Range	42–60	105–120	120–160	17–20	29–3

Table 141
Skull measurements (mm) of *Sekeetamys calurus calurus* (Figure 149)

	Males (10)			Females (8)		
	Mean	Range	SD	Mean	Range	SD
GTL	36.4	34.3–38.9	1.5	36.0	35.0–38.4	1.2
CBL	31.0	29.3–32.9	1.0	31.7	29.7–36.0	2.0
ZB (n = 8)	18.4	17.1–19.3	0.7	18.5 (n = 6)	17.6–19.2	0.7
BB	16.3	15.4–18.0	0.9	16.4	15.1–18.7	1.1
IC	5.7	5.0–6.2	0.4	5.8	5.2–6.3	0.3
UT	5.0	4.8–5.4	0.2	4.9	4.6–5.1	0.2
LT	5.0	4.7–5.3	0.2	4.8 (n = 7)	4.5–5.0	0.2
M	20.4	18.8–21.6	0.8	20.5 (n = 7)	19.6–22.1	0.8
TB	13.9	12.9–14.8	0.6	14.0 (n = 7)	13.4–14.8	0.5

Karyotype: 2N = 38 for specimens from Israel (Wahrman *et al.*, 1988).

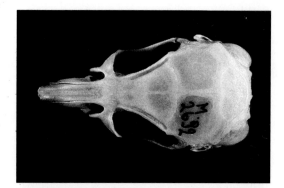

Figure 149: *Sekeetamys calurus calurus.* Dorsal, ventral and lateral views of the cranium, and a view of the mandible.

1 cm

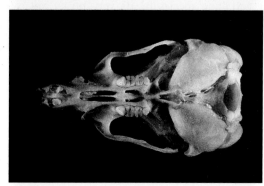

Figure 150: Distribution map of *Sekeetamys calurus calurus.*

● – museum records.

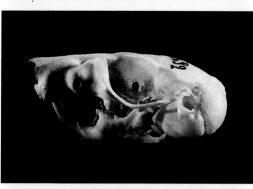

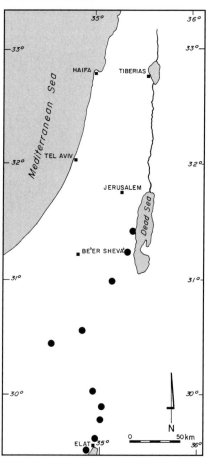

326

reflects this: "kalos" — beautiful, "oura" — tail. It is covered along its whole length with long hairs, giving this gerbil a squirrel-like appearance. Sometimes the tip of the black tail is white. The body fur is yellowish-brown, with black-tipped hairs. Its colour is darker than other desert gerbils. There are no significant differences in size between the sexes.

Sekeetamys calurus calurus is the subspecies present in Israel (Harrison and Bates, 1991).

Distribution: This jird is confined to Arabia, Israel, Jordan, Sinai and eastern Egypt. In Israel it occurs on steep rocky slopes in the Negev and southern Judean Deserts east and south of the 100 mm isohyet, as far north as 'En Gedi (Figure 150).

Habits: The bushy-tailed jird occurs on cliffs and steep, rocky slopes. It is a very agile climber, even on vertical rock surfaces and can jump well. It is solitary, digging its burrows under rocks. In captivity it behaves like a squirrel, climbs on nets, branches and even runs along the roof of the cage, while hanging upside down. The basal metabolic rate (BMR) of this species is 47% of the expected, and its minimal metabolic conductance is 46% higher than the expected. Both characters are considered adaptations to its hot desert habitat (Haim and Borut, 1986). Its efficient kidneys enable it to survive on water with a salt concentration double that of sea water, and to feed on halophytic plants.

Food: It feeds on seeds and leaves, including of halophytic plants, as well as on invertebrates. Similarly to squirrels and hamsters, it tends to collect large stocks of seeds to a greater extent than other gerbillids.

Reproduction: Gestation lasts 24 days. Litter size ranges from 2–7, and in the Tel Aviv University Research Zoo mean litter size of six litters was four. In captivity it breeds all year round, and may live for up to 6 years.

Parasites: This species is affected by fleas (*Xenopsylla conformis, X. dipodilli, Nosopsyllus theodori*), ticks (Ixodidae — *Rhipicephalus sanguineus, Hyalomma* sp.), mites (*Androlaelaps androgynus, Laelaps acomydis*) (Theodor and Costa, 1967).

MERIONES Illiger, 1811

Meriones Illiger, 1811. *Prodromus Syst. Mamm. et Avium*, p. 82.

The genus *Meriones* includes several species that are relatively large, about the size of a small rat. They range from North Africa to Mongolia and India. Their colour is yellow to brown, according to the habitat. The tuft at the tip of the tail is black. They feed on seeds and also on much more green food than the smaller gerbillids. This genus is represented in Israel by three, largely parapatric, species: *M. tristrami* inhabits the Mediterranean region; *M. sacramenti* is found on sandy soils in the Northern Negev and the Coastal Plain south of the Yarqon River and *M. crassus* occurs on other soils in desert areas (Figure 151). The distribution areas of these species partly overlap in the Be'er Sheva' area.

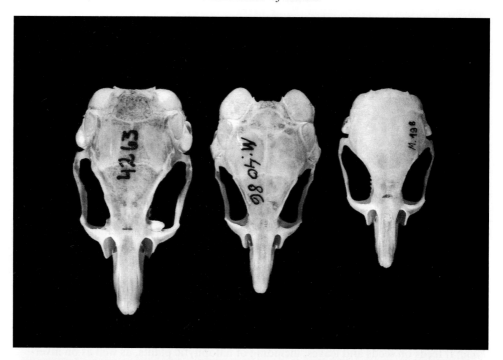

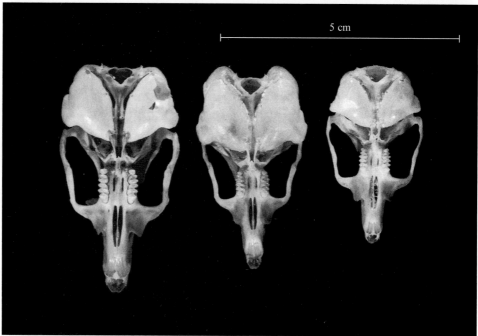

Figure 151: Skulls of *Meriones tristrami tristrami* (smallest), *M. crassus crassus* (middle) and *M. sacramenti sacramenti* (largest). Note the large bullae tympanica of *M. crassus*, which extend beyond the occipital region.

Meriones (Meriones) tristrami Thomas, 1892   Tristram's Jird

Figs 151–153; Plate XXIX:1   מריון מצוי

Meriones tristrami Thomas, 1892. *Annals Mag. nat. Hist.*, (6),9: 148. Dead Sea Region, Palestine.

Synonymy in Harrison and Bates (1991).

Its colour is generally yellowish-brown; specimens from the south (Mamshit area) are lighter in colour, almost as light as *Meriones crassus*. Specimens found in a very humid area with dense vegetation, near a brook (Naḥal Poleg, Coastal Plain) were very dark, blackish-brown. When kept in captivity, they molted after several

Table 142
Body measurements (g and mm) of *Meriones tristrami tristrami*

	Body		Tail	Ear	Foot
	Mass	Length	Tail	Ear	Foot
	Males (12)				
Mean	77	136	138	18	31
SD	17	13	7	4	1
Range	51–104	115–155	125–146	16–21	29–33
	Females (12)				
Mean	62	135	131	18	30
SD	10	12	12	2	3
Range	48–63	120–155	113–150	16–21	22–34

Table 143
Skull measurements (mm) of *Meriones tristrami tristrami* (Figure 152)

	Males (13)			Females (10)		
	Mean	Range	SD	Mean	Range	SD
GTL	37.7	34.9–40.0	1.2	36.0	33.3–39.0	2.0
CBL	33.8	31.6–35.3	1.1	32.3	29.7–34.7	1.8
ZB	20.3	18.6–22.0	1.25	19.6	17.9–20.7	1.0
ZB (male n = 12 female n = 9)						
BB	15.7	15.0–16.5	0.4	15.4	14.8–16.6	0.5
IC	6.05	5.5–6.7	0.4	5.8	5.5–6.2	0.2
UT	5.4	5.1–5.9	0.2	5.3	4.9–5.8	0.3
LT	5.4	5.0–5.9	0.3	5.4	5.1–5.9	0.2
LT (male n = 11)						
M (male n = 11)	22.7	21–24.1	1.0	21.7	19.5–23.7	1.5
TB (M n = 13)	12.4	11.2–13.3	0.5	12.6	11.6–13.6	0.7

Karyotype: 2N = 72, FN = 76–78, for specimens for Israel (Wahrman *et al.*, 1988).

months and then developed the normal colour, showing that the dark coloration was environmentally influenced.

There is no north-south body size gradient or high correlation with climatic factors, but there is a shift in body size when sympatric with one of its congeners in Israel (Chetboun and Tchernov, 1983).

Meriones tristrami tristrami is the subspecies present in Israel (Harrison and Bates, 1991).

Distribution: It occurs in Asia Minor and east to Iraq and north western Iran, and south to Lebanon and Israel.

In Israel it is abundant over most of the Mediterranean zone and penetrates into the Negev as far south as the central mountain range near 'Avedat north of the 100 mm isohyet. It occurs along the Jordan River as south as Jericho, and on Mount Ḥermon, up to 1,600 m above sea level. It inhabits various soils, including sand, loess, rendzina and terra-rossa, but always well-drained ones (Figure 153).

Fossil record: The earliest remains of this species found in Israel are from the Late Acheulian period (about 250,000 years BCE (Tchernov, 1988).

Habits: This species is normally restricted to non-sandy habitats, but in coastal areas north of Mount Carmel, where *Gerbillus andersoni* does not occur, *M. tristrami* inhabits dunes (Abramsky and Sellah, 1982). It also occurs on stabilized sandy soil such as between dunes, where there is dense vegetation. The population density south of Mount Carmel fluctuated between 1–4 individuals per 1000 m^2 (Abramsky and Sellah, 1982).

It is a solitary species, constructing its burrows to a depth of 30–40 cm and more. Each burrow has several openings, which sometimes it plugs with soil from the inside. It constructs a resting chamber with a nest, made of dry grass, at the end of one of the tunnels. It is primarily nocturnal.

The young produce very high frequency calls, which occasionally also are produced by adults. When alarmed they drum with their hind feet.

This species shows remarkable resistance to adverse conditions of hunger, thirst and temperature (Bodenheimer, 1949). Specimens which were kept at -15°C survived for 4 days by huddling together.

Food: It feeds on grain and leaves. During peak years it hoards large quantities of grain in its burrows, thus causing damage to crops.

Reproduction: The gestation period lasts 21–22 days and litter size ranges between 1–8. In captivity young are born all year round, but in nature it breeds in spring and summer (March–September). Litter size is 1–7, mean 3.6 (Bodenheimer, 1949). In captivity one female produced 13 litters in 523 days (Naftali and Wolf, 1954). Newborn weigh four g and are blind and hairless. When 10 days old they are covered with fur, and when 10–12 days old they open their eyes. They are weaned when three weeks old (Z. Zook-Rimon, pers, comm.). Only the female takes care of the young. Sexual maturity is reached at three months. In captivity they may live up to five years, but under natural conditions males do not normally survive for more than 2.5 years, and females even less.

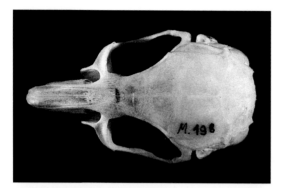

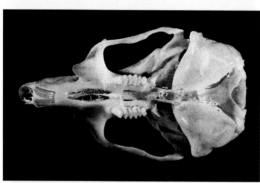

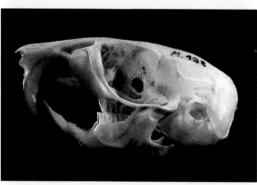

Figure 152: *Meriones tristrami tristrami.* Dorsal, ventral and lateral views of the cranium, and a view of the mandible.

1 cm

Figure 153: Distribution map of *Meriones tristrami tristrami.*

● – museum records.

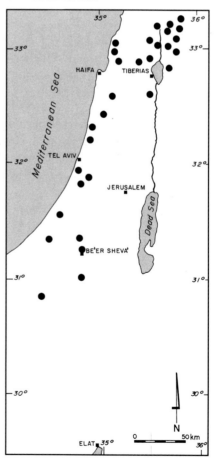

331

Parasites : This species is affected by fleas (*Nosopsyllus iranus, N. sincerus, N. henleyi, Xenopsylla ramesis, X. cheopis, Synosternus cleopatrae, Stenoponia tripectinata, Leptopsylla algira*) and ticks (lxodidae — *Rhipicephalus sanguineus, Hyalomma* sp., *Haemaphysalis erinacei*), mites (*Eulaelaps stabularis, Androlaelaps ovalis, Laelaps acomydis*), Protozoa (*Nuttallia danii, Grahamella* sp.) and spirochaetes (Theodor and Costa, 1967).

Relations with man: This species may become a minor agricultural pest in years when its population size is large (Bodenheimer, 1949). As it feeds on seeds as well as on green food, it is easy to control with poison grain bait. A bait composed of 10% wheat grain soaked with thallium sulfate and mixed with 90% non-poisoned wheat controls the jirds successfully, but prevents secondary poisoning of raptors and carnivores (Z. Zook-Rimon, pers. comm.).

Meriones (Meriones) sacramenti Thomas, 1922 Buxton's Jird
Figs 151,154–155; Plate XXIX:2 מריון החולות

Meriones sacramenti Thomas, 1922. *Annals Mag. nat. Hist.*, (9), 10:552. 10 miles south of
Beersheba, Palestine.
Synonymy in Harrison and Bates (1991).

This is the largest of its genus and of all the gerbillids in Israel. Its colour is light earth-brown in the Coastal Plain, lighter in the Negev. The tuft of the tail is black. Specimens from the Coastal Plain are on average larger than those of the Negev.

Meriones sacramenti sacramenti is the subspecies present in Israel (Harrison and Bates, 1991).

Distribution: It is the only endemic mammal in Israel, occurring in sandy and sand-loess areas in the western Negev and along the Coastal Plain as far north as the Yarqon River (Figure 155).

Fossil record: The earliest remains of this species found in Israel are from the Natufian period (Tchernov, 1988).

Habits: Buxton's jird is psammophilous, dwelling in burrows it digs in the soil.

Food : Species of the genus *Meriones* generally prefer green food, and the present species is much more herbivorous than granivorous. It feeds mainly on green food, but also on seeds and insects. It feeds also on young shoots of *Retama raetam* bushes, under which it often digs its burrows. In a study in the western Negev (Bar *et al.*, 1984) these foods comprised 29%, 68% and 3% of the annual diet, respectively. During winter green food comprised 97% and in spring seeds comprised 57% of the diet. The preferred food species were Gramineae, *Erodium* and *Plantago*.

Reproduction: In nature it breeds during spring and autumn, while in the Tel Aviv University Research Zoo young were born between February–June and in September. Gestation lasts about 24–25 days, and litter size ranges from 2–8 (mean of 17 litters in captivity was 5).

Table 144
Body measurements (g and mm) of *Meriones sacramenti sacramenti*

	Body		Tail	Ear	Foot
	Mass	Length			
	Males (11)				
Mean	136	161	132	19	36
SD	22	16	27	3	3
Range	110–175	140–190	110–168	13–22	33–40
	Females (10)				
			n = 9		
Mean	113	153	153	19	38
SD	23	11	9.5	1	2.5
Range	72–142	135–172	135–165	17–21	34–41

The Negev specimens are on average smaller than those of the Coastal Plain. Size seems to be influenced by environmental factors, as demonstrated by the following observation: Specimens from the Coastal Plain that were sent to the zoo in Zurich, Switzerland, bred there and their offspring grew to a much larger size than their parents, reaching the size of brown rats (*Rattus norvegicus*).

Table 145
Skull measurements (mm) of *Meriones sacramenti sacramenti* (Figure 154)

	Males (10)			Females (8)		
	Mean	Range	SD	Mean	Range	SD
GTL	43.4	41.1–45.3	1.5	42.2	41.0–43.4	0.9
CBL	38.9	36.7–40.8	1.5	37.9	36.3–39.7	1.3
ZB	24.0	23.2–25.4	0.8	22.9	21.9–23.7	0.7
BB	18.9	18.1–19.6	0.5	18.9	18.1–19.7	0.6
IC	6.8	6.2–7.5	0.4	6.7	5.9–7.5	0.5
UT	6.5	6.0–7.0	0.3	6.8	6.5–7.4	0.4
LT	6.5	6.1–7.0	0.3	6.5	6.2–6.9	0.2
M	26.0	24.5–28.3	1.3	25.2	24.2–26.1	0.7
TB	16.6	15.9–17.4	0.4	16.2	15.4–17.4	0.6

Karyotype: 2N = 46 for specimens from Israel (Wahrman *et al.*, 1988).

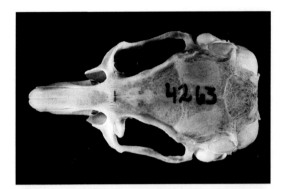

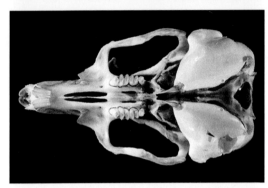

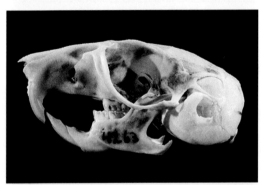

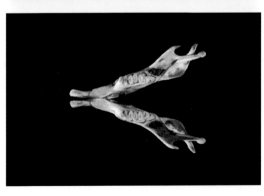

Figure 154: *Meriones sacramenti sacramenti.* Dorsal, ventral and lateral views of the cranium, and a view of the mandible.

|—— 1 cm ——|

Figure 155: Distribution map of *Meriones sacramenti sacramenti.*

● – museum records.

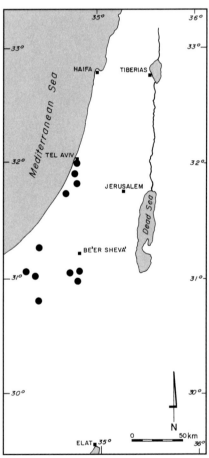

In one case, at Tel Aviv University Research Zoo, the female of a pair died when the pups were one week old. The male stayed with the pups, warmed and groomed them, retrieved them when they were removed from the nest and kept them alive until they starved after a week. In captivity, the male may also take care of the young. In captivity longevity is up to five years.

Parasites: This species is affected by fleas (*Synosternus cleopatrae, Stenoponia tripectinata*), ticks (Ixodidae — *Rhipicephalus sanguineus, Hyalomma* sp.), mites (*Androlaelaps centrocarpus, A. insculptus, A. marshalli*) and Protozoa (*Nuttallia* sp., *Grahamella* sp.) (Theodor and Costa, 1967).

Meriones (Meriones) crassus Sundevall, 1842   Sundevall's Jird
Figs 151, 156–157; Plate XXIX:3–4   מריון המדבר

Meriones crassus Sundevall, 1842. *Svenska Vet. Akad.*, ser. 3:233. Fons Moses (Ain Musa), Sinai.
Synonymy in Harrison and Bates (1991).

This jird is light-brown to greyish-yellow in colour, always lighter than *M. tristrami*, with which it may be sympatric in some areas. The big, bushy tuft of the tail is conspicuously black. Its heavy build is reflected in its specific name, "crassus", meaning "heavy" in Latin.

Meriones crassus crassus is the subspecies present in Israel (Harrison and Bates, 1991).

Distribution: It occurs in desert areas in the Sahara from Algeria to Egypt, and through Arabia and Iran to Afghanistan.

In Israel it occurs in the Negev and Judean Deserts, in wide wadis, loess and sandy-loess plains and hamadas, as far north as Yeruḥam and Kurnub (Mamshit) in the central Negev (Figure 157). In the loess-sand areas of Kurnub it is sympatric with *Meriones tristrami*.

Habits: The biology of this species was studied by Krasnov *et al.* (1996) in Makhtesh Ramon. It inhabits various habitats such as sand dunes, wadis and gravel plains. Its distribution is patchy and is related to shrub cover in summer. During October–November its population density is high in open plains and slopes and low in wadis, and the reverse occurs in January. This is a result of relocation of individuals from one habitat to another. Population density also varies between years and seasons and ranges between 0.17–4.0 individuals per hectare. Mean distance between consecutive capture locations is 25 m, but there have been occasional recaptures up to 190 m from the previous location. A study of radio-implanted individuals indicated that an individual remained 3–7 nights in one locality and then moved an average 47 m (range 5–140 m) to another location where it stayed for a similar period. Jirds spend most of their above ground activity below bushes. The studied population was apparently composed of solitary individuals, each living in a small home range. However, according to Harrison (1972) and Haim and Tchernov (1974) it

Table 146
Body measurements (g and mm) of *Meriones crassus crassus*

| | Body | | Tail | Ear | Foot |
	Mass	Length			
			Males (10)		
Mean	86	134	133	15	32
SD	13	11	11	2	1
Range	63–100	120–155	120–150	11–20	30–33
			Females (10)		
	n = 9		**n = 8**		
Mean	72	139	133	15	31
SD	19	17	12	1	1
Range	51–110	105–170	110–145	12–17	29–33

Field data show that males are generally heavier than females, and that environmental conditions, such as low ambient temperatures and scarcity of rainfall, negatively affect body mass (Krasnov *et al.*, 1996).

Table 147
Skull measurements (mm) of *Meriones crassus crassus* (Figure 156)

| | Males (10) | | | Females (10) | | |
	Mean	Range	SD	Mean	Range	SD
GTL	39.9	38.3–42.0	1.5	40.0	38.3–42.1	1.2
CBL	34.8	32.1–36.5	1.5	34.5	32.7–35.7	1.1
ZB	20.5	18.7–21.6	1.0	20.6	19.7–21.7	0.7
BB	16.5	15.5–17.0	0.5	16.5	16.0–17.3	0.5
IC	6.3	5.9–6.8	0.3	6.2	5.9–6.7	0.3
UT	5.4	5.2–5.7	0.2	5.6	5.2–6.0	0.2
LT	5.6	5.1–6.2	0.4	5.7	5.4–6.2	0.3
M	22.0	20.5–23.5	1.1	22.2	16.1–18.1	0.7
TB	16.9	16.2–17.6	0.4	16.8	16.1–18.1	0.3

Karyotype: $2N = 60$, $FN = 70$, for specimens from Israel (Wahrman *et al.*, 1988).

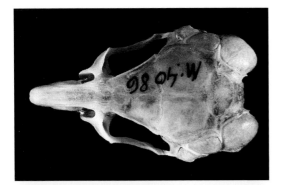

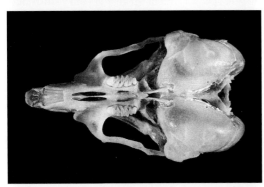

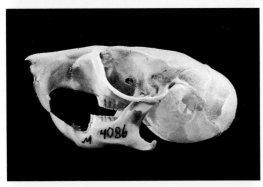

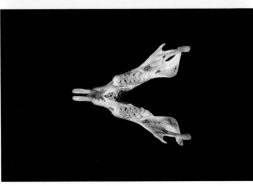

Figure 156: *Meriones crassus crassus*. Dorsal, ventral and lateral views of the cranium, and a view of the mandible.

├──────┤ 1 cm

Figure 157: Distribution map of *Meriones crassus crassus*.

● – museum records.

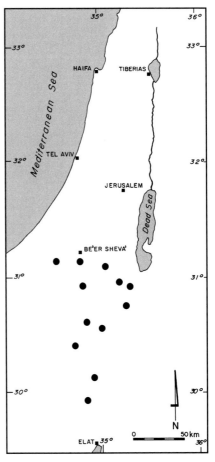

337

is a colonial species, and 8–10 jirds may live in one hillock. Such an aggregation may, however, comprise a pair with a litter of grown young, before they disperse. They construct extensive burrow systems, where they often hoard seeds. This increases the water content of the seeds, as they absorb water vapor from the high air humidity in the burrow. In Israel several individuals were found in one burrow system, but their relationship is not known.

Food: It feeds on leaves, seeds and insects, and in captivity, under suitable ambient conditions, it can survive for months on dry seeds without access to water. In captivity they breed only if green, succulent food is available.

Reproduction: Reproduction of this species was studied in Makhtesh Ramon by Krasnov *et al.* (1996). They found that this species breeds from February until September, and during March–May all adults were in breeding condition, i.e. females were either pregnant or lactating, or had either open vagina or vaginal plug. However, young born during late summer did not survive. In captivity it breeds throughout the year. The gestation period is 18–24 days (Misonne, 1959, Krasnov *et al.*, 1996) and litter size ranges from 3–7 (Harrison and Bates, 1991), average of 16 litters was 4.9 (Krasnov *et al.*, 1996). However, autopsied females captured in the field had 1–4 pups (Lay, 1967, Le Berre, 1990). Hair appears when the pups are 10–15 days old, eyes open at 17–20 days and the young start eating solid food when 17–23 days old. Sex ratio is 1:1.

Among individuals recorded for more than one occasion 70% were recorded for six months, 20% for one year and 10% for 1.5 years. In captivity individuals caught as adults lived more than three years (Krasnov *et al.*, 1996).

Parasites: This species is affected by fleas (*Xenopsylla conformis, X. dipodilli, X. ramesis, Synosternus cleopatrae, Nosopsyllus theodori, N. pumilionis, Stenoponia tripectinata, Coptopsylla africana*), Anoplura (*Polyplax paradoxa*), ticks (Ixodidae — *Rhipicephalus sanguineus, Hyalommaa savignyi*), mites (*Androlaelaps longipes*), Protozoa (*Nuttallia* sp., *Grahamella* sp.) and spirochaetes (*crocidurae* group) (Theodor and Costa, 1967; and Krasnov *et al.*, 1997).

PSAMMOMYS Cretzschmar, 1828

Psammomys Cretzschmar, 1828. In: Rüppell, *Atlas Reise Nordl. Afr. Saugeth*, p. 58, pl. 22.

This genus is represented in Israel by one species, *P. obesus*.

Psammomys obesus Cretzschmar, 1828   Sand Rat
Figs 158–159; Plate XXX:1–2   פסמון

Psammomys obesus Cretzschmar, 1828. In: Rüppell, *Atlas Reise Nordl. Afr. Saugeth*, p. 58,
  pl. 22. Near Alexandria, Egypt.
Synonymy in Harrison and Bates (1991).

The sand rat has large black eyes and relatively short ears. Its fur is light-brown, with greyish-white underparts. The relatively short tail has a well developed, black tuft. Its skin and claws are black. Black skin also characterizes *Acomys russatus*, another diurnal desert rodent, presumably as a protection against ultraviolet solar radiation. The upper incisors are smooth and do not have the median longitudinal groove that is characteristic for the front upper incisors of *Meriones* and other gerbillids. Its scientific name reflects that it may live on sands ("psamos" — sand, "mys" — mouse in Greek) and its heavy build ("obesus" — fat in Latin).

Psammomys obesus terrasanctae Thomas, 1902 is the subspecies present in Israel (Harrison and Bates, 1991).

Fossil record: The earliest remains of this species found in Israel are from the Late Acheulian period, about 250,000 years BCE (Tchernov, 1988).

Distribution: It occurs in North Africa from Mauritania to Egypt and Sudan, and east to Arabia and Syria.

In Israel it occurs in the Negev and Judean Deserts, in wadi beds, saline and saline-marsh areas and loess plains (Figure 159).

Habits : It is diurnal and its activity depends on ambient temperature: during winter it is active outside the burrow for about five hours daily during mid-day, while during summer activity is restricted to the early hours of the morning and the afternoon. In the lower Jordan Valley during summer it is also active during night, apparently due to high ambient temperature in this region. During winter its activity begins with basking in the sun, flattening the body and spreading the limbs in order to absorb solar radiation (Ilan and Yom-Tov, 1990).

It digs its burrows under the bushes from which it gets its food. The sand rat constructs an extensive burrow system, which is a well developed structure with 3–4 levels the lowest of which is as deep as 75 cm below surface, and spread over several dozen m^2. Young specimens dig a simple burrow, which is progressively extended. The burrow system of one adult pair, without young, which was completely excavated extended over an area of 12 × 7 m and went down to 80 cm depth. In another case the burrow system was flooded and the inhabitants fled from an opening on the

Table 148
Body measurements (g and mm) of *Psammomys obesus terrasanctae*

| | Body | | Tail | Ear | Foot |
	Mass	Length			
Males (11)					
Mean	156	171	125	13	36
SD	31	17.5	21	2	2
Range	125–208	125–190	95–150	10–18	33–39
Females (5)					
		n = 4			
Mean	177	173	138	16	38
SD	23	12.5	26	3.5	3
Range	146–207	155–190	110–160	13–22	34–40

Table 149
Skull measurements (mm) of *Psammomys obesus terrasanctae* (Figure 158)

| | Males (10) | | | Females (6) | | |
	Mean	Range	SD	Mean	Range	SD
GTL	45.4	42.9–48.1	2.0	44.4	42.4–45.8	1.4
CBL	42.6	39.2–47.8	3.3	41.3	39.3–43.0	1.5
ZB (n = 8)	26.1	24.3–28.2	1.3	25.1	24.2–25.8	0.6
BB (n = 9)	21.5	19.6–24.1	1.6	20.5	18.1–23.2	2.1
IC	7.3	6.9–7.9	0.3	7.3	6.7–7.7	0.3
UT	7.3	6.8–7.8	0.5	7.3 (n = 4)	7.0–7.6	0.3
LT	7.1	6.2–7.8	0.6	7.2	6.8–7.6	0.3
M	29.5	27.1–32.5	1.9	28.6	27.6–30.0	0.9
TB	16.2	15.4–17.5	0.7	15.3	14.6–15.9	0.5

Ilan (1984) found that in the field body mass differs between years, probably as an outcome of food availability: mean body mass was 221 g in 1981–1982, which was a rainy year, and 187 g in 1983–1984, which was a drier year. In his study males were significantly heavier than females (mean body mass of males and females was 208 g and 192 g, respectively). These data indicate that the museum specimens are lighter than those weighed in the field.

Karyotype: 2N = 48 for specimens from Israel (Wahrman *et al.*, 1988).

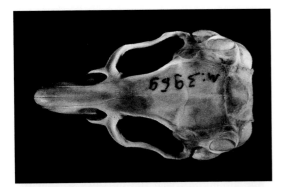

Figure 158: *Psammomys obesus terrasanctae*. Dorsal, ventral and lateral views of the cranium, and a view of the mandible.

1 cm

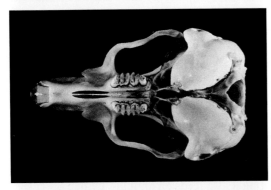

Figure 159: Distribution map of *Psammomys obesus terrasanctae*.

● – museum records.

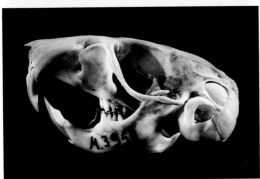

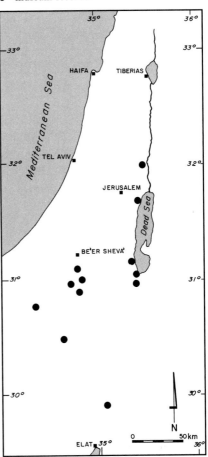

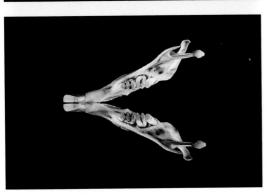

341

other side of a 12 m wide road, a distance of 20 m from the flooded opening. It constructs chambers for food storage, a toilet chamber and nests for sleeping. The young are delivered in one of the sleeping nests (Or, 1974). Much of the faeces are deposited outside the burrows.

It is colonial, but each burrow system is occupied by one adult. However, during the breeding season a pair or a family can often be found in one system. It climbs well on the shrubs on which it feeds and uses permanent, well marked trails.

Daily activity starts with feeding, followed by cleaning the burrows by removing food remains and faeces. The sand rat hoards large amounts of food plants in its burrows, not all of which are eaten, and the wilted remains are removed. The large amount of plant material raises the air humidity in the burrows so that the inhabitants lose less water by evaporation.

Population densities fluctuate greatly according to environmental conditions. *Psammomys* is a favourite prey of many animals, including snakes (i.e. *Spalerosophis*) and raptors, and at times when it is also nocturnal it is preyed on by owls. Near Qalya, north of the Dead Sea, jackals (*Canis aureus*) increased due to food availability from garbage dumps, and reduced the local *Psammomys* population density considerably. Sex ratio is 1:1.

This species communicates by drumming with its hind legs on the ground and by squeaking.

Food: It feeds only on leaves of salt bushes, particularly *Atriplex halimus* and other species of Chenopodiaceae. During spring it also feeds on annuals. Unlike other species of gerbillines, it does not eat seeds. It cuts a small branch with its teeth, grasping it with both front paws. The branch is then moved sideways in the mouth, similar to humans eating corn cobs, and the leaves are removed and eaten. The daily intake of dry matter ranged from about 25% (summer) to 32% of body mass (Degen et al., 1991). As the halophytes on which it feeds have a low energy value, it has to ingest large amounts of food and produces accordingly large amounts of excrement. Its diet contains large quantities of electrolytes (NaCl) and oxalates (about 40% ash of dry matter) which can cause osmotic stress. The animal scrapes off the salt from the leaves with its paws, sometimes using its incisors. The amount of leaf scraping is related to the water content of the *Atriplex*: When offered *Atriplex* with 84% water content, they scraped only 0.8% of the dry matter, and when offered *Atriplex* with 69% water content they scraped 14.3% of dry matter. Scraping lowers the ash and electrolyte content of the food considerably (Degen, 1988). Under natural conditions it does not drink, but in captivity it does so when water is offered and it is capable of drinking saline water (9% NaCl). It has very efficient kidneys, which excrete highly concentrated urine, about 18 times as concentrated as in humans. In captivity, when fed on high caloric food such as laboratory rodent pellets, it develops diabetes.

Reproduction: Gestation lasts 24 days. Litter size ranges from 1–7, and the average in captivity is 3–5 (Shaham, 1977). In nature, the young are born between December–April, but in captivity they breed all year round. The start of the breeding season depends on food availability, which in turn depends on the amount and distribution of rain. During drought periods they almost cease breeding (Ilan, 1984). The young are born blind and hairless. They open their eyes within a week, and suckle for up to 3–4 weeks. A female may give birth 2–4 times per season, and has an oestrus cycle of four days. Post-partum oestrus occurs in captivity, and possibly also in nature. Under such conditions pregnancy is extended by four days to 28 days. Every additional suckling pup adds two days to the length of the next pregnancy (Shaham, 1977).

In captivity sexual maturity is reached at three months (females) or four months (males), but in nature at a minimum of seven months of age (Shaham, 1977, Ilan, 1984). Only the females take care of the young. In a study at Sede Boqer it was found that individuals do not live for more than 14 months and thus breed for only one season (Ilan, 1984). In captivity they may live for 3–4 years.

Parasites: This species is affected by fleas (*Xenopsylla conformis, Synosternus cleopatra, Stenoponia tripectinata* (Theodor and Costa, 1967; and Krasnov *et al.*, 1997).

Relations with man: In the lower Jordan Valley this species is a vector of Leishmaniasis, that is transferred to humans by the sandfly *Phlebotomus*. (Schlein *et al.*, 1982).

MURIDAE   Rats and Mice

This family is represented in Israel by five genera: *Apodemus, Rattus, Mus, Nesokia* and *Acomys*.

Key to the Species of Muridae in Israel
(After Harrison and Bates, 1991)

1. Dorsal pelage with bristly spines at least over the posterior back.   2
  – Dorsal pelage normal, not modified to form bristles or spines.   3
2. Feet, ears and snout black. Spiny armature extends forwards to the occipital region.

Acomys russatus

  – Feet pale. Spiny armature does not extend in front of the mid-dorsal region.

Acomy cahirinus

3. Crowns of cheekteeth composed of simple transverse laminae, m^1 without trace of cusps in adults. Palatal foramina reduced, not extending between the toothrows.

Nesokia indica

  – Crowns of cheekteeth cuspidate. m^1 always with cusps in unworn condition.   4
4. m^1 and m^2 with a well developed postero-internal cusp when unworn.   5
  – m^1 and m^2 without a well developed postero-internal cusp when unworn.   6
5. Size largest, total length 196–269 mm; greatest length of skull 27.8–31.9 mm. Dorsal colour usually grey, underparts pure white, lacking any chest spot.

Apodemus mystacinus

343

–   Size smaller, total length 153–210 mm; greatest length of skull 22.3–27.6 mm. Dorsal colour brown, a yellow spot on the chest between the front legs   **Apodemus sylvaticus**

6.   Small mice, greatest length of skull 17.6–23.8 mm. m^1 enlarged, exceeding length of m^2 and m^3. Tail is equal to body length.   **Mus musculus**

–   Large rats, greatest length of skull 37.5–55.5 mm. m^1 not enlarged and not exceeding the combined length of m^2 and m^3.   7

7.   Tail distinctly shorter than head and body. m^1 without a clearly defined antero-external cusp. The ear, when folded forwards, reaches the eye, but does not cover it.

Rattus norvegicus

–   Tail longer than head and body, m^1 with a clearly defined antero-external cusp. The ear, when folded forwards, covers the eye.   **Rattus rattus**

APODEMUS Kaup, 1829

Apodemus Kaup, 1829. *Skizzirte Entw.-gesch. Naturl. Syst. Europ. Thierwelt*, 1:154.

Mice of the genus *Apodemus* have long limbs and tail and large ears. They are found in Europe, except in the extreme north, and throughout temperate Asia to eastern Asia. Most species are agile, climb and jump well. They are omnivorous, feeding on seeds, nuts and acorns, and on invertebrates, but also on small vertebrates and bird eggs. The fact that this genus is not commensal with man is reflected in its scientific (Greek) name: "apodem" — away from home, "mys" — mouse.

This genus is represented in Israel by two, possibly three, species. Harrison and Bates (1991) summarized the data available to them and suggested that *A. mystacinus* and *A. sylvaticus* are abundant in Israel while *A. flavicollis* occurs only on the southern slopes of Mount Ḥermon and the northern Ḥula Valley. On the other hand Fillippucci *et al.* (1989) suggested that *A. mystacinus* and *A. flavicollis* are the common species occurring in Israel and that *A. sylvaticus* does not occur in Israel. They also described a new species (*A. hermonensis*) from Mount Ḥermon, which is very close to *A. flavicollis*, from which it can be distinguished by one locus. Examining Fillippucci *et al.* (1989) data we notice that among the five populations of the sub-genus *Sylvaemus* they examined there is a clear trend of increase in body size from cold to warmer areas: the largest specimens are from the northern Ḥula Valley and the smallest are from Mount Ḥermon, while the other three populations are intermediates. A similar trend was found for European *Apodemus sylvaticus* (Alcantara, 1991). Hence, in our opinion there are two species of *Apodemus* in Israel and the population at high altitude on Mount Ḥermon belongs to *A. sylvaticus*. The local small *Apodemus*, considered as *A. sylvaticus*, also have, however, the yellow-throat spot in the same shape as European *A. sylvaticus* and not as European *A. flavicollis*, that have a collar-shaped throat spot. The ecological relationships between the two species are discussed below.

Apodemus mystacinus (Danford & Alston, 1877)   Broad-toothed Field Mouse

Figs 160–161; Plate XXX:3   יערון גדול

Mus mystacinus Danford & Alston, 1877. *Proceedings zool. Soc. Lond*, p. 279. Zebil, Bugar
    Dagh, Asia Minor.
Synonymy in Harrison and Bates (1991).

This is the largest species of its genus in Israel. Its large body makes it superficially
similar to a small roof rat. Its specific name reflects its long moustache ("mystac" in
Greek).

Apodemus mystacinus mystacinus is the subspecies present in Israel (Harrison and
Bates, 1991).

Distribution: This field mouse occurs in the Balkans, Asia Minor, western Transcaucasia, Syria, Lebanon, Jordan.

In Israel it occurs on Mount Hermon and along the central mountain range from
Upper and Lower Galilee, Mount Carmel, and south to Samaria and Judea south
of Jerusalem (Figure 161).

Fossil record: The earliest remains of this species found in Israel are from the Late
Acheulian period, about 250,000 years BCE (Tchernov, 1988).

Habits: It occurs in rocky areas with Mediterranean scrub forest, or in open areas
which were covered by such forest in the past. On Mount Meron, where it is sympatric with *Apodemus sylvaticus*, it occurs in dense forest while *A. sylvaticus* occurs in
shrub covered areas within the forest (Granot, 1978). A similar phenomenon was
observed at Tel Dan, where this species occurred in dense vegetation while *A. sylvaticus* occurred in drier areas with less vegetation. Haim and Rubal (1992) showed
that on Mount Carmel and in the Upper Galilee *A. mystacinus* was dominant in
Quercus calliprinos plots in mixed woodland, while *A. sylvaticus* was dominant in
Pinus halepensis forest. In the Lower Galilee *A. mystacinus* was again dominant in
Quercus calliprinos plots in mixed woodland, while *A. sylvaticus* was dominant in
Q. ithaburensis woodland. These observations, as well as a removal experiment carried out by Granot (1978) suggest that *A. mystacinus* exclude the smaller *A. sylvaticus* from preferred mesic habitats to drier ones. This conclusion is also supported by
physiological data which show that the osmoregulatory capability of *A. mystacinus*
is inferior to that of *A. sylvaticus* (Haim and Rubal, 1992).

It is nocturnal. A study near Jerusalem (Ritte, 1964) revealed two activity peaks, one
after sunset, the other between 2–3 a.m. There was almost no activity at 11 p.m. On
Mount Meron (Granot, 1978) they become active after sunset and continue until
after midnight, after which there is little activity.

The average radius of a home range varied from 8 m in young females to 15 m in
adult males, with considerable overlap between individual home ranges. No overlap
in home range between individuals is found only during periods of low population
density. About 40% of the population was captured only once, and it would appear
to have been transient. Maximum population size is reached at the end of winter and

Table 150
Body measurements (g and mm) of *Apodemus mystacinus mystacinus*

	Mass	Length	Tail	Ear	Foot
		Body			
	Males (12)				
Mean	37	112	122	19	25
SD	10	9	11	2	2
Range	29–57	100–125	105–140	16–22	19–26
	Females (12)				
Mean	33	115	120	19	26
SD	7	15	8	2	3
Range	20–43	90–120	110–135	16–21	23–35

Mean body mass in a population in the Galilee was 39 and 38 g for males and females, respectively (Granot, 1978). According to Ritte (1964), pregnant females may weigh up to 50 g.

Table 151
Skull measurements (mm) of *Apodemus mystacinus mystacinus* (Figure 160)

	Males (10)			Females (10)		
	Mean	Range	SD	Mean	Range	SD
GTL	29.9	29.1–30.7	0.7	30.2	29.2–30.6	0.7
CBL	26.6	25.6–27.8	0.7	26.9	25.6–27.3	0.8
ZB	15.0	14.1–15.9	0.5	15.1	14.5–15.5	0.3
BB	13.3	12.9–13.9	0.3	13.3	12.9–13.8	0.3
IC	4.6	4.4–4.9	0.2	4.6	4.4–4.8	0.2
UT	4.6	4.4–4.8	0.1	4.7	4.3–5.0	0.2
LT	4.7	4.4–4.9	0.2	4.8	4.4–5.0	0.2
M	18.5	17.6–19.1	0.5	18.7	17.7–20.0	0.6

Karyotype: 2N = 48, FN = 50, for specimens from Europe (Zima and Kral, 1984).

346

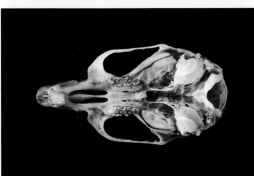

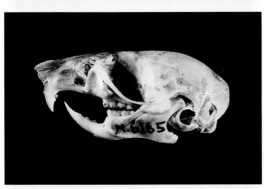

Figure 160: *Apodemus mystacinus mystacinus*. Dorsal, ventral and lateral views of the cranium, and a view of the mandible.

1 cm

Figure 161: Distribution map of *Apodemus mystacinus mystacinus*.
● – museum records.

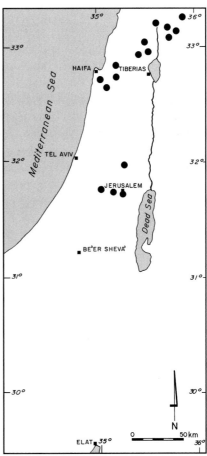

347

the beginning of spring, and minimum in autumn. There are no indications for large fluctuations in population size between years. Yahav and Haim (1980) noted a marked reduction in activity on Mount Carmel during September–January in comparison to February–June, and attributed this to low ambient temperatures. Individuals from Mount Hermon were found to be cold resistant in comparison to their conspecifics from Mount Carmel (Haim *et al.*, 1993).

It climbs well. Its relatively long tail and large sole pads (in relation to body length) are adaptations for climbing. It inhabits natural burrows in which it constructs a lined nest, but at times it also digs its own burrow. In captivity they seem to form hierarchical groups, sometimes occupying the same burrow; however, this might be the consequence of living in captivity, and in nature they may be solitary.

Food: Although they are more omnivorous than voles and gerbils, their main food is acorns (*Quercus*), but they also feed on pods of carob (*Ceratonia siliqua*), on seeds such as *Styrax officinalis*, *Rosa canina* and legumes, bark, snails, many insects and other invertebrates. They also take bird eggs and nestlings. The food is carried to permanent feeding sites in the territory ("tables"), where it is consumed and the inedible parts are discarded. "Tables" are found in deserted bird nests, and in holes in trees and rocks.

Reproduction: Pregnancy lasts 23–26 days and litter size ranges from 1–5. In the Judean hills females give birth between the end of October until April (Ritte, 1964), while on mount Meron they start earlier, from August, and breed until April (Granot, 1978). On Mount Carmel deliveries occur between November and March (Nizan, 1997). In a study near Jerusalem there were 1–4 peaks of births in one season, indicating that females may give birth up to 4 times annually. Males born by the end of the winter reach sexual maturity the following August–September, and there have been several cases of females born at the beginning of winter which gave birth in the following summer. The young are born blind and hairless, and have a very short tail. Within a week they double their size and are covered with grey fur. They leave the burrow when about one month old, at which time they weigh about 12 g. They reach adult mass when about 4 months old. Population density peaks in early spring, when the young leave their maternal nests, and is at a minimum in autumn. Life expectancy in nature is 7–8 months. Most individuals die before reaching one year of age, but there have been females which lived for 2.5 years and males which survived for two years (Ritte, 1964).

Parasites: This species is infected by fleas (*Leptopsylla segnis*, *Nosopsyllus sincerus*, *Xenopsylla cheopis*), Anoplura (*Polyplax (?) serrata*), and ticks (Ixodidae — *Rhipicephalus sanguineus*, *R. turanicus*, *Ixodes redikorzevi*, *Haemaphysalia erinacei*), mites (*Eulaelaps stabularis*, *Haemogamasus horridus*, *Ornithonyssus bacoti*) and Protozoa (*Grahamella* sp., *Eperythrozoon* sp.) (Granot, 1978; Theodor and Costa, 1967).

Apodemus sylvaticus (Linnaeus, 1758)   Common Field Mouse
Figs 162–163; Plate XXX:4   יערון קטן

Mus sylvaticus Linnaeus, 1758. *Systema Naturae*, 10th ed., 1:62. Upsala, Sweden.
Synonymy in Harrison and Bates (1991).

This species is a little larger than the house mouse, and has a longer tail, larger ears and longer legs. The colour is reddish to dark brown. The underside is white and there is generally a yellowish-brown spot between the fore legs. The local small *Apodemus*, considered as *A. sylvaticus*, have, however, the same longitudinal-shaped yellow throat spot as the European *A. sylvaticus* and not like the European *A. flavicollis*, which have a collar-shaped throat spot.
The subspecies found in Israel was referred by Ellerman and Morrison-Scott (1951) to *Apodemus sylvaticus tauricus* Barrett-Hamilton, 1900, but Harrison and Bates (1991) include it with *A.s. arianus* Blanford, 1881.
Fillippucci *et al.* (1989) suggested that this species is *A. flavicollis* and that *A. sylvaticus* does not occur in Israel. They also described a new species (*A. hermonensis*) from Mount Ḥermon.
Distribution: This species is widespread in the western Palaearctic throughout Europe to the Pamir and the Himalayas, as well as in North Africa.
In Israel it occurs in the forests and denuded forest areas of the Galilee and Mount Carmel (Figure 163).
Fossil record: The earliest remains of this species found in Israel are from the Late Acheulian period, about 250,000 years BCE (Tchernov, 1988).
Habits: It occurs in areas covered with bushes, but also in dense forest ("sylvaticus" — growing among trees in Latin). On Mount Ḥermon it also lives in rocky areas with little perennial cover. It inhabits burrows which it digs itself or uses burrows of other rodents (mole rats or voles), with branching tunnel systems (Granot, 1978). It jumps and climbs well.
It is nocturnal. A study on Mount Meron revealed two activity peaks, one after sunset, the other after midnight. Ambient temperature does not affect this activity pattern. The average radius of a home range is about eight metres, with little difference between the sexes and adults and young animals (Granot, 1978).
The population on Mount Ḥermon (which is referred to as a separate species *A. hermonensis* by Fillippucci *et al.*, 1989) has adapted to its environment by increasing the resting metabolic rate (up to 35% above the expected value from their body mass. Haim *et al.*, 1993).
The ecological relationships between *A. sylaticus* and *A. mystacinus* are discussed above (see *Apodemus mystacinus*).
Food: These field mice feed on acorns which seem to be their preferred food, but also on seeds, mainly of legumes, on fruits and other vegetable matter, as well as on insects and snails. They bring their food to permanent feeding sites ("tables") where remains of food can be found. They hoard acorns and seeds in their burrows,

349

Table 152
Body measurements (g and mm) of *Apodemus sylvaticus arianus*

	Mass	Length	Tail	Ear	Foot
		Body			
			Males (11)		
Mean	19	85	99	15	22
SD	2	6	6	2	2
Range	16–23	75–93	85–115	11–17	19–27
			Females (12)		
Mean	20	88	95	16	21
SD	3	6	11	3	2
Range	16–25	82–100	75–118	14–20	19–23

Mean body mass in a population in the Galilee was about 26 g for both sexes, 20% heavier than that of the museum specimens (Granot, 1978).

Table 153
Skull measurements (mm) of *Apodemus sylvaticus arianus* (Figure 162)

	Males (10)			Females (10)		
	Mean	Range	SD	Mean	Range	SD
GTL	26.1	24.8–27.6	1.1	25.6	24.2–26.8	1.0
CBL	23.7	22.0–25.0	1.2	23.2	21.5–24.6	0.9
ZB	13.4	12.3–14.6	0.7	13.1	12.1–13.8	0.6
BB	11.6	11.3–12.3	0.3	11.5	11.0–11.7	0.2
IC	4.1	3.8–4.3	0.2	4.0	3.8–4.1	0.1
UT	3.8	3.5–4.1	0.2	3.9	3.7–4.1	0.1
LT	3.8	3.4–4.0	0.2	3.9	3.5–4.2	0.2
M	16.3	15.4–17.3	0.7	16.1	15.2–16.8	0.6

Karyotype: 2N = 48 for specimens from Europe (Zima and Kral, 1984).

and on Mount Ḥermon this stored food is consumed during periods when the ground is snow covered.

Reproduction: Pregnancy lasts 23–26 days. On Mount Meron females give birth between September and May (Granot, 1978). On Mount Carmel deliveries occur mainly between January and March (Nizan, 1997). Females appear to give birth 3–4 times annually. Litter size ranges from 4–7 pups that are suckled for three weeks. The young open their eyes when two weeks old, and attain sexual maturity when two months old. Population density peaks in spring, when the young leave their maternal nests (Granot, 1978). In captivity they may live up to four years.

Parasites: This species is infected by fleas (*Leptopsylla segnis, Nosopsyllus inanus, Typhloceras poppei*), Anoplura (*Hoplopleura affinis*), and ticks (Ixodidae — *Ixodes redikorzevi, Rhipicephalus turanicus, R. sanguineus*), mites (*Haemoga masus, Eulaelaps*

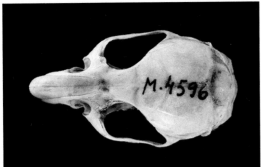

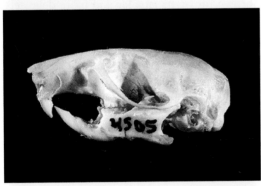

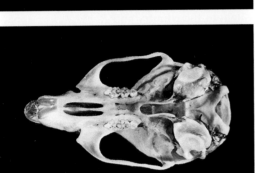

Figure 162: *Apodemus sylvaticus arianus.* Dorsal, ventral and lateral views of the cranium, and a view of the mandible.

|—— 1 cm ——|

Figure 163: Distribution map of *Apodemus sylvaticus arianus.*

● – museum records.

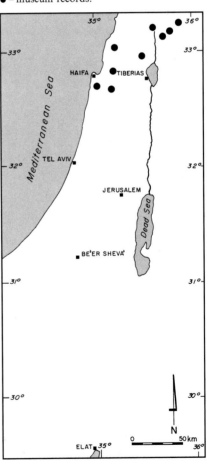

351

stabularis, Laelaps agilis, Hylomma sp.) and Protozoa (*Grahamella* sp.) (Granot, 1978, Theodor and Costa, 1967).

RATTUS Fischer, 1803

Rattus Fischer, 1803. *Das National-mus. Naturg. Paris*, 2:128.

This genus comprises larger murids with a body length of more than 15 cm. The genus includes about 55 species, most of which live in southern, south-eastern and eastern Asia. In Israel it is represented by two species, *R. rattus* and *R. norvegicus*. *R. rattus* is widespread throughout the country and *R. norvegicus* apparently depending on high air humidity, inhabits only the Mediterranean Coastal Plain, where it is commensal with man. Both species are serious pests and carriers of several diseases that can infect humans and domestic animals.

Rattus rattus (Linnaeus, 1758)　　　　　　　House Rat, Black Rat, Roof Rat
Figs 164–165; Plate XXXI:1–3　　　　　　　　　　　　　　חולדה מצויה

Mus rattus Linnaeus, 1758. *Systema Naturae*, 10th ed., 1:61. Sweden.
Synonymy in Harrison and Bates (1991).

This species is distinguished from the brown rat by its smaller and more delicately built body, a tail which is longer than the combined length of the head and body, (with the brown rat the tail equals or is shorter than the body), and larger ears which cover the eye when folded forward.
Different morphs, earlier described as subspecies, are found in Israel. The black, domestic morph, common in Europe, has only very rarely been found in Israel. The common morph is grey-brown, with white or grey, sometimes dark-grey, occasionally yellow-underside. In open habitats the long-haired frugivorous morph is occasionally found.
Rattus rattus rattus is the subspecies present in Israel (Harrison and Bates, 1991).
Distribution: This species is widespread throughout most of the world as a result of commensalism with man.
In Israel it is very common and widespread in the Mediterranean zone in natural habitats such as scrub forest, but also in human settlements and agricultural areas, even in the desert. It is the most common, often the only, mammal in the planted monocultures of Aleppo pine. It lives in the Ḥula swamp, and is also common in oases in the Negev and Judean Deserts (Figure 165). It has been found in mangrove areas on the shores of Sinai. After tourists from Israel began to frequent the area, rat populations increased, encouraged by the large amount of garbage left behind.
In many places in the high humidity coastal areas it has been displaced by the Norway rat. The following observation demonstrates this process. A wooden, one storey

building in Tel Aviv was overrun by roof rats that moved freely everywhere. In 1936 Norway rats appeared. These, however, were only active on the floor. When specimens of both species met, the Norway rat would attack the roof rat. If the roof rat did not succeed in escaping at once, and ascending a wall, the Norway rat would catch it and the squeals, uttered during the ensuing fight would attract more Norway rats. The roof rat was generally quickly killed and the Norway rats would begin to eat it. After some time the partly eaten corpse would be pulled into one of the burrows. The roof rats, deprived of the food sources on the floor and threatened with continued attacks, disappeared within half a year. In recent years the roof rat has become quite rare in Tel Aviv.

Fossil record: The earliest remains of this species found in Israel are from the Kebaran period (Tchernov, 1988).

Habits: This species is very euryoecous and occupies many habitats. Its nests are found on trees in the Mediterranean shrub forest, ruins, inhabited buildings, even high-rise buildings in which it climbs up elevator shafts or the rough surface of the building walls. Inside buildings it tends to construct nests in high places — below roofs and in attics, particularly when sympatric with the Norway Rat. It also constructs nests in climbing vines on the exteriors of buildings, up to 10 m above ground. It rarely digs burrows or uses available underground burrows and hides.

The roof rat is nocturnal. It is very social, living in large, territorial groups with one dominant male, some subordinate males and several females. Both males and females mark their territory with excretions from the chin and abdominal glands. Within the territory they use trails, usually along walls, which they mark with faeces and urine, that stick to the soles of their feet. These trails are dark in colour and conspicuous. This rat is a very agile climber and jumps well.

Food: It is omnivorous, but prefers herbivorous foods such as fruits. It is also a predator of various smaller animals, but less so than the Norway Rat. In the Ḥula it is known to feed on birds eggs and chicks, but also on the fruit of *Nuphar* and in the mangrove areas in southern Sinai they feed on crabs (*Uca*) which are often eaten on egret (*Egretta gularis*) nests, that are used as tables. In Israel it is rather common in the uniform, planted pine forests, where it feeds on pine seeds. These rats have developed a special technique for opening the pine cones: the rat disconnects a cone from the branch by gnawing at its base and carries it to a feeding site on the tree, where the scales are gnawed off with the teeth, beginning at the base of the cone and continuing along the spiral arrangement of scales. *Apodemus* ssp. also open cones in this way. Young rats learn this technique from their mothers, and those which do not acquire it usually do not survive in the pine plantations, where pine seeds are the sole source of food for rats (Aisner and Terkel, 1992). Below the feeding sites piles of bare cone shafts accumulate.

Reproduction: The oestrus cycle lasts 4–6 days and pregnancy is 21–22 days. The roof rat breeds all year round, and females give birth 3–6 times annually. Litter size ranges from 5–10. The young are born blind and their eyes open when two

Table 154
Body measurements (g and mm) of *Rattus rattus rattus*

| | Body | | Tail | Ear | Foot |
	Mass	Length	Tail	Ear	Foot
	Males (18)				
Mean	141	177	214	23	35
SD	23	20	16	2	3
Range	100–175	145–200	199–250	20–26	27–40
	Females (9)				
Mean	152	177	215	23	35
SD	31	15	16	2	2
Range	100–185	155–200	195–240	20–25	30–36

Table 155
Skull measurements (mm) of *Rattus rattus rattus* (Figure 164)

| | Males (15) | | | Females (6) | | |
	Mean	Range	SD	Mean	Range	SD
GTL	41.5	37.5–44.3	2.0	41.9	40.3–44.7	1.7
CBL	38.2	34.3–40.9	2.1	39.1	36.7–41.8	1.9
ZB	20.0	18.6–21.3	0.8	20.1	19.7–20.8	0.4
BB	16.4	15.9–17.4	0.5	16.3	15.8–16.6	0.3
IC	5.9	5.5–6.2	0.3	5.8	5.6–5.9	0.2
UT	6.8	6.1–7.8	0.4	6.5	6.0–6.8	0.3
LT	6.5	6.1–7.0	0.3	6.4	6.1–6.6	0.2
M	26.3	22.2–29.8	2.2	26.9	25.9–28.4	1.1

Karyotype: $2N = 38$, $FN = 58$, for specimens from Europe (Zima and Kral, 1984).

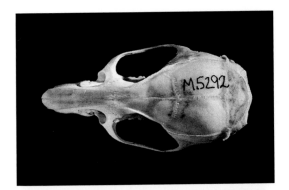

Figure 164: *Rattus rattus rattus*. Dorsal, ventral and lateral views of the cranium, and a view of the mandible.

1 cm

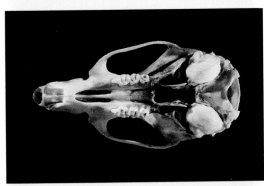

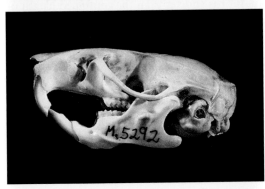

Figure 165: Distribution map of *Rattus rattus rattus*.

● – museum records.

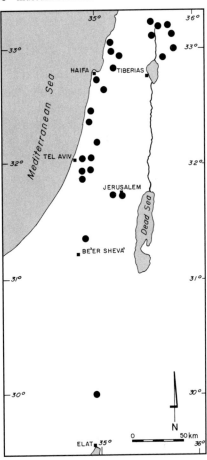

weeks old. They are nursed for three weeks. They become independent when six weeks old, and reach sexual maturity at three months. Life expectancy is 3–4 years.

Parasites: This species is infected by fleas (*Leptopsylla segnis, L. algira, Ctenocephalides felis, Echidnophaga murina, Xenopsylla cheopis, Nosopsyllus sincerus, N. londiniensis*), Anoplura (*Polyplax spinulosa, Hoplopleura oenomydis*), mites (*Echidnolaelaps echidninus, Laelaps nuttalli, Ornithonyssus bacoti*), ticks (*Rhipicephalus sanguineus, Ixodes redikorzevi, Hyalomma sp., Haemaphysalis erinacei*) and Protozoa (*Trypanosoma lewisi., Hepatozoon sp.) (Theodor and Costa, 1967).

Relations with man: Due to its abundance in human habitations, its omnivorous diet and its relative intelligence, this species is a major pest, causing damage to various crops and stored food products and in chicken coops and cowsheds. It is also a vector of disease, including the plague. It is easier to control than the Norway Rat, which is more cautious when presented with new food, poison bait or traps.

Rattus norvegicus (Berkenhout, 1769) Norway Rat, Brown Rat

Figs 166–167 חולדת החוף

Mus norvegicus Berkenhout, 1769. *Outlines Nat. Hist. Great Britain and Ireland*, 1:5. Great Britain.

This species is bigger than the roof rat, more clumsy and less agile, and its ears and tail are much shorter. Its tail length equals or is shorter than the combined length of its head and body. Its colour is dirty grey-brown with light underparts. There are no colour variations, unlike the roof rat.

Rattus norvegicus norvegicus is the subspecies present in Israel (Harrison and Bates, 1991).

Distribution: This species apparently originated from east Asia, probably Korea and Manchuria, where wild populations have been found, but currently is widespread throughout the world as a result of commensalism with man.

In Israel it occurs in the Mediterranean Coastal Plain, where it arrived during the 1930s (Figure 167). It did not spread to other areas of the country, possibly being dependent on high air humidity and easy availability of water.

Habits: It is normally nocturnal, but occasionally and in places of high population density it is active during the day, mainly in the late afternoon. It is very gregarious, dwelling in large polygamous, territorial groups which have an elaborate social hierarchy. It swims and dives well, but jumps and climbs less well than the roof rat. It digs wide, ramified burrow systems, in which it spends the day and builds nests. Nests are rarely built in hides above the ground, and never high up, unlike the roof rat. In towns it often lives in the sewer systems and it seems that generations may live there without ever leaving this habitat.

Food: The brown rat is omnivorous, feeding on a large variety of food, including matter not eaten by man such as soap and shoe polish. It is more aggressive than the roof rat and tends to be more carnivorous. It may attack and kill small or

356

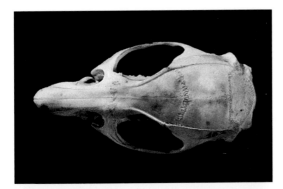

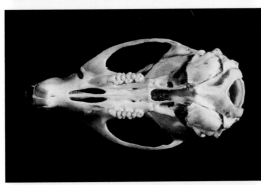

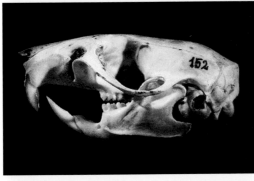

Figure 166: *Rattus norvegicus norvegicus.* Dorsal, ventral and lateral views of the cranium, and a view of the mandible.

├─────┤ 1 cm

Figure 167: Distribution map of *Rattus norvegicus norvegicus.*

● – museum records.

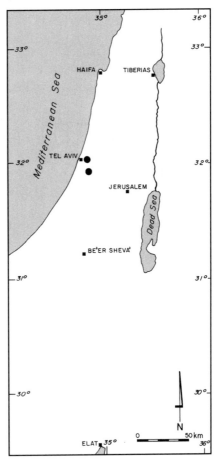

357

Table 156
Body measurements (g and mm) of *Rattus norvegicus norvegicus*

| | Body | | Tail | Ear | Foot |
	Mass	Length			
	Males (4)				
		n = 2	n = 2	n = 2	n = 2
Mean	296	228.5	191.5	20	42
SD	103	5	11	1	3
Range	220–441	225–232	184–199	19–21	40–44
	Females (5)				
Mean	419	225	195	21	42
SD	69	17	14	1	2
Range	315–508	205–250	172–205	19–22	40–45

One of us (H. Mendelssohn) weighed a pregnant female of 650 g.

Table 157
Skull measurements (mm) of *Rattus norvegicus norvegicus* (Figure 166)

| | Males (4) | | | Females (10) | | |
	Mean	Range	SD	Mean	Range	SD
GTL	50.6	47.9–53.8	2.6	48.0	45.4–51.6	2.1
CBL	48.0	45.1–51.4	2.6	45.7	42.6–49.2	2.1
ZB	22.9	20.9–24.0	1.4	23.3	21.7–25.8	1.5
BB	17.3	16.9–17.9	0.5	16.6	15.8–17.3	0.5
IC	7.1	6.6–7.5	0.4	6.7	6.4–7.2	0.2
UT	7.2	6.9–7.5	0.3	7.1	6.7–7.5	0.3
LT	6.9	6.7–7.3	0.3	7.1	6.7–7.7	0.3
M	33.9	31.2–36.1	2.2	32.2	29.7–34.2	1.5

Karyotype: 2N = 42, FN = 62, for specimens from Europe (Zima and Kral, 1984).

weak animals, or animals that are confined in cramped quarters and cannot escape or defend themselves, such as piglets, lambs, chickens, rabbits and geese. Cases are known of brown rats that have badly wounded chained-up elephants by gnawing deep holes into their feet.

Reproduction: It breeds all year round. Females give birth in nests in underground burrows or low shelters. The oestrus cycle lasts 4–6 days, pregnancy is 21–22 days, and litter size is 6–12. A female may give birth 3–6 times annually. The young weigh 7 g at birth. Their eyes are opened at two weeks old, at which stage they are covered with hair and are capable of leaving the maternal burrow. They are nursed for three weeks, and become independent when six weeks old, but sometime stay with their mother until two months old. Sexual maturity is reached at three months. Life expectancy is 3–4 years.

Parasites: This species is infected by fleas (*Leptopsylla segnis, Ctenocephalides felis, C. canis, Echidnophaga murina, Xenopsylla cheopis, Nosopsyllus sincerus, N. fasciatus, Stenoponia tripectinata*), Anoplura (*Polyplax spinulosa, Hoplopleura oenomydis*), mites (*Echidnolaelaps echidninus, Laelaps nuttalli, Ornithonyssus bacoti*), ticks (*Rhipicephalus sanguineus*) and Protozoa (*Trypanosoma lewisi., Hepatozoon* sp., *Grahamella* sp.) (Theodor and Costa, 1967).

Relations with man: Due to its abundance in human habitations, its omnivorous diet and relative intelligence, this species is a major pest, causing damage to various crops and stored food and in chicken houses and cowsheds. It is also a vector of diseases, including jaundice. It is more difficult to control than the roof rat, because it is more cautious with unknown food, such as poisoned bait. Entire populations easily develop poison shyness, if some specimens die from poisoned bait. The slow-working anticoagulant poisons are quite effective in controlling brown rat populations, but some populations have developed resistance against anticoagulant baits. The white laboratory rat was developed from this species.

MUS Linnaeus, 1758

Mus Linnaeus, 1758. *Systema Naturae*, 10th ed., 1:59.

According to Harrison and Bates (1991), this genus is represented in Israel by a single species, *M. musculus*. However, Auffray *et al.* (1990a, b) claimed that two species inhabit Israel. *M. macedonicus* has a shorter tail than *M. musculus* (44.5% and 49.7% of body length, respectively). They described the presence of *Mus spretoides* (currently named *M. macedonicus*; Auffray *et al.*, 1990a, b) in Israel and suggested that this species is restricted to Mediterranean environments whereas *M. musculus* also occurs in semi-desert areas.

Mus musculus Linnaeus, 1758   House Mouse
Figs 168–169; Plate XXXI:4   עכבר הבית

Mus musculus Linnaeus, 1758. *Systema Naturae*, 10th ed., 1:62. Upsala, Sweden.
Synonymy in Harrison and Bates (1991).

A small, agile, omnivorous mouse with very variable coloration. The underside is
white or grey. Domestic and feral specimens in the Mediterranean part of Israel
are dark grey-brown with light grey or white underside. Feral specimens in the
desert, however, are light yellowish-brown, similar to some gerbils. The black
domestic form common in Europe has not been found in Israel.
Mus musculus praetextus Brants, 1827 is the subspecies present in Israel (Harrison
and Bates, 1991).

Table 158
Body measurements (g and mm) of *Mus musculus praetextus*

	Body		Tail	Ear	Foot
	Mass	Length	Tail	Ear	Foot
Males (12)					
Mean	14	79	70	13	17
SD	6	6	22	3	2
Range	11–22	74–90	60–89	11–19	12–19
Females (12)					
Mean	14	72	73	13	16
SD	8	23	9	2	2
Range	8–28	65–92	60–85	10–17	11–19

Table 159
Skull measurements (mm) of *Mus musculus praetextus* (Figure 168)

	Males (10)			Females (10)		
	Mean	Range	SD	Mean	Range	SD
GTL	22.2	21.5–23.2	0.5	21.8	21.0–23.4	0.8
CBL	20.6	19.8–21.5	0.5	20.4	19.1–21.8	1.0
ZB	11.5	11.0–12.1	0.35	11.3	10.8–12.1	0.45
BB	9.7	9.3–10.1	0.3	9.6	9.3–10.1	0.3
IC	3.5	3.4–3.6	0.1	3.5	3.4–3.6	0.1
UT	3.4	3.2–3.5	0.1	3.35	3–3.7	0.2
LT	3.05	2.9–3.4	0.2	3.1	2.6–3.5	0.25
M	13.8	12.8–14.5	0.4	13.7	12.9–14.9	0.7

Karyotype: 2N = 40, based on four males from Mount Carmel (Ivanitskaya *et al.*,
1996b).

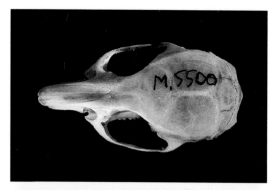

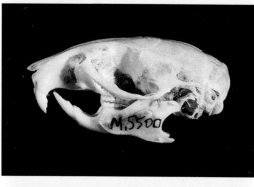

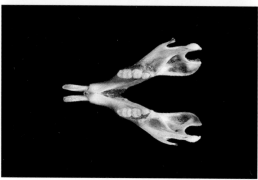

Figure 168: *Mus musculus praetextus.* Dorsal, ventral and lateral views of the cranium, and a view of the mandible.

1 cm

Figure 169: Distribution map of *Mus musculus praetextus.*

● – museum records.

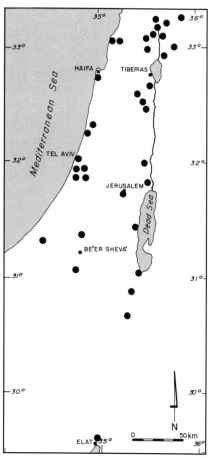

Distribution: This species has a worldwide distribution due to commensalism with man.

In Israel it occurs throughout the country, even in desert areas, but it is particularly common in and around human settlements and agricultural areas (Figure 169).

Fossil record: The earliest remains of this species found in Israel are from the Late Acheulian period, about 150,000 years BCE (Tchernov, 1988).

Habits: It is nocturnal, but is also occasionally active during the day. It uses self-dug burrows or hides below rocks and other cover. Because of its small size it can use narrow crevices in the ground or in houses. In Israel it is commensal with man as well as an inhabitant of open fields, including agricultural fields, also in desert areas and desert oases. Groups of mice occasionally inhabit wheat sheaves and breed in these nutritious and well protected places. It climbs well, making use of its tail.

Food: It is omnivorous, feeding mainly on seeds but also on invertebrates and green food. The domestic form eats anything available in human habitations and does much damage by gnawing paper, books, etc. Its small size enables it to penetrate almost everywhere.

Reproduction: Pregnancy lasts 20–21 days and litter size ranges from 6–8. There is often a post-partum oestrus. The female rears the young alone. They open their eyes when 13 days old and are weaned when 18 days old. Sexual maturity is reached at two months. Females give birth all year round, and may do so 10 times annually.

Parasites: This species is infected by fleas (*Leptopsylla segnis, L. algira, Synosternus cleopatrae, S. tripectinata, Nosopsyllus sincerus*), Anoplura (*Polyplax serrata, Hoplopleura captiosa*), mites (*Androlaelaps hirstionyssoides, Laelaps algericus*), ticks (*Rhipicephalus sanguineus, Isodes redikorzevi*) and Protozoa (*Eperythrozoon* sp., *Grahamella* sp.) (Theodor and Costa, 1967).

Relations with man: Populations in agricultural areas occasionally undergo a mass increase, concomitantly with *Microtus guentheri* and *Meriones tristrami*, causing damage to grain crops. A special kind of damage was found in the 'Arava Valley: house mice would gnaw holes in green peppers and eat the seeds inside, leaving behind their excrement, spoiling the peppers, which were grown for export. Several strains of domestic and laboratory mice have been developed from this species, the most common one being the albino laboratory mouse.

NESOKIA Gray, 1842

Nesokia Gray, 1842. *Annals Mag. nat. Hist.* (7), 10:264.

This genus is represented in Israel by a single species, *N. indica*.

Nesokia indica (Gray & Hardwick, 1832)   Short-tailed Bandicoot Rat
Figs 170–171; Plate XXXII:4   נזוקיה

Arvicola indica Gray & Hardwick, 1832. *Illustrated Indian Zool.*, 1:pl. xi. "India".
Synonymy in Harrison and Bates (1991).

This is a big, clumsy rodent, somewhat similar to a brown rat, but with heavier head and shorter tail. In many specimens the colour is similar to that of a brown rat, but many others are reddish-brown. Many specimens have a white spot on the breast.

Nesokia indica bacheri Thomas, 1919 is the subspecies present in Israel (Harrison and Bates, 1991).

Distribution: This species occurs in the southern Palaearctic from Egypt to north-western India.

In Israel it occurs in the lower Jordan and the 'Arava valleys, from the Jericho area to Elat, as well as in oases in the Negev, such as 'En Ziq (Figure 171). The Israeli population in the 'Arava has increased due to an increase in irrigated agriculture in this area.

Habits: It inhabits salty marshes and other humid areas. The various populations along the 'Arava and in desert oases appear to be isolated, but it is possible that wandering individuals migrate through the desert from one place to another. One pregnant female was trapped on a hammada (stony desert flat), far from any suitable habitat.

It is nocturnal, digging burrows in the soil, particularly in wet soils. The burrow system is extensive and spreads over a large area (Z. Zook-Rimon, pers. comm.).

Food: It is herbivorous, feeding on vegetation, including roots. In the 'Arava its main food are rhizomes of reeds (*Phragmites*), but it also feeds on the soft, inner part of date palm trees, and especially the young shoots, thus causing damage to agriculture.

Reproduction: A captive colony was kept in Tel Aviv University Zoo. Litter size was 4–8 and pregnancy lasted three weeks. Weaning takes place when the young are one month old. A pregnant female, with seven large embryos, was caught near Jericho on 15 July, 1973. There were several cases of post-partum oestrus (Z. Zook-Rimon, pers. comm.).

Parasites: This species is infected by fleas (*Xenopsylla astia, Nosopsylla londiniensis*) and mites (*Androlaelaps longipes*) (Theodor and Costa, 1967).

Table 160
Body measurements (g and mm) of *Nesokia indica bacheri*

| | | Body | | | |
	Mass	Length	Tail	Ear	Foot
			Males (4)		
	n = 3				
Mean	274	205	130	19	40
SD	33	26	19	4	4
Range	237–300	180–230	105–145	13–21	35–44
			Females (3)		
Mean	326	208	129	18	40
SD	40	18	16	2	5
Range	280–350	190–225	112–145	15–20	35–45

Table 161
Skull measurements (mm) of *Nesokia indica bacheri* (Figure 170)

| | Males (1) | Females (4) | | |
	Mean	Mean	Range	SD
GTL	46.6	45.9	40.5–52.8	5.1
CBL	44.3	45.0	39.2–52.7	5.8
ZB		27.5	24.8–29.5	2.0
BB	20.2	17.6	16.3–18.4	0.9
IC	6.7	6.7	6.4–7.4	0.5
UT	9.9	9.6	8.8–10.3	0.6
LT	10.2	8.9	8.5–9.2	0.3
M	34.5	36.0	32.8–41.0	3.5
TB		9.5	8.7–10.3	1.1

Karyotype: 2N = 42, FN = 66, for specimens from Iran (Kamali, 1975).

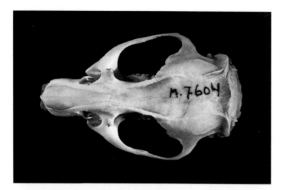

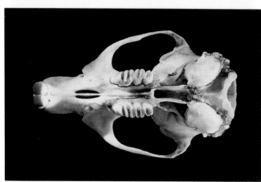

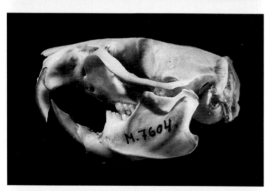

Figure 170: *Nesokia indica bacheri*. Dorsal, ventral and lateral views of the cranium, and a view of the mandible.

1 cm

Figure 171: Distribution map of *Nesokia indica bacheri*.

● – museum records.

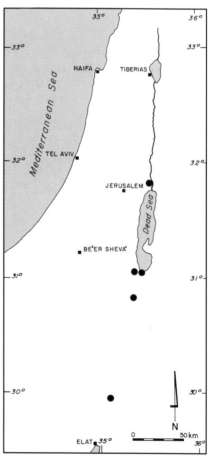

365

ACOMYS I. Geoffroy, 1838

Acomys I. Geoffroy, 1838. *Annales des Sci. Nat Zool.*, (2), 10:126.

This genus is represented in Israel by two species, *A. cahirinus* and *A. russatus*. Its back is covered with spiny bristles, hence its scientific name: "ac" — a needle, "mys" — a mouse in Greek. The skin is very thin. Both species are generally nocturnal, but when sympatric with *A. cahirinus*, as is the situation in most of the Negev and Judean Desert areas, *A. russatus* is diurnal. When this competition is experimentally removed it becomes active at night as well (Shkolnik, 1971).

Acomys cahirinus (Desmarest, 1819)   Egyptian Spiny Mouse
Figs 172–173; Plate XXXII:1–2   קוצן מצוי

Mus cahirinus Desmarest, 1819. *Nouveau Dictionnaire d'Hist. Nat.*, 29:70. Cairo, Egypt. Synonymy in Harrison and Bates (1991).

The back is covered with grey-brown bristles. Fur colour varies greatly: individuals from the Galilee are dark grey-brown and much darker than the ones from the Negev which have yellowish-grey fur. The darkest of all are those from basaltic areas (i.e. in the Golan), which are dark-grey to blackish.

Acomys cahirinus dimidiatus (Cretzschmar, 1826) is the subspecies present in Israel (Harrison and Bates, 1991).

Distribution: The Egyptian spiny mouse occurs throughout North Africa from Mauritania to Egypt, in Cyprus and Crete (that it probably reached as a commensal of humans), and in western Asia from Sinai, Israel and Turkey to Pakistan.

In Israel it is common in rocky areas throughout the country, and in the 'Arava and the Jordan Valley it is also found in non-rocky habitats, where it lives in burrows dug by other rodents (Figure 173).

Fossil record: The earliest remains of this species found in Israel are from the Kebaran period (Tchernov, 1988).

Habits: It is nocturnal but active mainly during the early hours of the night. During the day it hides in natural holes, crevices in and below rocks and in stone piles. It does not dig burrows of its own and does not build nests, unlike many other rodents. However, food remains are often found in front of their hiding places. In Jerusalem this species is commensal with man, as it is in Sinai (Haim and Tchernov, 1974) and in Egypt.

In the Judean hills the home range radius is 15 and 12 m for adult males and females, respectively, and 10 m for young animals of both sexes (Ritte, 1964). At 'En Gedi population density is 2–4 individuals per dunam (1000 m^2), but this is an overestimate because enlarging the study area did not reveal more individuals in it (Shargal, 1997). Hence, in reality, population density at 'En Gedi is apparently much lower. In 'En Gedi mean home ranges of males is 0.9 dunam and 0.6 dunam for

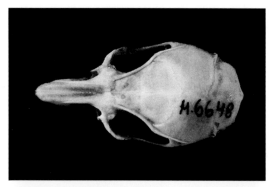

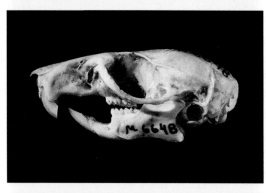

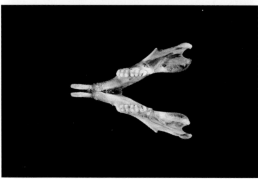

Figure 172: *Acomys cahirinus dimidiatus.* Dorsal, ventral and lateral views of the cranium, and a view of the mandible.

1 cm

Figure 173: Distribution map of *Acomys cahirinus dimidiatus.*

● – museum records.

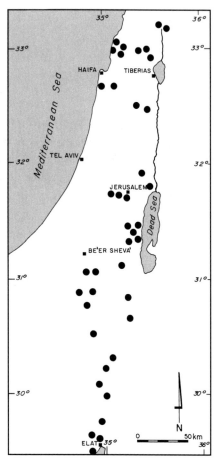

367

Table 162
Body measurements (g and mm) of *Acomys cahirinus dimidiatus*

	Body		Tail	Ear	Foot
	Mass	Length			
Males (12)					
Mean	38	114	93	18	20
SD	10	11	31	3	2
Range	26–48	85–120	75–113	14–22	13–21
Females (12)					
Mean	44	113	101	18	19
SD	8	5	13	4	3
Range	29–57	108–120	85–123	16–23	19–21

Table 163
Skull measurements (mm) of *Acomys cahirinus dimidiatus* (Figure 172)

	Males (10)			Females (10)		
	Mean	Range	SD	Mean	Range	SD
GTL	31.1	29.9–32.4	1.2	31.1	29.0–32.8	1.1
CBL	28.2	25.5–29.7	1.3	28.1	26.3–29.6	1.0
ZB	14.9	13.0–15.8	0.8	14.7	14.0–15.8	0.5
BB	13.0	12.7–13.5	0.2	12.9	14.0–15.8	0.3
IC	5.1	4.6–5.4	0.2	4.9	4.6–5.1	0.2
UT	4.8	4.6–5.0	0.2	4.7	4.6–5.0	0.1
LT	4.4	4.3–4.5	0.1	4.4	4.2–4.5	0.1
M	19.2	16.8–20.3	1.1	19.1	17.6–20.6	0.8

In Israel, in accordance with Bergmann's rule, there is a north-south decline in body length and mass, while tail, ear and foot length increase in the same direction in accordance with Allen's rule (Nevo, 1989). However, no latitudinal trend in skull dimensions could be detected in our material.

Karyotype: $2N = 36$, for specimens from Israel, 38 from eastern Sinai. $FN = 68$ (Wahrman and Goitein, 1972).

females, and there is a considerable overlap in home range between individuals of both sexes (Shargal, 1997). There is a considerable overlap in home range of individuals, and spiny mice tend to live together even at times of low population density. Population size reaches a peak in autumn and is low in spring. There is much movement of individuals within the population, and many adults arrive before the breeding season starts. In the desert spiny mice often travel distances of up to 300 m during the night (Y. Yom-Tov, pers. observ., Mann, 1986). They are easily caught in traps.

In captivity they are very social and form large groups which cohabit. Females take interest in the process of delivery of other females, and may lick the newborn of other females and eat their placentas. When population density becomes too high in captivity, they bite each others tails. The infected wounds cause a gradual loss of the tail, so that in such dense populations many specimens are tailless or have short tails.

Spiny mice are resistant to heat and drought, but less so than the golden spiny mice. The basal metabolic rate of this species is 23% lower than the expected value (Degen, 1994). Cutaneous water loss is high due to the thin skin (Shkolnik and Borut, 1966). However, the kidneys are very efficient in concentrating electrolytes and urea (up to 4700 ml/M). This is an adaptation for a protein-rich food, such as snails and insects, and explains their ability to balance their water economy in spite of the high rate of cutaneous water loss. They are therefore able to fed on succulent halophytes and to survive for a long time on a diet of only dry seeds. They are also able to survive in low temperatures (around freezing point) for a short time. Specimens from 'En Gedi are relatively resistant to the venom of *Echis colorata*; their LD 50 is 38 times higher than that of laboratory mice (Weisenberg *et al.* 1996). *A. cahirinus* produces chirping noises, but it is not vociferous.

Food: It is omnivorous, feeding on seeds, green food, including halophytes, beetles and other insects, isopods and other invertebrates. Snails are a preferred food in the desert, where they are a source of water, protein and energy. They are collected from a wide area, carried to the entrance of the hide and gnawed at from the apex where the shell is thin, and the soft contents are extracted from the opening. Piles of hundreds of gnawed shells accumulate in front of the crevices inhabited by spiny mice in the desert.

Reproduction: In the Judean hills young are born between April–October (Ritte, 1964) and at 'En Gedi between April–September (Mann, 1986; Shargal, 1997). Pregnancy lasts 35–38 days, a relatively long period for a rodent of this size. During the birth the female does not lie on the side, unlike most other mice, but remains standing and after the pup is born she turns around and licks it. Litter size ranges from 1–5, mostly two or three. In a study area near Jerusalem litter size ranged between 2–3 (Ritte, 1964), and mean of 11 litters at the Tel Aviv University Research Zoo was two. The young are precocious and are able to walk soon after birth. They are born covered with short, soft, grey fur which is replaced at the age of two months by the adult colour fur with spines. The brown colour appears first in the middle of the back and gradually spreads to cover the entire back. Their eyes and ears are open upon birth, or several hours later. On the second day they are already able to move around. In their second week they start to feed on solid food, and are weaned when three weeks old. In a study area near Jerusalem young were caught in traps when they were about a month old and weighed 15–20 g (Ritte, 1964). Sexual maturity is reached at two months. In captivity they breed all year round, but in nature mainly between March–September.

In captivity they normally live three years, but some survive for five years (Bodenheimer, 1949). In the Judean hills mean longevity was 6–7 months, but some lived more than 2.5 years (Ritte, 1964). In 'En Gedi individuals were caught for up to 22 months after their initial capture (Shargal, 1997). Females tend to live longer than males.

Parasites: This species is infected by fleas (*Parapulex chefrenis*, *Leptopsylla segnis*, *Xenopsylla cheopis*, *X. nubica*, *X. conformi*, *X. dipodilli*, *Nosopsyllus sincerus*, *N. theodori*, *Stenoponia tripectinata*), Anoplura (*Polyplax brachyrrhyncha*, *P. oxyrrhyncha*), ticks (Ixodidae — *Rhipicephalus sanguineus*, *Hyalomma* sp., *Hyalommina rhipicephaloides*, *Ixodes redikorzevi*, *Haemaphysalis erinacei*), mites (*Trombicula acomys*, *Trombicula palestinensis*, *Laelaps acomydis*, *Ornithonyssus bacoti*, *Allodermanyssus sanguineus*, *A. aegyptiacus*), Protozoa (*Grahamella* sp. , *Eperythrozoon* sp.) and Spirochaetes (Theodor and Costa, 1967; Krasnov *et al.*, 1997).

Acomys russatus (Wagner, 1840)   Golden Spiny Mouse
Figs 174–175; Plate XXXII:3   קוצן זהוב

Mus russatus Wagner, 1840. *Abhandlungen Akad. Wiss. Munich*, 3:195, pl. 3, fig. 2. Sinai. Synonymy in Harrison and Bates (1991).

The back and the nape are covered with brown-orange coloured spiny bristles ("russus" — reddish in Latin). The area covered with bristles is larger than in *A. cahirinus*. The skin is black, and its colour is easily seen where the fur is thin, as it is on the ears, nose, tail and feet. The tail is shorter and thinner and its ears are shorter than those of *A. cahirinus*.

Acomys russatus russatus is the subspecies present in Israel (Harrison and Bates, 1991).

Distribution: This species has a limited distribution in north-eastern Egypt, Sinai, southern Israel and Jordan.

In Israel it occurs in rocky areas in the Negev and Judean Deserts as far north as 'Ein Fashkha, east and south of the 100 mm isohyet (Figure 175).

Habits: Population density in 'En Gedi is 3–6.5 individuals per dunam (1000 m^2) (Shargal, 1997). However, this is an overestimate because enlarging the study area did not reveal other individuals. During inactivity the golden spiny mouse hides in natural burrows, rock crevices below rocks and in stone piles. It does not dig burrows of its own and does not line them, unlike many other rodents. 'En Gedi individuals travelled up to 350 m in their home range (Mann, 1986). Its activity period is determined by the presence of *A. cahirinus* and by ambient temperature. When sympatric with *A. cahirinus* it is diurnal, but when this competition is experimentally removed it becomes active at night as well (Shkolnik, 1971). In the morning it emerges from its hiding place, and in cold ambient temperatures it warms up by basking and absorbing solar radiation, like reptiles. When ambient temperature rises above 32°C it retreats to its hiding place. Its thermoregulation is not efficient, and under laboratory conditions with temperatures below 18°C it becomes sluggish. However, one population, inhabiting the high mountains of Sinai, possesses an efficient non-shivering thermogenetic mechanism and is much more cold resistant; these mice line their nests, hoard food and accumulate fat for the winter (Haim and Borut, 1975; Haim, 1991). The basal metabolic rate of this species is 43% lower than the expected value (Degen, 1994).

It is more resistant to heat and drought than *A. cahirinus* and ambient lethal temperature is 42.5°C. Cutaneous water loss is high due to its thin skin. However, its kidneys are very efficient in concentrating electrolytes and urea (up to 4700 ml/M). This is an adaptation for protein-rich food, such as snails, and for electrolyte-rich food such as halophytes, and explains its ability to balance its water economy despite the high rate of cutaneous water loss (Shkolnik and Borut, 1969). 'En Gedi mean home ranges of males is 700 m^2 and 500 m^2 for females, and there is a considerable overlap in home range between individuals of both sexes (Shargal, 1997). The preferred habitat of this species are rocks and boulders.

Food: The golden spiny mouse is omnivorous, feeding on seeds, green food, insects, isopods and other invertebrates. Snails are a preferred food in the desert, as a source of water, protein and energy. They are collected from a wide area, carried to the hide and the shell is gnawed until the soft parts can be extracted. Piles of hundreds of emptied shells mark the burrows of many spiny mice in the desert.

Reproduction: Pregnancy lasts 38–44 days, a relatively long period for the size of this rodent. Litter size ranges from 1–5 (mean of 17 litters at the Tel Aviv University Research Zoo was two). The young are precocial and are born covered with soft, grey fur which at the age of two months is replaced by the adult coloured fur with spiny bristles. Their eyes and ears are open at birth, or several hours later. On their first day they are already able to move around and are weaned when 12 days old. Females reach sexual maturity at about 31 days (Searight, 1987b). In captivity breeding occurs all year round, but in nature mainly between March–August. At 'En Gedi young are born between April–July (Shargal, 1997). In captivity they tend to become obese and such specimens do not breed. Hence, if breeding is required they must be kept on a restricted diet. In captivity they may live up to six years. 'En Gedi individuals were caught for up to two years after their initial capture (Shargal, 1997).

Parasites: This species is infected by fleas (*Parapulex chefrenis, Nosopsyllus theodori, Xenopsylla dipodilli, Leptopsylla algira*), Anoplura (*Polyplax brachyrrhyncha, Polyplax oxyrrhyncha*), ticks (Ixodidae — *Rhipicephalus sanguineus*) and mites (*Allodermanyssus aegyptiacus*) (Theodor and Costa, 1967; and Krasnov *et al.*, 1997).

Table 164
Body measurements (g and mm) of *Acomys russatus russatus*

| | Body | | Tail | Ear | Foot |
	Mass	Length			
		Males (9)			
	n = 8	**n = 4**			
Mean	48	112	69	17	18
SD	10	11	3	2	1
Range	37–64	98–130	66–73	16–21	17–20
		Females (10)			
		n = 8			
Mean	59	120	67	18	19
SD	19	29	17	2	1
Range	31–94	85–188	40–99	15–20	18–20

Table 165
Skull measurements (mm) of *Acomys russatus russatus* (Figure 174)

| | Males (9) | | | Females (10) | | |
	Mean	Range	SD	Mean	Range	SD
GTL	29.5	28.1–30.4	0.7	30.1	28.7–31.3	0.9
CBL	27.4	25.9–29.3	1.1	28.4	26.6–30.6	1.5
ZB (M n = 7)	14.7	14.1–15.4	0.5	15.2	14.5–16.5	0.7
BB (F n = 9)	13.1	12.3–14.1	0.5	13.1	12.2–13.9	0.5
IC	5.0	4.6–5.3	0.2	5.0	4.5–5.5	0.3
UT	4.9	4.7–5.3	0.2	4.9	4.5–5.2	0.2
LT	4.7	4.3–5.2	0.3	4.8	4.3–5.4	0.3
M (F n = 9)	18.7	17.8–19.9	0.7	19.5	18.2–22.7	1.4

Karyotype: 2N = 66, FN = 76, for specimens from Israel (Wahrman and Goitein, 1972).

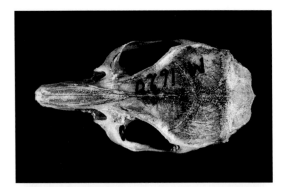

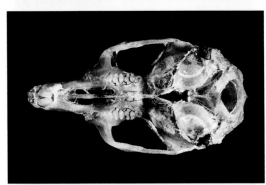

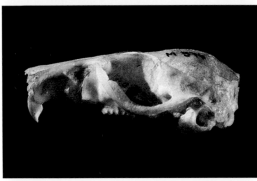

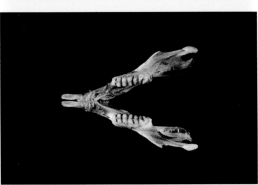

Figure 174: *Acomys russatus russatus*. Dorsal, ventral and lateral views of the cranium, and a view of the mandible.

|— 1 cm —|

Figure 175: Distribution map of *Acomys russatus russatus*.

● – museum recordss.

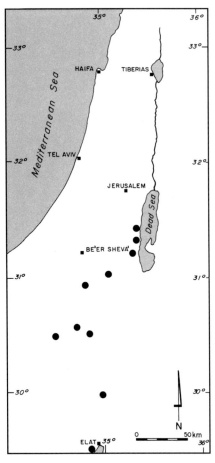

373

SPALACIDAE   Mole Rats   חולדיים

This family is represented in Israel by a single, highly variable, species, *Spalax leucodon*.

SPALAX Gueldenstaedt, 1770

Spalax Gueldenstaedt, 1770. *Nova Comm. Acad. Sci. Petrop.*, ser. 14, 1:40.

Spalax leucodon (Nordmann, 1840)   Lesser Mole Rat
Figs 176–177; Plate XXXIII:1–2   חולד

Spalax typhlus leucodon Nordmann, 1840. *Observations sur la faune Pontique*, 3:34. Near
   Odessa, Russia.
Synonymy in Harrison and Bates (1991).

A fossorial, blind and tailless rodent, without external ears, with short legs, cylindrical body and soft fur. The short, soft hairs are not directed from the front to the rear, as with most mammals, but are bi-directional, creating a velvet-like fur that enables the mole rat to move easily in its tunnels, backwards as well as forwards. Its colour is grey-brown all over and some specimens have a white spot on the head and the abdomen.

Males are about 15% larger than females. There are marked differences between the various populations occurring in Israel (see below). In accordance with Bergmann's rule there is a north — south cline in body size. Average body mass of Negev specimens is about 100 g whereas those living on Mount Hermon weigh 240 g. Negev specimens are also lighter in colour than northern ones (Nevo *et al.*, 1986). There are also large differences in body mass within the same area, which are probably an outcome of food availability. For example, near Tel Aviv, mole rats living in heavy soil, on which *Oxalis pes-caprae* is very common, are much larger than those living in hamra soil that has fewer geophytes.

Spalax leucodon ehrenbergi Nehring, 1848 is the subspecies present in Israel (Harrison and Bates, 1991).

Distribution: It occurs from the Balkans through Turkey to Transcaucasia, and from Libya to Jordan.

In Israel it occurs throughout the Mediterranean zone and south to the Northern Negev (Figure 177).

Fossil record: The earliest remains of this species found in Israel are from the Late Acheulian period, about 250,000 years BCE (Tchernov, 1988).

Habits: There are marked physiological, morphological, ecological and genetic differences between the four chromosomal forms. Basic metabolic rates decrease from north to south. The $2N = 60$ race are the best thermoregulators of the four forms.

Genetic variability increases from south to north, and this phenomenon is explained as an adaptation to an increase in climatological predictability from south to north. Population density increases and territory size decreases from south to north. Maximal population density in the Galilee is 170 per km^2, and it is less than 100 in the Northern Negev (Nevo *et al.*, 1982).

The mole rat is fossorial. Each individual digs its own tunnel system to its own width. Using its front paws and incisors, it excavates the soil and pushes it either back with the hind legs or above ground through an opening, with the head, thus forming the typical mounds. The density and distribution of the mounds above the ground does not reflect the shape of individual territories (Zuri and Terkel, 1996). The depth of the tunnel system depends on soil type, moisture and climate, being about 10–40 cm in winter and as deep as 1.5 metre during summer. At these depths temperature and humidity are relatively constant during the respective seasons. This difference in depth appears to be mainly due to two factors: during winter oxygen penetration through the wet soil is lower, and during summer temperature of the upper layer of the soil is above optimum. When a tunnel is opened the animal plugs it within a short time, generally minutes. Each tunnel system includes chambers for food storing, for deposition of faeces and for a nest.

This is a solitary, territorial and aggressive rodent. In captivity they attempt to kill each other if kept together, and in nature they vigorously guard their territories against intrusion. Territories undergo constant change in use and location throughout the year, with tunnel lengths averaging about 19 metres in the dry season and 39 metres in the rainy season. The basic pattern of the tunnel system consists of a main tunnel with several secondary tunnels branching off it. In each tunnel there is a nest, which is connected to the tunnel system by 2–3 exits. During one 24 hour period, a mole-rat travelled about 17 times the length of its territory (Zuri and Terkel, 1996). Experimental removal of an individual is followed by the occupation of its territory by another mole rat within hours or at most a few days. Territorial disputes may end with the expelling of the owner from its territory and its death due to wounds inflicted by the invader. Expelled individuals may escape above ground, where they are exposed to predation (Zuri and Terkel, 1996). Laboratory study suggests that mole rats scent-mark their territorial boundaries (underground) with urine and faeces (I. Zuri pers. comm.). Mole rats are active about 50% of the time all year round, mainly during the day, and always return to the same nest for resting periods (Zuri and Terkel, 1996).

While still in the maternal tunnel system the young communicate with their mother by producing various calls, which are transmitted through the air in the tunnels. Following dispersal, adult mole rats communicate by drumming the tunnel roof with their heads at frequencies ranging between 50–100 Hz. Neighbouring mole rats perceive this seismic signalling by pressing their lower jaw to the tunnel wall, and sensing the vibrations which are transmitted to the ear (Rado *et al.*, 1992).

Table 166
Body measurements (g and mm) of *Spalax leucodon ehrenbergi*

| | Body | | Foot |
	Mass	Length	
	Males (10)		
Mean	195	201	22
SD	35	17	3
Range	118–240	180–230	20–25
	Females (16)		
Mean	153	172	22
SD	36	11	1
Range	102–217	151–187	20–24

Table 167
Skull measurements (mm) of *Spalax leucodon ehrenbergi* (Figure 176)

| | Males (10) | | | Females (10) | | |
	Mean	Range	SD	Mean	Range	SD
GTL (M n = 9)	45.9	40.9–52.2	3.6	44.8	41.2–48.6	2.6
CBL	43.0	37.8–48.7	3.2	42.0	38.7–45.7	2.5
ZB	33.2	28.9–37.4	2.5	31.9	28.2–34.3	2.1
BB	23.0	21.8–24.9	1.0	22.2	20.7–24.1	1.1
IC	6.9	6.3–7.5	0.4	6.5	6.1–6.9	0.3
UT	7.35	6.9–8	0.3	7.3	6.9–7.6	0.3
LT	7.2	6.7–7.7	0.3	7.1	6.6–7.4	0.3
M	35.0	30.6–38.8	2.65	34.4	31.9–38.5	2.2

Dental formula is $\frac{1.0.0.3}{1.0.0.3} = 16$.

Karyotype: There are four chromosomal forms in Israel: 2N = 54 in the Golan, 52 in the Upper Galilee, 58 in the Lower Galilee, Yizre'el Valley, Mount Carmel and the Coastal Plain, and 60 in Judea, Samaria and the Northern Negev. In transition zones between chromosomal forms there are hybrids in which 2N equals a mean between two forms. The width of the transition zones ranges from several hundred metres to a few kilometres. Generally the width of hybrid zones is greater in the south than in the north (3000 m and 300 m, respectively, Nevo, 1969).

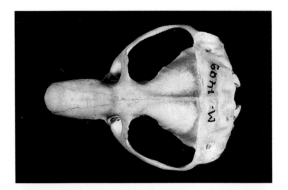

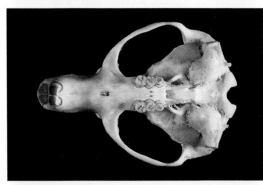

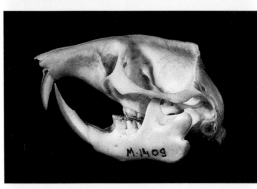

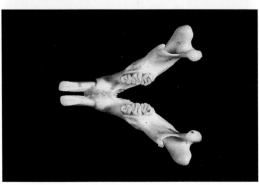

Figure 176: *Spalax leucodon ehrenbergi.* Dorsal, ventral and lateral views of the cranium, and a view of the mandible.

2 cm

Figure 177: Distribution map of *Spalax leucodon ehrenbergi.*

● – museum records.

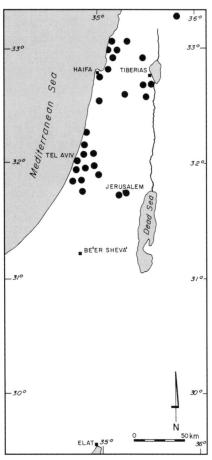

The mole rat is well adapted to high CO_2 and low O_2 pressures (similar to O_2 at 9 km height) which are found in its tunnels. Its basal metabolic rate is about 30% of the rate expected from its body mass.

Food: It is herbivorous, feeding mainly on underground parts of plants — bulbs, tubers and thick roots. Mole rats store large quantities of food in special underground chambers. In one case 25 kgs of potatoes and carrots were found in one burrow. Generally each food item is hoarded in a different chamber. Mole rats also feed on the green parts of plants above ground, mainly alfalfa, lettuce and Gramineae, which they cut below the surface and drag down into their tunnels. Long leaves are often only partially eaten, and their distal parts are left above ground. Only occasionally a mole rat may emerge from its tunnel, cut a plant and drag it to its burrow. Its food contains a large percentage of water, so that it does not require drinking water. A preferred food is the bulbs of the weed *Oxalis pes-caprae* (introduced from South Africa). By hoarding the bulbs and carrying them over large distances the mole rat disseminates this weed, that in Israel does not produce seeds (Galil, 1967).

Reproduction: A female selects its mate after a courting process in which potential partners approach one another cautiously, with some initial aggression, which gradually diminishes, while making various calls. Mating observed in captivity included 60 copulations within 90 minutes, and was followed by the partners licking one another. Females give birth once annually between January–March. There is a possibility that some females may occasionally produce a second litter, especially if the winter is long and rainy. Pregnancy lasts 28–36 days and litter size ranges from 2–6 (usually 3–4). In light soil the female gives birth in an underground, grass lined chamber, while in heavy soil she does so in a similar chamber located in a mound above ground. This mound can be up to one metre high and one metre in diameter. The young are born hairless and weigh about five grams. By two weeks old they are fully covered with hair, and start eating solid food.

The young disperse when two months old, during March–April, by digging lateral extension tunnels, and establishing their own territories near their mother's territory. Male young leave the maternal burrow before females. The dispersion process lasts 4–5 weeks, during which a connection to the maternal tunnel is maintained until the extension tunnel is about 4 m long, and at the end of this stage the offspring seal the connection with the mother's tunnel. In a laboratory study male offspring do not enter the maternal territory after about 11–12 weeks old, while female offspring return to their mother's territory even when five months old (I. Gazit and J. Terkel, pers. comm.). When there is insufficient space for all the offspring to establish territories around that of their mother, the young compete for tunneling space, and young mole rats can be seen above ground (Rado *et al.*, 1992), where they fall easy prey to nocturnal (*Bubo bubo*, *Tyto alba*, *Strix aluco* and *Asio* spp.) and diurnal (*Neophron percnopterus*, *Milvus migrans*) birds of prey and mammalian carnivores. In captivity many live up to four and some even to ten years of age.

GLIRIDAE Dormice נמנמניים

This family is represented in Israel by two genera, each with one species: *Eliomys melanurus* and *Dryomys nitedula*.

Dental formula is $\dfrac{1.0.0.3}{1.0.0.3} = 16$.

Key to the Species of Gliridae in Israel
(After Harrison and Bates, 1991)

1. Large species with a total length of 222–270 mm. The external ears are large (24–31 mm). The distal two-thirds of the tail is black. Greatest length of the skull is 33.5–37.1 mm. **Eliomys melanurus**

– Small species with a total length of about 190 mm. The external ears are small, (12–14 mm). The tail is uniformly coloured. Greatest length of the skull is 26.1–29.7 mm. **Dryomys nitedula**

ELIOMYS Wagner, 1840

Eliomys Wagner, 1840. *Abhandlungen bayer. Akad. Wiss.*, 3:176.

Eliomys melanurus (Wagner, 1840)   Black-tailed Dormouse
Figs 178–179; Plate XXXIII:3   נמנמן הסלעים

Eliomys (Myoxus) melanurus Wagner, 1840. *Abhandlungen bayer. Akad. Wiss.*, 3:176, pl. 3, fig. 1. Sinai.

The soft hair has a beautiful ash-grey colour, the underside is white, a black ring surrounds the eye, continuing as a black band to below the ear and separating the grey upper part of the head from the white cheeks. The bushy tail is black with grey at its base, hence its specific name ("melas" — black, "oura" — tail in Greek). The large ears are naked.
Eliomys melanurus melanurus is the subspecies present in Israel (Harrison and Bates, 1991).
Distribution: This dormouse occurs in North Africa from Morocco to Egypt, and in western Arabia, Israel, Jordan, Lebanon, Syria.
In Israel there are two populations, one in the central Negev Desert, from the Makhtesh Ramon to Sede Boqer and near Revivim (U. Paz, pers. comm.) and another in the Golan Heights and Mount Hermon. It has also recently been found in 'En Yahav in the 'Arava Valley (Figure 179). It is possible that the two populations are connected by a series of semi-isolated populations along the Rift Valley (Filippucci *et al.*, 1988).

379

Table 168
Body measurements (g and mm) of *Eliomys melanurus melanurus*

| | Body | | Tail | Ear | Foot |
	Mass	Length			
			Males (17)		
Mean	48	117	110	29	25
SD	9	6	9	2	1
Range	32–65	106–128	97–120	25–31	23–28
			Females (17)		
Mean	54	123	109	27	25
SD	10	10	16	2	1
Range	35–68	108–140	75–130	24–30	22–27

Table 169
Skull measurements (mm) of *Eliomys melanurus melanurus* (Figure 178)

| | Males (10) | | | Females (10) | | |
	Mean	Range	SD	Mean	Range	SD
GTL	34.9	33.7–36.0	0.7	35.3	33.5–36.7	0.9
CBL	31.1	30.1–32.2	0.6	31.7	30.1–32.6	0.8
ZB	19.9	18.9–20.5	0.5	19.9	19.0–20.4	0.5
BB	15.6	15.2–16.6	0.4	15.8	14.5–16.5	0.6
IC	4.7	4.4–5.0	0.2	4.7	4.5–5.8	0.9
UT	5.1	4.7–5.5	0.2	5.2	5.1–5.3	0.1
LT	5.2	4.6–5.5	0.3	5.2	4.9–5.4	0.2
M	20.7	19.9–21.4	0.4	20.7	19.3–21.8	0.7
TB	13.2	12.3–13.6	0.4	13.3	13.0–13.7	0.2

Karyotype: 2N = 48, FN = 86 for specimens from Israel (Filippucci *et al.*, 1988).

Rodentia: Eliomys

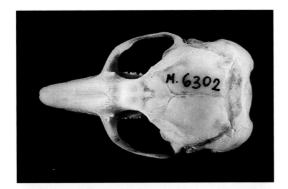

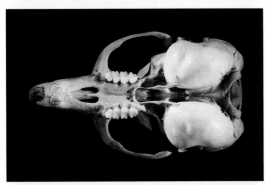

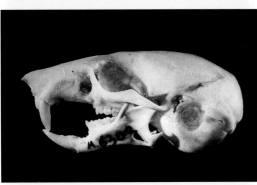

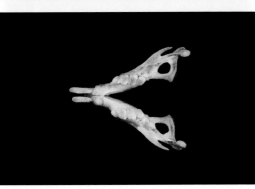

Figure 178: *Eliomys melanurus melanurus.* Dorsal, ventral and lateral views of the cranium, and a view of the mandible.

1 cm

Figure 179: Distribution map of *Eliomys melanurus melanurus.*

● – museum records, ○ – verified field records.

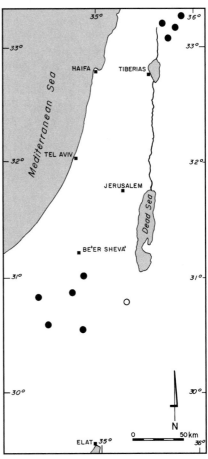

381

Fossil record: The earliest remains of this species found in Israel are from the Natufian period (Tchernov, 1988).

Habits: It is nocturnal, living mainly in rocky habitats, where it uses natural holes in the ground, crevices and holes in trees and rocks in which it constructs nests similar to those of birds. In the central Negev they prefer holes in *Pistacia atlantica* trees, which are the only trees there.

In the Negev it tends to appear in a locality, stays there for several months and disappears (Y. Yom-Tov, pers. observ.).

Before winter it accumulates much fat. Specimens caught during autumn in the Negev tend to fall asleep in the traps and wake very slowly, as if from hibernation. In captivity, this species may enter daily periods of torpor which can save up to 65% of the average daily energy expenditure of normothermic individual under the same condition (Haim and Rubal, 1995).

Food: It feeds on various invertebrates, mainly snails and insects, small vertebrates, including eggs and chicks of birds, seeds, mainly oily ones, and fruit.

Reproduction: Pregnancy lasts 22 days. Litter size ranges from 3–7 (mean of 19 litters at the Tel Aviv University Research Zoo was four). A female with seven pups was found in July in a nest in 'En Yahav (Y. Cenaani, pers. comm.). In captivity young are born between April–August. In nature females apparently give birth twice annually, but more often in captivity. In captivity they may live up to six years.

Parasites: This species is infected by fleas: *Myoxopsylla laverani* and *Xenopsylla ramesis* (Krasnov *et al.*, 1997).

DRYOMYS Thomas, 1906

Dryomys Thomas, 1906. *Proceedings zool. Soc. Lond.*, 1905 (2): 348.

Dryomys nitedula (Pallas, 1779)   Forest Dormouse
Figs 180–181; Plate XXXIII:4; Plate XXXIV:1   נמנמן העצים

Mus nitedula Pallas, 1779. *Nova Spec. Quad. Glir. Ord.*, p. 88. Region of Lower Volga, Russia. Synonymy in Harrison and Bates (1991).

Its colour is brown, the underparts are white and the bushy tail is dark grey. It has a black face pattern similar to that of the black-tailed dormouse. On the hind foot the first and fifth toes are directed sidewards at almost 90° to the direction of the other three toes. This may be an adaptation for more efficient grasp of branches. This arrangement is not found in the black-tailed dormouse, that lives mainly on rocky slopes.

Dryomys nitedula phrygius Thomas, 1907 is the subspecies present in Israel (Harrison and Bates, 1991).

Distribution: It occurs from the eastern Alps and the Balkan, through Russia east to the Tien Shan mountains and south to Asia Minor, Syria and Israel.

In Israel it is found in the forests of the Upper Galilee (Figure 181).

Fossil record: The earliest remains of this species found in Israel are from the Late Aurignacian period (Tchernov, 1988).

Habits: It is nocturnal. It builds nests, mainly on *Quercus calliprinos* trees, preferably in tangles of climbing vines (*Smilax aspera, Lonicera, Clematis*). This is reflected in the scientific name of the genus: "dry" — a tree, mainly oak, "mys" — a mouse in Greek. The nests are bell-shaped with an entrance from the side. They are 10–12 cm in diameter, built of soft material, dry weeds, moss and bark and disintegrate easily. The nests are found in groups of 4–6 on adjacent trees at a height of 1–6 m at distances of up to 100 m between each group (Nevo and Amir, 1961). During the breeding season either a single male or a pregnant female or a female with pups can be found in the nest. Sometimes a male and a female can be found in neighbouring nests.

In Israel it has a short period of hibernation, apparently lasting only a few days, during which its body temperature drops to 11.5°C. During hibernation the body is curled up in a tight ball with the tail wrapped around it and the eyes closed (Nevo and Amir, 1961). Hibernation appears to take place in nests built in holes in the ground or among rocks.

Food: It is omnivorous, feeding on oily seeds, fruits (acorns of *Quercus calliprinos*, fruits of *Rosa canina, Arbutus andrachne* and *Styrax officinalis*) and insects (Nevo and Amir, 1961). Soft fruit are a preferred food.

Reproduction: In captivity it breeds 2–3 times annually, between March-October. Gestation lasts 23–25 days. Litter size is 1–4 (mean of 11 litters at the Tel Aviv University Research Zoo was 3). A female can produce 2 -3 litters a year. The young weigh 1.3 g at birth (Nevo and Amir, 1961).

Table 170
Body measurements (g and mm) of *Dryomys nitedula phrygius*

		Body			
	Mass	Length	Tail	Ear	Foot
		Males (3)			
Mean	30	100	82	13	23
Range	22–37	85–110	65–100	12–14	20–30
		Females (5)			
Including three individuals from captivity					
Mean	22	85	90	13	20
SD	7	7	15	2	4
Range	13–28	76–92	74–105	10–15	13–22

Table 171
Skull measurements (mm) of *Dryomys nitedula phrygius* (Figure 180)

	Males (3)		Females (3)	
	Mean	Range	Mean	Range
GTL	28.1	27.5–28.9	27.1	26.1–28.0
CBL	24.5	23.9–25.4	24.2	23.3–24.7
ZB	16.5	(n = 1)	15.6	14.9–15.9
BB	12.7	12.3–13.1	12.7	11.8–13.3
IC	4.1	4.0–4.2	4.0	3.8–4.1
UT	3.8	3.7–3.9	3.9	3.8–4.0
LT	4.2	4.1–4.2	4.0	3.9–4.0
M	16.4	15.9–17.3	16.4	16.3–16.5
TB	9.6	9.3–10.1	8.6	8.0–9.2

Karyotype: 2N = 48, FN = 92 for specimens from Europe (Zima and Kral, 1984).

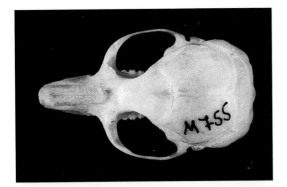

Figure 180: *Dryomys nitedula phrygius.* Dorsal, ventral and lateral views of the cranium, and a view of the mandible.

1 cm

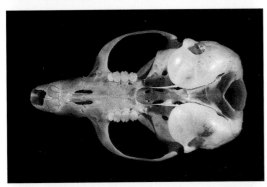

Figure 181: Distribution map of *Dryomys nitedula phrygius.*

● – museum records.

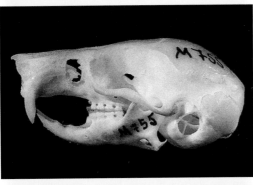

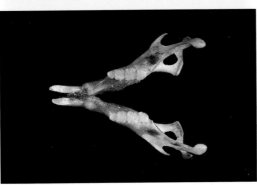

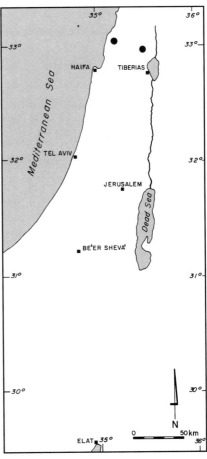

385

DIPODIDAE   Jerboas   ירבועיים

This family is represented in Israel by two species of the genus *Jaculus*: *J. jaculus* and *J. orientalis*. The scientific name reflects on its habit of jumping: "iacta" — to throw in Latin.

Dental formula is $\frac{1.0.0.3}{1.0.0.3} = 16$.

Key to the Species of Dipodidae in Israel
(After Harrison and Bates, 1991)

1.   Size large; hind foot more than 68 mm; greatest length of skull more than 38.0 mm. Glans penis with a pair of long spines attached to the baculum.   **Jaculus orientalis**

–   Size small; hind foot less than 70 mm; greatest length of skull less than 36.5 mm. Glans penis without long spines.   **Jaculus jaculus**

JACULUS Erxleben, 1777

Jaculus Erxleben, 1777. *Systema regni animales, Classis 1, Mammalia*, p. 404.

Jaculus jaculus (Linnaeus, 1758)   Lesser Jerboa
Figs 182–183; Plate XXXIV:2; Plate XXXV:1–4   ירבוע מצוי

Mus jaculus Linnaeus, 1758. *Systema Naturae*, 10th ed,. 1:63. Giza Pyramids, Egypt.
Synonymy in Harrison and Bates (1991).

Jerboas are characterized by their well developed hindlegs and much smaller forelegs. The soles of the three toes of the hindlegs are covered with long, strong hairs, that prevent sinking in the sand. The head is wide. The fur is light yellowish-grey, the underside is white and a whitish stripe extends along the thigh. The vibrissae are well developed and, when extended forward, form a fan that touches the soil in front of the jerboa and probably transfers information. One bristle on each side is longer than the body. When the jerboa moves, these long bristles are extended forwards-sidewards and transfer information on obstacles. The well developed tail tuft is black with a conspicuous white tip.

Two subspecies are recognized in Israel: *J. j vocator* Thomas, 1921 in the deserts and around the Dead Sea, including the lower Jordan Valley, and *J. j. schlueteri* Nehring, 1901 along the sand dunes of the southern Coastal Plain and extending into Sinai. Eastwards it reaches Be'er Sheva'. Harrison and Bates (1991) state that both subspecies intergrade in the central Negev. However, both in fact occur side by side in an area south-west of Be'er Sheva' and no intergrades have been found. In captivity mixed pairs did not breed whereas pure pairs of both subspecies reproduced. These facts would justify giving *J. schlueteri* the status of a species.

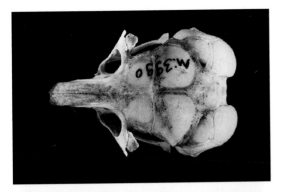

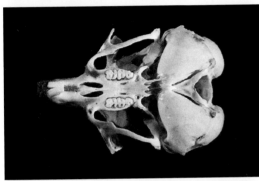

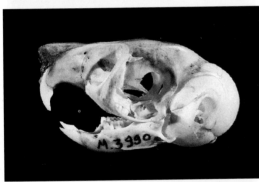

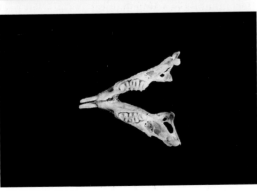

Figure 182: *Jaculus jaculus*. Dorsal, ventral and lateral views of the cranium, and a view of the mandible.

1 cm

Figure 183: Distribution map of *Jaculus jaculus*.

● – museum records.

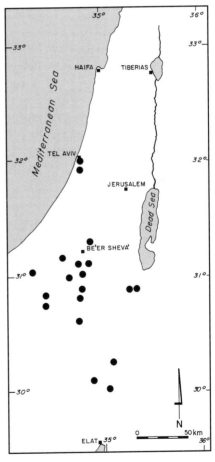

Table 172
Body measurements (g and mm) of *Jaculus jaculus vocator*

	Body		Tail	Ear	Foot
	Mass	Length			
Males (12)					
Mean	52	114	171	22	62
SD	10	6	15	2	5
Range	33–66	100–120	130–190	20–25	55–70
Females (12)					
Mean	61	116	189	23	58
SD	15	11	15	2	5
Range	42–91	105–145	165–215	18–25	50–63

Table 173
Skull measurements (mm) of *Jaculus jaculus vocator* (Figure 182)

	Males (10)			Females (10)		
	Mean	Range	SD	Mean	Range	SD
GTL	34.0	32.8–35.9	0.9	34.5	32.3–35.9	1.3
CBL	29.6	28.5–31.0	0.8	30.1	28.5–31.1	1.0
ZB	22.5	21.1–23.3	0.7	22.4	20.9–24.3	1.2
BB	23.8	23.0–25.2	0.7	23.5	21.7–25	1.0
IC	12.6	12.0–13.5	0.5	12.4	11.6–13.7	0.6
UT	5.2	4.8–5.5	0.2	5.2	4.8–5.5	0.2
LT	5.2	4.9–5.5	0.2	5.2	4.7–5.5	0.3
M	20.2	19.5–20.9	0.5	20.8	19.2–22.8	0.9
TB	15.6	15.1–16.0	0.3	15.3	13.9–16.2	0.7

J. j. schlueteri is somewhat larger than *J. j. vocator*, and the ears are less elongated and conspicuously rounded. The black colour of the tail tuft does not surround the tail, unlike *J. j. vocator*, but is separated on the underside by a white area. The general colour is sandy yellow. *J. j. schlueteri* inhabits the same habitat as *Gerbillus pyramidum* (and both species have similar colour), loose sand with sparse vegetation. *J. j. schlueteri* is currently seriously endangered in Israel by destruction of its habitat, and in Sinai by extreme overgrazing, that destroys the plant cover so that little food remains for rodents.

Table 174
Body measurements (g and mm) of *Jaculus jaculus schlueteri*

	Mass	Body length	Tail	Ear	Foot
		Males (8)			
Mean	56	112	173	20	66
SD	14	11	8	2	1
Range	43–85	98–130	160–185	18–23	65–68
		Females (12)			
Mean	49	113	176	18	66
SD	9	11	12	6	5
Range	40–60	97–128	152–190	17–23	61–68

Table 175
Skull measurements (mm) of *Jaculus jaculus schlueteri*

	Males (4)			Females (3)		
	Mean	Range	SD	Mean	Range	SD
GTL	33.7	33.2–34.2	0.4	33.5	32.0–34.9	(n=2)
CBL	29.4	28.7–30.6	0.9	28.9	28.1–29.6	(n=2)
ZB	23.0	22.5–23.2	0.3	20.5		(n=1)
BB	23.3	23–24	0.5	23.5	21.3–24.8	
IC	12.9	12.5–13.2	0.3	12.6	11.4–13.4	
UT	5.1	4.9–5.4	0.3	5.1	5.0–5.2	
LT	5.1	4.9–5.1	0.1	5.2	4.9–5.4	
M	20.0	19.6–20.4	0.3	20.3	19.4–21.0	
TB	14.6	13.8–15.2	0.6	14.5	13.7–14.9	

Karyotype: 2N=48, FN=92, for specimens from Saudi Arabia (Al-Sabeh and Khan, 1984).

Distribution: Jaculus jaculus occurs throughout North Africa from Mauritania to Egypt, and in western Asia from Sinai and Israel and Arabia to south-western Iran. In Israel it occurs in the Negev and Judean Deserts, with two penetrations north: one in the southern Jordan Valley (*J. j. vocator*), and the other (*J. j. schlueteri*) along the sand dunes of the Coastal Plain as far north as the Yarqon River (Figure 183). However, the development of human settlements south of the Yarqon in Tel Aviv, Holon and Bat Yam as well as in Ashdod, resulted in a considerable decrease in population size in and near these settlements due to habitat change and predation by feral cats.

Habits: It is nocturnal. During the day it hides in burrows which are dug with the front feet, while the hind legs push the excavated soil backwards. Each burrow system has several openings. An escape branch of the main tunnel is often dug towards the surface without an outside opening, but in case of danger the animal pushes quickly through the remaining soil and escapes through this opening. A tunnel system can be several metres long and about one metre deep. When the animal is inside its burrow, it closes most openings with a plug of soil, which it pushes forward with its head, apparently in order to preserve a favourable microclimate within the burrow system. In the burrow there is a resting chamber lined with dry weeds and plant wool; in winter this lining is larger and thicker than in summer, becoming ball shaped.

Its nocturnal activity begins in the evening with sand bathing and grooming of the fur with paws and teeth. Sand bathing is crucial to remove excess fat from the fur and to keep it in good condition. In captivity it is essential to enable jerboas to sand bath regularly.

When resting, the jerboa sits on its long hind legs and is supported by the curved tail. When moving slowly, for instance while foraging, it walks with small, alternate steps, with its small forelegs contracted under its throat, hidden by the fur and almost invisible (hence the name *Dipus* meaning "two legs"). The normal gait is by hopping on the hindlegs. The hops are 40–50 cm long and the tracks reveal that one foot alights a little in front of the other. When fleeing, the hops become jumps 80–100 cm long. Generally the jumps are quite flat, not more than 20 cm high, but when a jerboa is suddenly alarmed. it can instantly jump to one m high. During all these movements the front paws are hidden below the throat. They are used only when sifting the soil for food, handling food, climbing plants or grooming the fur.

Food: It feeds on seeds which it finds by filtering sand with its front feet. It also eats leaves, which it collects by climbing bushes. Normally jerboas do not drink water, but when they do, they dip their hands in the water and lick them, unlike the direct drinking of other animals.

Reproduction: Mating behaviour in captivity was observed by H. Mendelssohn. During mating the male approaches the female by jumping in front of her, sometimes while chattering its teeth, blocking her way, raising and lowering his head in front of her and drumming with his paws on her head. He then approaches from the rear, touches the female's rear with his paws and tries to copulate. The male reclines with his front legs on the female's back and copulates many times in succession. After copulation the female squats and brushes the vulva on the ground. In captivity births occurred all year round. Gestation lasts 27 days. Several days before giving birth the female builds a large nest lined with soft material, and aggressively attacks approaching conspecifics. Litter size in captivity was 1–4 (mean of eight litters at the Tel Aviv University Research Zoo was three).

The newborn resemble other rodents rather than jerboas, with the hind legs being shorter than the front ones, which are mainly used while crawling in the nest. During the first weeks the legs are moved alternately. After the rear legs begin to grow, they are moved synchronously, and the pups push themselves along in the burrow system. The head is not as wide as that of the adult. The young develop slowly. Fur starts growing at 18 days and the eyes open at 30 days, they leave the nest for short periods at 35 days and nurse until 42 days of age. Mating behaviour (without copulation) was already observed at 46 days. In captivity some lived to 6 years of age.

Parasites: This species is infected by the flea *Xenopsylla ramesis* (Krasnov *et al.*, 1997)

Relations with man: The Bedouin catch jerboas by digging them out of their burrows in order to eat them, which they do not do with other rodents. After skinning and removing the entrails they tie them together by their tails and roast them.

Jerboas are easy to keep in captivity, provided they are given plenty of sand which is necessary for sand-bathing and digging, as well as a shelter. Several specimens can be kept together in the same cage, but they will breed only if kept in pairs.

Jaculus orientalis Erxleben, 1777   Greater Egyptian Jerboa
Figs 184–185; Plate XXXIV:2–4   ירבוע גדול

Jaculus orientalis Erxleben, 1777. *Systema regni animales, Classis 1, Mammalia*, p. 404, Egypt (mountains separating Egypt from Arabia, G. Allen).

Jaculus orientalis is considerably larger than *J. jaculus*. Its legs and tail are relatively shorter, and its ears are larger and longer relative to the lesser jerboa, making them somewhat similar to the long ears of the genus *Alactaga*. Its postures are similar to those of *J. jaculus*, but it tends to keep its body in a more horizontal posture. Its colour is light brown, the underparts are white and there is a white band on the thigh. The tail tuft is black with a large, white tip. A conspicuous difference to *J. jaculus* is in its movements. Normally it walks or runs bipedally, and only when alarmed and fleeing does it resort to jumping.

Jaculus orientalis orientalis is the subspecies present in Israel (Harrison and Bates, 1991).

Distribution: It occurs in North Africa from Morocco to Egypt, and east to Sinai and Israel.

In Israel it is found mainly on plains of loess soil in the Northern Negev and western Judean Desert, mainly near 'Arad, where the average precipitation is 200–300 mm (Figure 185). This species in endangered in Israel, as the loess soils, its only habitat, have been taken over by agriculture and no suitable reserves exist.

Habits: Little is known about its biology in Israel. It is nocturnal. This species tends more to walk or run with alternate steps, rather than to jump, unlike the smaller jerboas. When alarmed, however, they flee, using metre wide jumps.

Food: It feeds on seeds and leaves of which the latter form a larger proportion of its diet than in the Lesser Jerboa.

Table 176
Body measurements (g and mm) of *Jaculus orientalis orientalis*

| | Body | | Tail | Ear | Foot |
	Mass	Length			
Males (7)					
Mean	140	155	219	32	74
SD	30	17	7	2	4
Range	112–175	134–180	210–230	30–35	68–78
Females (4)					
Mean	164	156	237	34	78
SD	18	6	10	2	2
Range	140–182	150–165	228–250	32–35	76–80

There are no significant differences in body size between the sexes.

Table 177
Skull measurements (mm) of *Jaculus orientalis orientalis* (Figure 184)

| | Males (4) | | | Females (1) |
	Mean	Range	SD	Mean
GTL	40.2	38.5–41.2	1.2	41.3
CBL	36.4	35.4–37.3	1.0	37.4
ZB	28.2	26.8–29.0	1.0	27.3
BB	25.2	24.5–26.6	0.9	26.0
IC	15.4	14.4–16.3	0.8	15.0
UT	6.6	6.6–6.7	0.1	6.7
LT	6.3	6.0–6.5	0.2	6.6
M	25.4	24.6–25.8	0.5	26.8
TB	16.4	15.7–16.9	0.6	17.1

Reproduction: A pregnant female caught near 'Arad on 11 April, 1983 and kept at Tel Aviv University Research Zoo gave birth to five young in a large lined nest she built, and another gave birth to three young in March.

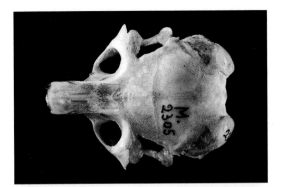

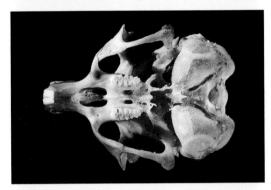

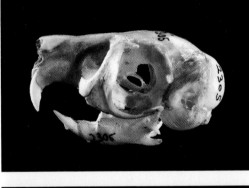

Figure 184: *Jaculus orientalis orientalis.* Dorsal, ventral and lateral views of the cranium, and a view of the mandible.

1 cm

Figure 185: Distribution map of *Jaculus orientalis orientalis.*

● – museum records.

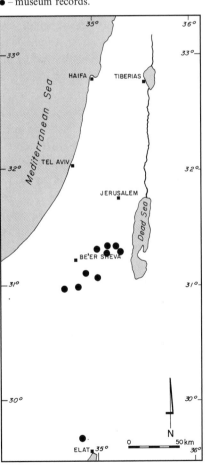

393

HYSTRICIDAE Porcupines דרבנ״ים

This family is represented in Israel by one species, *Hystrix indica*.

HYSTRIX Linnaeus, 1758

Hystrix Linnaeus, 1758. *Systema Naturae*, 10th ed., 1:56.

Hystrix indica (Kerr, 1792)   Indian Crested Porcupine
Figs 186–187; Plate XXXVI:1–4; Plate XXXVII:1   דרבן

Hystrix cristata var. *indica* Kerr, 1792. *Animal Kingdom*, p. 213. Based on Smellie's Buffon, 1781, 7:pl. 206. India.
Synonymy in Harrison and Bates (1991).

This species is the largest rodent in Israel, its weight ranges between 10–15 kg, but an occasional specimen may weigh 18.5 kg. The body is covered with coarse, bristly hair, among which are small, short spines on the neck, the sides of the body and the upper part of the legs. On its head and neck is a mane of long bristles, up to 30 cm long. The back is covered with long (up to 40 cm long), thin, flexible quills, coloured in alternate black and white bands, and shorter, thicker spines, with sharp points, also coloured black and white, that are very effective weapons. In case of alarm or excitement the mane and dorsal quills and spines are erected and the porcupine appears much bigger and more impressive than in normal posture. Bristles, quills and spines are also erected in case of positive excitement, for instance when eating a very tasty food or when copulating.

On the lower back and the tail are white, non-erectile spines, the white colour of which perhaps serves as a visual mark for the copulating male. On the tail there are special quills, about 12 cm long with a thin base: these are hollow and open at the distal end. If startled it wags its tail swiftly, causing the hollow quills to emit a rattling noise. This noise, together with stomping its hind feet, can serve to warn conspecifics. The general overall colour of *Hystrix indica* is blackish, with white throat collar and the already-mentioned black and white quills and spines. The tail and the lower back are white. The legs are short and strong, and the soles are naked. The forefeet have four digits with strong claws for digging, and the hindfeet have five digits.

Hystrix indica indica is the subspecies present in Israel (Harrison and Bates, 1991).
Distribution: It occurs from Asia Minor and south Turkestan east to Kashmir and India, south of latitude 44°N. Alkon and Saltz (1988a) suggested that the northern limit of its distribution is determined by the minimum amount of time available for night foraging, which is about 7 hours.

In Israel it is found throughout the country, but is more abundant in the Mediterranean zone than in the desert (Figure 187).

Fossil record: The earliest remains of this species found in Israel are from the Late Acheulian period, about 150,000 years BCE (Tchernov, 1988).

Habits: *Hystrix indica* has been studied by Sever and Mendelssohn (1991) near Tel Aviv and in captivity, and by Alkon and Saltz (1988a, b, 1989) near Sede Boqer, central Negev. During the day these porcupines live in permanent burrows they dig in the soil on slopes, often under rocks which serve as roofs, or in natural caves. Each tunnel has one or more openings and is up to 15 m or more long. The opening is about 30 cm high and 40 cm wide, fitting the diameter of the body of the owner, and the tunnel has a wider chamber at its far end. No lining or food is brought into it. In addition to the permanent burrow porcupines have several alternative burrows in their territories which are used in times of danger.

Porcupines generally live in pairs, but there are also solitary males and trios of a male and two females. These arrangements are temporary, for porcupines prefer to establish a pair bond. After the breeding season a burrow system may be occupied by a pair and its one or two young. Burrow systems which are branched (mostly natural burrows) may be occupied by other animals, such as mongooses, badgers, caracals and owls. Thick vegetation cover is used mainly during spring and summer (dry seasons in Israel), and often by solitary males (Sever and Mendelssohn, 1991). Porcupines emerge from their burrows about 20 minutes after sunset during summer and up to two hours after sunset in winter. They return at about sunrise in summer and at 2–3 a.m. during winter; hence daily activity is about 1–2 hours longer in summer, probably due to the relative food shortage and harder soil at this time. Average activity time in one night is 9.2 hours during summer and 6.7 hours during winter (Alkon and Saltz, 1988b). In Sede Boqer activity is shorter in agricultural than natural areas, and there is a tendency for shorter activity during moonlit nights. Upon emerging from their den porcupines stand for a few minutes at the entrance of the burrow and leave only if there is no disturbance. After emerging from their dens porcupines walk for one km or longer (two hours or longer) to a particular foraging area, where they spend most of the night, walking around a distance of about 100 m for several hours, and towards sunrise they return more quickly to their dens. Walking speed ranges from 9–37 m/min. Return speed is higher than outgoing speed, and after returning from their nightly activity they stand near their burrow for up to 30 minutes, apparently to cool themselves after walking back and before entering the narrow burrow, in which dissipation of body heat would be difficult. Ambient temperature does not seem to affect their activity, but on nights with a combination of high humidity and low temperature they tend not be active.

Porcupines mark their trails with excretion from glands around and between the anus and the genitals, which they smear on various objects such as stones and branches.

Mean home range of six individuals near Sede Boqer was 1.5 km², and the animals travelled on an average 2.8 km per night. Animals which depended on natural forage had a larger home range than those which foraged on crops. There was a considerable overlap in home range between individuals. During moonlit periods the porcu-

Table 178
Body measurements (kg and mm) of *Hystrix indica indica*

	Body		Tail	Ear	Foot
	Mass	Length			
Males (7)					
				n = 6	**n = 6**
Mean	8.2	666	134	40	100
SD	2	146	16	4	7
Range	5.9–11.3	390–850	120–165	35–46	94–113
Females (5)					
	n = 4				
Mean	9.9	675	141	40	106
SD	1.9	72	19	6	12
Range	7–11	560–713	81–170	32–48	74–120

Table 179
Skull measurements (mm) of *Hystrix indica indica* (Figure 186)

	Males (5)			Females (5)		
	Mean	Range	SD	Mean	Range	SD
GTL	144.8	136.9–155	8.8	144.1	135.2–147.3	5.1
CBL	134.3	130.7–138.5	3.1	136	130.3–139.6	3.8
ZB	80.3	80–84.3	2.8	78.5	70.9–84	5.7
BB	46.9	44.9–48.4	1.3	45.7	41.8–48.2	2.6
IC	51.6	46.3–56.9	4.2	52.1	48.7–54.8	2.4
UT	32	30–34.7	2.0	31	28.9–32.4	1.7
LT	33.8	31–37.9	2.6	33.3	31.5–64.7	1.2
M	104.1	103.5–114.3	11.9	106.6	104–109.9	2.4
TB	18.9	17–21.3	1.8	19	18–21.1	1.2

The normal body mass of adult porcupines is no less than 10 kg, and specimens lighter than this mass were emaciated. The heaviest specimen weighed in Israel was a male of 18.5 kg (Z. Sever, pers. comm.). There are no significant differences in size between the sexes.

Dental formula is $\frac{1.0.1.3}{1.0.1.3} = 20$.

Karyotype: $2N = 60$, $FN = 114$ for specimens from Europe (Renzoni, 1967).

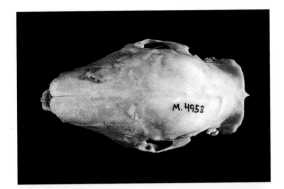

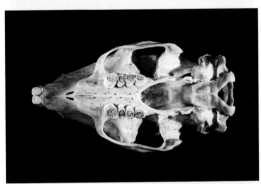

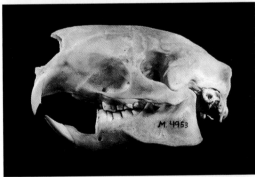

Figure 186: *Hystrix indica indica.* Dorsal, ventral and lateral views of the cranium, and a view of the mandible.

5 cm

Figure 187: Distribution map of *Hystrix indica indica.*

● – museum records.

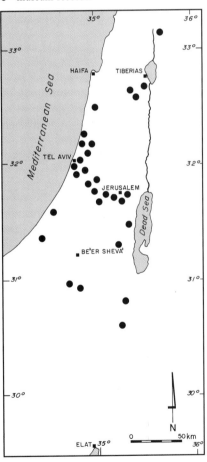

397

pines tended to remain closer to their dens (Saltz and Alkon, 1989). Near Tel Aviv, in the Mediterranean zone of Israel, the home range of two pairs in an agricultural area was about 0.43 km^2 compared with 1.2 km^2 for a pair which foraged on natural food. Single males have a much larger home range, up to 4.2 km^2, probably because they try to locate females. Average distance covered per night was 1.6 km for pairs, and 2.3 km for solitary males, but in one case a solitary male walked 7.7 km in one night. Moonlight did not affect foraging range in this study (Sever and Mendelssohn, 1991).

When threatened porcupines run quickly while erecting their spines and barking huskily. When cornered they erect their spines like a fan and jump backwards or sidewards towards the attacker, or suddenly stop running, so that the attacker runs into their spines. They are able to inflict serious injuries in this way. The embedded spines can cause pain and infections, even to leopards.

Porcupines also erect their spines when excited while mating, meeting conspecifics under other circumstances or feeding on favourite foods. While fighting among themselves they do not stab each other, but bite the back spines and skin of their opponent's back. When afraid they stamp on the ground with the hind legs, producing quite a loud noise. They also produce a rattling noise by quivering the hollow, open tail spines that have a thin base.

Leopards lie in wait for adult porcupines near their trails and try to kill them with a quick slap of the powerful forefoot. This method is not always successful, as shown by an adult male leopard, that was shot by a Beduin not far from Sede Boqer. It was quite emaciated and had porcupine spines stuck in its head and forefoot. It had apparently been unable to extract them and they prevented it from hunting. Young porcupines are eaten by other, smaller predators such as wolves and jackals. Chewed porcupine spines were found embedded under skins of leopards and in their faeces, in the Galilee and in the Judean Desert.

Food: Most of the year porcupines feed mainly on geophytes — bulbs, roots and rhizomes which they dig out with their front paws. In the 'Arava Valley they feed on roots of the camel thorn (*Alhagi maurorum*). The dig has the distinctive shape of a triangle with a pile of soil on its wider end, and these depressions serve as good germination spots for seeds in the desert, as moisture accumulates there. The porcupine grasps the vegetation between its front paws, frequently while lying down. In spring they feed mainly on green vegetation. The food is ground mainly by the molar teeth, but also by the hard, horny plates on the palate and the tongue. Sometimes they gnaw bones, but this behaviour is not as common in Israel with its lime rich soils as it is in South Africa where lime is lacking in some soils. Porcupines may drink water, but normally get all the water they need from their food.

The faeces are long and narrow, resembling small unpeeled peanuts, and are found in piles of about 20 on porcupine trails.

Reproduction: In the Mediterranean zone porcupine pairs maintain permanent contact, walking and foraging at a distance of 10 m or less from one another. They appear to be monogamous. A study on captive porcupines (Sever and Mendelssohn,

1988) indicated that the members of the pair copulate on an average eight times per night every night during the year, even during pregnancy and lactation. These copulations, only a few of which serve for reproduction, seem to function to maintain the pair bond. The female solicitates copulations. The male carefully put its paws on the females back behind the erected long quills and copulates without its body touching the female's back and barely even her raised tail. Copulation without injury from the spines on the back is made possible by the length of the penis, which is about nine cm.

In a sample of 113 porcupines caught between June–November in the Judean hills, sex ratio was male biased (58/42), 15% of the females were pregnant and 9% in oestrus (Eitan Bezer, pers. comm.). In the Tel Aviv area females become oestrous from January, pregnancy lasts 3 months, and young are born between March-September. Females may give birth twice yearly, in spring and autumn. Litter size ranges from 1–4, but mostly two. Litter size at the Tel Aviv University Research Zoo was 1–3 (mostly one young; n = 14), and young were born between February–August. Young are born in the burrow. The newborn weigh about 350 g, have open eyes, are able to walk, and are covered with soft spines which harden within days. When born they have only incisors, but within several days they start to nibble at food, and are weaned when three months old (Sever and Mendelssohn, 1988). The female has four teats, two on each side, that are situated on the sides of the body behind the front legs. However, females with one or three teats on one side were observed (Z. Sever, pers. comm.). During lactation the female lies flat on her abdomen, with her front legs extended forwards and the young suckle while lying on the abdomen at a right angle to the female. In all litters of three born at the Tel Aviv University Research Zoo, the two larger newborns bit the third to death, without the parents intervening (Z. Sever, pers. comm.). Since only two teats produce milk, and each is adopted and used by a certain pup, this behaviour has some biological function.

Sexual maturity is reached at about one year, and the young weigh about 10 kg at this time. One individual was kept for 14 years at the Tel Aviv University Research Zoo.

Relations with man: Porcupines cause damage to agriculture by gnawing on tree trunks and low branches. They also eat melons, water melons, potatoes and various vegetables, and fruits from low branches.

Porcupine meat is regarded as a delicacy and is considered by the Druze to be a powerful aphrodisiac. They are favoured game animals in Israel. The Bedouin believe in the therapeutic nature of various parts of the porcupine body: they use the blood to treat asthma, the liver for curing female sterility, and the spines are used in back acupuncture to treat high blood pressure.

CAPROMYIDAE נוטריים

This family is represented in Israel by one, introduced species, *Myocastor coypu*.

Myocastor coypu Kerr, 1792   Nutria, Coypu
Figs 188–189; Plate XXXVII:2–3   נוטריה

Mus coypu Molina, 1782. 287. Rio Maipo, Santiago province, Chile.
Synonymy in Wilson and Reeder (1993).

This is a large rodent (body mass can be up to eight kg). The hindfeet are large and propel the nutria while swimming, while the forefeet are used for digging and manipulating food. The toes, apart from the fifth, which is free, are connected by large webs. The rat-like tail takes no part in swimming. The body is covered with coarse brown fur, with dense, woolly, soft, water-proof underfur. This underfur, if the coarse guard hair is removed, is a popular fashion fur. As nutrias are easy to keep and breed, they are raised in fur farms in many parts of the world.

Table 180
Body measurements (g and mm) of *Myocastor coypu*

	Body		Tail	Ear	Foot
	Mass	Length			
	Males (2)				
Mean	4350 (n=1)	518	392	31	127
Range		485–550	385–440	25–37	124–130
	Females (1)				
Mean	5800	550	425	25	140

Table 181
Skull measurements (mm) of *Myocastor coypu* (Figure 188)

	Males (1)	Females (1)		Sex unknown (3)
	Mean	Mean	Range	
GTL	114.6	112.6	128.5	120.8–140.1
CBL	100.1	102.6	112.7	105.6–117.7
ZB	65.9	66.6	73.4	68.3–80.3
BB	36.2	34.8	37.4	36.1–39.2
IC	30.7	27.8	32.1	28.9–35.0
UT	29.7		27.9	26.6–28.6
LT	33.6	26.7	29.9	29.8–30.0
M	86.0	87.2	99.7	91.8–107.6

Dental formula is $\dfrac{1.0.1.3}{1.0.1.3} = 20$.

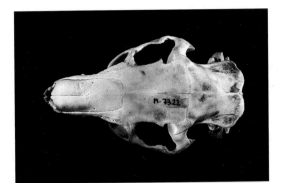

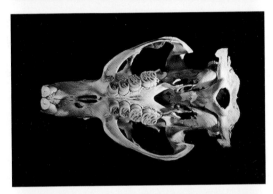

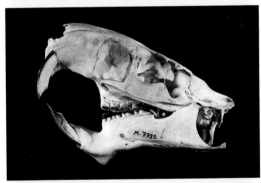

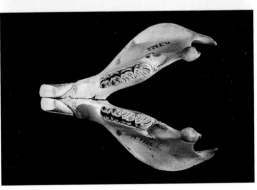

Figure 188: *Myocastor coypu*. Dorsal, ventral and lateral views of the cranium, and a view of the mandible.

5 cm

Figure 189: Distribution map of *Myocastor coypu*.

● – museum records.

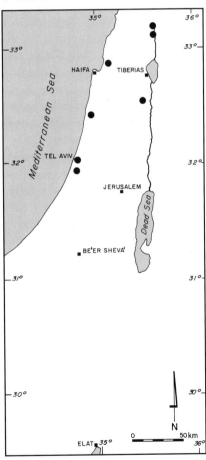

Distribution: The coypu is widely distributed in southern South America and was introduced as a fur animal to various countries in continental Europe and England, Kenya and several of the states of the USA (Woods *et al.*, 1992). In the early fifties nutria were imported to Israel to be bred for their fur. However, in the hot climate it produced a worthless fur, and the nutria farmers released their animals, to join those which had escaped earlier. A new species was thereby added to the faunal list of Israel. The nutria is now distributed throughout the Mediterranean areas of Israel wherever there are streams, fish ponds and water reservoirs in the Galilee, the northern valleys, the Jordan River and the Mediterranean Coastal Plain as far south as Ashqelon (Figure 189). Even a small ditch is sufficient to support a nutria population.

Habits: The nutria is mainly nocturnal, but is also active during the day, particularly where population density is high. It digs its own burrow along the shores of various water bodies. The burrow is several metres long with an opening of about 20 cm diameter. In Israel they generally live in burrows. Occasionally, in areas with rich swamp vegetation, they build huge nests of reeds, etc., in which an entire family lives.

The nutria spends most of its time near and in water, but may occasionally be found several hundred metres from water. It is a slow swimmer, using its hindlegs alternately to propel itself while the forelegs are held close to the body. It is a social animal, living in groups composed of a dominant pair and several of their female offspring.

Food: The nutria is herbivorous and uses its paws to manipulate its food. It feeds on a wide variety of plants which grow near and in the water. Occasionally it eats snails and other invertebrates.

Reproduction: Pregnancy lasts about 130 days and litter size ranges from 3–6 young. The female can have a post-partum oestrus and can produce three litters within a year. The young are suckled for 7–8 weeks, but are capable of feeding on solid food and of following their mother from birth. They are precocial and born with open eyes and covered with fur.

Relations with man: In Israel, it was believed for some time that the nutria might be useful for controlling reeds in fish ponds. This advantage was, however, offset by the damage they did by digging their burrows into the fish pond dikes. They also do much damage to vegetable fields and by digging and eating peanuts. Due to their high reproductive potential and because the small local carnivores do not prey on the strong and aggressive adult nutria, they spread quickly over most of the plains area of the country, south to the north-western Negev. As they eat almost any vegetation, food is apparently not a limiting factor. Carnivores (jackals, foxes, mongooses and jungle cats) gradually learned that young nutria are easy prey, thereby preventing a population explosion. The effect of predation is evidenced by a rampant increase of nutria which followed the 1964 poisoning of predators in Israel. This population explosion subsided after the mongoose and jungle cat populations recovered.

ADDENDA

Introduced Ungulates

Since the completion of this work in January 1996 four species of ungulates have been introduced to Israel by the Nature Reserves Authority (NRA). These are the Persian fallow deer (*Dama dama mesopotamica*), the roe deer (*Capreolus capreolus*), the wild goat (*Capra hircus*) and the Arabian oryx (*Oryx leucoryx*). The following is a short description of the history of introduction and the present status of these species in Israel. Much of the following information was provided by David Saltz of the NRA.

The introduced fallow deer in Israel originate from five females brought to Israel from Iran before the Islamic revolution there, and one pair that was brought from Germany, but originated from Iran. All these deer were kept in an enclosure on Mount Carmel (Ḥai Bar) where they multiplied to about 150 individuals. Between 1996 and 1998 about 50 individuals were released in several consecutive releases into a natural oak forest in Naḥal Beẓet in the western Galilee. Most of these animals live in three separate groups, each composed of several males and females, all within a distance of 1.5 kms of the site from which they were released. Individual home ranges vary between 86–365 hectares, with much overlap between the individuals of each of the three groups, and with no difference in the size of home range between the sexes. They browse and graze in the forest, from which they rarely come out. They prefer to live in moderately sloped areas rather on steep slopes. At times they penetrate into nearby orchards, where they do some damage to the trees. Fawns were observed, indicating that the released animals bred after being released.

The introduced roe deer originate from Hungary, Holland and France, although a population of *Capreolus capreolus coxi*, the subspecies that lived in Israel still exists in Turkey. They were kept in Ḥai Bar Carmel before being released. Nine individuals were released in 1996 into the scrub forest of the southern Carmel (Ramat HaNadiv area), and all were provided with radio transmitters on neck collars. Radio signals indicate that all, apart from one individual which was apparently hunted by a poacher, were alive in autumn 1998. Various observers reported sighting of fawns which were born after the release.

The introduced Arabian oryx originated from the herd of San Diego Zoo, a herd which multiplied from animals captured in the southern Arabian peninsula. Several oryx were imported to Israel and were kept in Ḥai Bar Yotvata, in the southern 'Arava. In March 1997, 21 individuals were released near 'En Shaḥak, in the eastern Negev Desert, and ten more were released there during spring 1998. All roam in an area of about 100 km^2 near the area where they were released, in groups whose size changes constantly. At least five calves were born since the initial release.

The introduced goats originate from a herd of wild goats which were brought to Israel from the island of Crete. As the wild goats of Crete could have reached the island only with the help of man, they probably descend from goats in the first stages of domestication. As they are, however, very similar to wild goats in Turkey, they can be considered as real wild goats. This herd was kept in Hai Bar Carmel, and several individuals, the number of which is not known, escaped and are now living on Mount Carmel.

Microchiroptera of the Dead Sea Area (Feldman, 1998)

Nycteris thebaica has a diet composed mainly of Lepidoptera, Diptera and Coleoptera (Scarabaeidae), whose remains occur in 67%, 56% and 44% of the faeces examined, respectively. There were also remains of Hymenoptera (Formicidae) and Hemiptera.

Rhinolophus clivosus. Measurements found were: mean body weight was maximal during spring (11.5 g), and females were heavier than males (11.1 and 8.9 g) and had longer forearms than males (49.6 and 46.3 mm). It emerges from its roosts immediately after sunset. In the Dead Sea area it feeds mainly on Coleoptera (4 families), Hymenoptera (Formicidae) and Lepidoptera whose remains occur in 71%, 6% and 47% of the faeces examined, respectively. There were also remains of Trichoptera, Diptera (2 families), Homoptera (Cicadellidae) and Neuroptera.

Rhinolophus hipposideros flies fast, often close to the ground and between vegetation. Nocturnal activity in the Dead Sea area starts about an hour after sunset. Its diet is composed mainly of Lepidoptera and of Diptera (Chironomidae) whose remains occur in 94% and 78% of the faeces examined, respectively. There were also remains of Trichoptera, Coleoptera (2 families), Hymenoptera, Hemiptera and Homoptera.

Asellia tridens emerges from its roosts immediately after sunset. Its diet in the Dead Sea area is composed mainly of Coleoptera (4 families), Hymenoptera (Formicidae) and Hemiptera (3 families) whose remains occur in 83%, 65% and 28% of the faeces examined, respectively. There were also remains of Lepidoptera, Trichoptera, Diptera (2 families), Homoptera (2 families) and Neuroptera.

Eptesicus bottae has a diet composed mainly of Hymenoptera, Lepidoptera (2 families) and Coleoptera (Scarabaeidae), whose remains occur in 73%, 46% and 36% of the faeces examined, respectively.

Pipistrellus kuhli is an opportunistic feeder in the Dead Sea area. Its diet is composed mainly of Diptera (2 families), Lepidoptera and Coleoptera (3 families), whose remains occur in 70%, 60% and 70% of the faeces examined, respectively. There were also remains of Hymenoptera (Formicidae), Hemiptera (Lygaeidae), Neuroptera and Homoptera (2 families).

Pipistrellus rueppelli. In the Dead Sea area its mean body weight is 6.3 g, heaviest in the autumn (7.4 g) and lightest in spring (5.9 g). It was caught more frequently during winter than during summer. It is an opportunistic feeder, and its diet is composed chiefly of Diptera (mainly Chironomidae) whose remains occur in 90% of the faeces examined, but also of Coleoptera (Scarabaeidae), Trichoptera and Lepidoptera. It hunts mainly above water, and its faeces are generally noticeably wet.

Pipistrellus bodenheimeri has a bi-modal activity pattern, with one activity peak after sunset until about 4 hours later, and another about an hour before dawn. In the Dead Sea area it was observed foraging near vegetation, rocky slopes, street lights and above water. It is an opportunistic feeder, and its diet is composed mainly of Diptera (4 families), chiefly during winter, Hymenoptera (Formicidae) and Coleoptera (5 families), whose remains occur in 70%, 70% and 60% of the faeces examined, respectively. There were also remains of Lepidoptera, Trichoptera, Hemiptera (4 families) and Homoptera (3 families).

Plecotus austriacus. Mean body weight was 7.5 g, and mean forearm length was 40.3 mm. The specimens were captured only after the second hour after sunset. Its diet is composed mainly of Lepidoptera, whose remains occur in 95% of the faeces examined, but also of Diptera, Trichoptera and Coleoptera.

Sekeetamys calurus in 'En Gedi

Shargal *et al.* (1998) found that mean body mass of females and males in 'En Gedi was 39.4 and 40.6 g, respectively, and there was no significant difference in body mass between the sexes. Population density ranged between 0–8 individuals per 0.5 hectares, with a mean of 2.6. Males were found in reproductive state in January, March, July, August and October, suggesting a prolonged reproductive season.

Degen *et al.* (1986) found that dry matter intake in 'En gedi during spring was 87%.

REFERENCES

Abadi, M. (1989) 'A small cat in the 'Arava', *Teva Va'Aretz*, 31:17–28 (in Hebrew).

Abramsky, Z. (1980) 'Ecological similarity of *Gerbillus allenbyi* and *Meriones tristrami*', *J. Arid Envir.*, 3:153–160.

Abramsky, Z. (1984) 'Population biology of *Gerbillus allenbyi* in northern Israel', *Mammalia*, 48:197–206.

Abramsky, Z. and Sellah, C. (1982) 'Competition and the role of habitat selection in *Gerbillus allenbyi* and *Meriones tristrami*: a removal experiment', *Ecology*, 63:1242–1247.

Abramsky, Z., Rosenzweig, M.L. and Brand, S. (1985) 'Habitat selection of Israel desert rodents: comparison of a traditional and a new method of analysis', *Oikos*, 45:79–88.

Afik, D. and Alkon, P.U. (1983) 'Movements of a radio-collared wolf (*Canis lupus pallipes*) in the Negev Highlands, Israel', *Israel J. Zool.*, 32:138–146.

Aharoni, I. (1917) 'Zum Vorkommen der Säugetiere in Palästina und Syrien', *Z. Mitteil. u. Nachr. Deutsch. Palästina, Ver.*, 40:235–242.

Ahlen, I. (1990) *Identification of Bats in Flight*. The Swedish Society for the Conservation of Nature.

Aisner, R. and Terkel, J. (1992) 'Cultural transmission of pine cone opening behaviour in the roof rat (*Rattus rattus*)', *Anim. Behav.*, 44:327–336.

Alcantara, M. (1991) 'Geographical size variation in body size of the wood mouse *Apodemus sylvaticus* L.', *Mammalian Rev.* 21:143–150.

Aldos, C.L. (1986) 'The use of tail and rump patch in the dorcas gazelle (*Gazella dorcas* L.)', *Mammalia*, 50:439–446.

Al-Khalili, A.D. and Delany, M.J. (1986) 'The post-embryonic development and reproductive strategies of two species of rodents in south-west Saudi Arabia', *Cimbebasia*, 8:175–185.

Al-Robaae, K. (1966) 'Untersuchugen der Lebensweise irakischer Fledermäuse', *Säugetierkundliche Mitt.*, 14:177–211.

Al-Robaae, K. (1968) 'Notes on the biology of the Tomb Bat, *Taphozous nudiventris magnus* v. Wettstein 1913, in Iraq', *Säugetierkundliche Mitt.*, 16(1):21–26.

Al-Saleh, A.A. and Khan, M.A. (1984) 'Cytological studies of certain desert mammals of Saudi Arabia. 1. The karyotype of *Jaculus jaculus*', *J. Coll. Sci. King Saud University*, 15:163–168.

Al-Saleh, A.A. and Khan, M.A. (1985) 'Cytological studies of certain desert mammals from Saudi Arabia. 3. The karyotype of *Paraechinus aethiopicus*', *Cytologia*, 50:507–512.

Alkon, P.U. and Saltz, D. (1988a) 'Foraging time and the northern range limits of Indian crested porcupines (*Hystrix indica* Kerr)', *J. Biogeogr.*, 15:403–408.

Alkon, P.U. and Saltz, D. (1988b) 'Influence of season and moonlight on temporal activity patterns of Indian crested porcupines (*Hystrix indica*)', *J. Mammal.*, 69:71–80.

Alkon, P.U. and Saltz, D. (1989) 'On the spatial behaviour of Indian crested porcupines (*Hystrix indica*)', *J. Zool. Lond.*, 217:255–266.

Aronson, L. (1982) *The Ibex — King of the Cliffs*. Massada Press, Tel Aviv. 87 pp. (in Hebrew).

Assa, T. (1990) 'The biology and biodynamics of the red fox (*Vulpes vulpes*) in the northern Arava Valley, Israel', M.Sc. thesis, Department of Zoology, Tel Aviv University, Israel (in Hebrew, with English summary).

Atallah, S.I. (1977) 'Mammals of the eastern Mediterranean region: their ecology, systematics and zoogeographical relationships', Part 1. *Säugetierkundliche Mitt.*, 25:241–320.

Auffray, J.C., Marshall, J.T., Thaler, L. and Bonhomme, F. (1990a) 'Focus on the nomenclature of European species of *Mus*', *Mouse Genome*, 88:7–8.

Auffray, J.C., Tchernov, E., Bonhomme, F., Heth, G., Simson, S. and Nevo, E. (1990b) 'Presence and ecological distribution of *Mus "spretoides"* and *Mus musculus domesticus* in Israel. Circum-Mediterranean vicariance in the genus *Mus*', *Z. Säugetierkunde*, 55:1–10.

Ayal, Y. (1990) 'The status of the leopard population in Israel: an evaluation based on data accumulated by G. Ilani between 1970–1989'. Report submitted to the Nature Reserves Authority of Israel. pp. 1–18 (in Hebrew).

Ayal, Y., and Baharav, D. (1985) 'Conservation and management plan for the mountain gazelle in Israel', in: Brown, S.L. and Robertson, S.F. (eds), *Game Harvest Management*. C. K. Wildlife Research, Atlanta, Georgia, pp. 269–278.

Baharav, D. (1980) 'Habitat utilization of the dorcas gazelle in a desert saline area', *J. Arid Envir.*, 3:161–167.

Baharav, D. (1982) 'Desert habitat partitioning by the dorcas gazelle', *J. Arid Envir.*, 5:323–335.

Baharav, D. (1983) 'Reproductive strategies in female mountain and dorcas gazelles (*Gazella gazella gazella* and *Gazella dorcas*)', *J. Zool. Lond.*, 200:445–453.

Baharav, D. and Meiboom, U. (1981) 'The status of the Nubian ibex *Capra ibex nubiana* in the Sinai Desert', *Biol. Conser.*, 20:91–97.

Baharav, D. and Meiboom, U. (1982) 'Winter thermoregulatory behaviour of the Nubian ibex in the southern Sinai Desert', *J. Arid. Envir.*, 5:295–298.

Baker, R.J., Davis, B.L., Gordon, R.G. and Binous, A. (1974) 'Karyotypic and morphometric studies of Tunisian mammals: 1. Bats', *Mammalia*, 38:695–710.

Bar, Y., Abramsky, Z. and Gutterman, Y. (1984) 'Diet of gerbilline rodents in the Israeli desert', *J. Arid Envir.*, 7:371–376.

Bar-Oz, G. (1996) 'The fauna of the Geometric-Kebaran site, Neve David: taphonomic, economic and ecological implications', M.Sc. thesis, Department of Zoology, Tel Aviv University (in Hebrew, with English summary).

Barak, Y. (1989) 'The biology of the Kuhl's bat *Pipistrellus kuhli* in Israel', M.Sc. thesis, Department of Zoology, Tel Aviv University, Israel (in Hebrew, with English summary).

Barak, Y. and Yom-Tov, Y. (1989) 'The advantage of group hunting by Kuhl's bat *Pipistrellus kuhli* (Microchiroptera)', *J. Zool. Lond.*, 219:670–675.

Barak, Y. and Yom-Tov, Y. (1991) 'The mating system of *Pipistrellus kuhli* (Microchiroptera) in Israel', *Mammalia*, 55:285–292.

References

Barzilai, E., and Tadmor, A. (1972) 'Experimental infection of the mountain gazelle (*Gazella gazella*) with bluetongue virus', *Refuah Veterinarit*, 29:45–50 (in Hebrew).

Bate, D.M.A. (1940) 'The fossil antelopes of Palestine in Natufian (Mesolithic) times, with description of new species', *Oecological Magazine*, 77:418–443.

Belecheva, R.G., Peshev, T.H. and Peshev, D.T. (1980) 'Chromosome C- and G-banding patterns in a Bulgarian population of *Microtus guentheri* Danford and Alson (Microtinae, Rodentia)', *Genetica*, 52/3:45–48.

Ben-David, M. (1988) 'The biology and ecology of the marbled polecat (*Vormela peregusna syriaca*) in Israel', M.Sc. thesis, Department of Zoology, Tel Aviv University, Israel (in Hebrew, with English summary).

Ben-David, M., Pellis, S.M. and Pellis, V.C. (1991) 'Feeding habits and predatory behaviour in the marbled polecat (*Vormela peregusna syriaca*): I. Killing methods in relation to prey size and prey behaviour', *Behaviour*, 118:127–143.

Ben-Yaacov, R. and Yom-Tov, Y. (1983) 'On the biology of the Egyptian mongoose, *Herpestes ichneumon*, in Israel', *Z. Säugetierkunde*, 48:34–45.

Bhatnager, A.N and El-Azawi, T.F. (1978) 'A karyotype study of chromosomes of two species of hedgehogs, *Hemiechinus auritus* and *Paraechinus aethiopicus* (Insectivora, Mammalia)', *Cytologia*, 43:53–59.

Bodenheimer, F.S. (1935) *Animal Life in Palestine*. L. Mayer, Jerusalem, 507 pp.

Bodenheimer, F.S. (1949) 'Ecological and physiological studies on some rodents', *Physiologia Comp. Oecol. Haag.*, 1:376–389.

Bodenheimer, F.S. (1953) *The Fauna of Eretz Israel*. Dvir Publication House, Tel Aviv (in Hebrew).

Bodenheimer, F.S. (1958) 'The present taxonomic status of the terrestrial mammals of Palestine', *Bull. Res. Counc. Israel*, 7B:165–190.

Booth, B.D. (1961) 'Breeding of the Sooty Falcon in the Libyan desert'. *Ibis*, 103A:129–130.

Bouskila, Y. (1983) 'The biology and behaviour of the striped hyaena, *Hyaena hyaena syriaca* in the Sede Boqer region', The Society for the Protection of Nature in Israel (in Hebrew).

Bouskila, Y. (1984) 'The foraging groups of the striped hyaena (*Hyaena hyaena syriaca*)', *Carnivore*, 7:2–12.

Bovey, R. (1949) 'Les chromosomes des Chiroptères et des Insectivores', *Revue suisse Zool.*, 56:371–460.

Brand, S. and Abramsky, Z. (1987) 'Body masses of gerbilline rodents in sandy habitats of Israel', *J. Arid Envir.*, 12:247–253.

Brosset, A. (1962) 'The bats of central and western India', *J. Bombay nat. Hist. Soc.*, 59:1–78.

Burgos, M., Jimenez, R. and Diaz de la Guardia, R. (1989) 'Comparative studies of G- and C-banded chromosomes of five species of Microtidae', *Genetica*, 78:3–12.

Carlisle, D.B. and Ghobrial, L.I. (1968) 'Food and water requirements of dorcas gazelles in the Sudan', *Mammalia*, 32:570–576.

Carruthers, D. (1909) 'Big game of Syria, Palestine and Sinai', *Field, London*, 114:1135.

Catzeflis, F., Maddalena, T., Hellwing, S. and Vogel, P. (1985) 'Unexpected findings on the taxonomic status of East Mediterranean *Crocidura russula* auct. (Mammalia, Insectivora)', *Z. Säugetierkunde*, 50:185–201.

Chetboun, R. and Tchernov, E. (1983) 'Temporal and spatial morphological variation in *Meriones tristrami* (Rodentia: Gerbillidae) from Israel', *Israel J. Zool.,* 32:63–90.

Chilcott, B. J., Phillips, M. R., Zwick, K. L., Ahmmad, Q., Khaled, Y. and Yousef, M. (1995) 'Dana Nature Reserve baseline wildlife surveys', Royal Society for the Conservation of Nature, Amman.

Cloudsley-Thompson, J.F. and Ghobrial, L.I. (1965) 'Water economy of the dorcas gazelle', *Nature,* 207:1313.

Clutton-Brock, J. (1979) 'The mammalian remains from the Jericho tell', *Proceedings of the Prehistory Society,* 45:135–157.

Cnaani, G. (1972) 'Studies on the ecology and the behaviour of the wild boar (*Sus scrofa lybicus* Gray, 1868) in the Mt. Meron Region', M.Sc. thesis, Department of Zoology, Tel Aviv University, Israel (in Hebrew, with English summary).

Cohen-Shlagman, L., Yom-Tov, Y. and Hellwing, S. (1984a) 'The biology of the Levant vole *Microtus guentheri* in Israel. I. Population dynamics in the field', *Z. Säugetierkunde,* 49:135–148.

Cohen-Shlagman, L., Hellwing, S. and Yom-Tov, Y. (1984b) 'The biology of the Levant vole *Microtus guentheri* in Israel. II. The reproduction and growth in captivity', *Z. Säugetierkunde,* 49:149–156.

Corbet, G.B. (1978) *The Mammals of the Palaearctic Region: A Taxonomic Review.* British Museum (Natural History), London.

Danin, A. (1988) 'Flora and vegetation of Israel and adjacent areas', in: Yom-Tov Y. and Tchernov E. (eds), *The Zoogeography of Israel.* W. Junk Publ., Dordrecht, pp. 129–157.

Danin, A. and Plitmann, U. (1986) 'Revision of the plant geographical territories of Israel and Sinai', *Plant Syst. Evol.,* 156:43–53.

Danin, A. and Yom-Tov, Y. (1990) 'Ant nests as primary habitats of *Silybum marianum* (Compositae)', *Plant Systematics and Evolution,* 169:209–217.

Davis, S. (1977) 'The ungulate remains from Kebara cave', *Eretz Israel,* 13:150–163.

Davis, S. (1977) 'Animal remains from the Kebaran site Ein Gev I, Jordan Valley, Israel', *Paleorient,* 212:453–462.

Davis, S. (1980) 'Late Pleistocene-Holocene gazelles of northern Israel', *Israel J. Zool.,* 29:135–140.

Dayan, T., Simberloff, D., Tchernov, E. and Yom-Tov, Y. (1989a) 'Inter- and intraspecific character displacement in mustelids', *Ecology,* 70:1526–1539.

Dayan, T., Tchernov, E., Yom-Tov, Y. and Simberloff, D. (1989b) 'Ecological character displacement in Saharo-Arabian *Vulpes*: Outfoxing Bergmann's rule', *Oikos,* 55:263–272.

Dayan, T., Simberloff, D., Tchernov, E. and Yom-Tov, Y. (1990) 'Feline canines: Community wide character displacement among the small cats of Israel', *Am. Nat.,* 136:39–60.

Dayan, T., Simberloff, D., Tchernov, E. and Yom-Tov, Y. (1991) 'Calibrating the paleothermometer: character displacement and the evolution of size', *Paleobiology,* 17:189–199.

Dayan, T., Simberloff, D., Tchernov, E. and Yom-Tov, Y. (1992) 'Canine carnassials: character displacement of the wolves, jackals and foxes of Israel', *Biol. J. Linn. Soc.,* 45:315–331.

References

DeBlase, A.F. (1972) '*Rhinolophus euryale* and *R. mehelyi* (Chiroptera, Rhinolophidae) in Egypt and southwest Asia', *Israel J. Zool.*, 21:1–12.

DeBlase, A.F. (1980) 'The bats of Iran: systematics, distribution, ecology', *Fieldiana, Zool.*, (N.S.) 4:1–424.

Degen, A.A. (1988) 'Ash and electrolyte intakes of the fat sand rat, *Psammomys obesus*, consuming saltbush, *Atriplex halimus*, containing different water content', *Physiol. Zool.*, 61:137–141.

Degen, A. (1994) 'Field metabolic rates of *Acomys russatus* and *A. cahirinus* and a comparison with other rodents', *Israel J. Zool.*, 40:127–134.

Degen, A.A., Hazan, A., Kam, M. and Nagy, K.A. (1991) 'Seasonal water influx and energy expenditure of free-living fat sand rats', *J. Mammal.*, 72:652–657.

Degen, A.A., Kam, M., Hazan, A. and Nagy, K.A. (1986) 'Energy expenditure and water flux in three sympatric desert rodents', *J. Anim. Ecol.*, 55:421–429.

Dittrich, L. (1972) 'Gestation periods and age of sexual maturity of some African antelopes', *International Zoo Yearbook*, 12:184–187.

Dmi'el, R. and Schwartz, M. (1984) 'Hibernation patterns and energy expenditure in hedgehogs from semi-desert and temperate habitats', *J. Comp. Physiol.*, B. 155:117–123.

Dor, M. (1947) 'Observations sur les micromammifères trouvés dans les pelotes de la chouette effraie (*Tyto alba*) en Palestine', *Mammalia*, 11:49–54.

Ducos, P. (1968) 'l'Origine des animaux domestiques en Palestine', Mémoire No. 6, Publications de l'Institut de Préhistoire de l'Université de Bordeaux (Delmas, Bordeaux), 191 pp.

Dulic, B. and Mrakovcic, M. (1980) 'Chromosomes of European freetailed bat, *Tadarida teniotis teniotis* (Rafinesque, 1814. Mammalia, Chiroptera, Molossidae)', *Biosistematika*, 6:109–112.

Dulic, B and Mutere, F.A. (1973) 'Comparative studies of the chromosomes of some molossid bats from East Africa', *Period. Biol.*, 75:61–65.

Dulic, H., Soldalovic, B. and Rimsa, D. (1967) 'La formule chromosomique de la Noctule, *Nyctalus noctula* (Mammalia, Chiroptera)', *Biol. Glasnik*, 19:65–97.

Effron, M., Bogart, M.H., Kumamoto, A.T. and Benirschke, K. (1976) 'Chromosome studies in the mammalian subfamily Antilopinae', *Genetica*, 46:419–444.

Eig, A. (1931–2) 'Les éléments et les groupes phytogéographiques auxiliaires dans la flore palestinienne. I and II', *Fedes Report. Spec. Nov. Reg. Veget. Beihf.*, 63.

Ellerman, J.R. and Morrison-Scott, T.C.S. (1951) *Checklist of Palaearctic and Indian Mammals, 1758–1946*. British Museum (Natural History), London, 810 pp.

Essghaier, M.F.A. and Johnson, D.R. (1981) 'Distribution and use of dung heaps by dorcas gazelle, *Gazella dorcas*, in western Libya', *Mammalia*, 45:153–156.

Fedyk, A. and Fedyk, S. (1971) 'Karyological analysis of representativs of the genus *Plecotus* Geoffroy 1818 (Mammalia: Chiroptera)', *Caryologia*, 24:483–492.

Feldman, R. (1998). 'The ecology of the insectivorous bats (Microchiptera) of the Dead Sea Area', M.Sc. Thesis, Department of Zoology, Tel Aviv University (in Hebrew, with English summary).

Fenton, M.B. (1975) 'Observations on the biology of some Rhodesian bats, including a key to the Chiroptera of Rhodesia', *Life Sci. Contr. R. Ont. Mus.*, 104:1–27.

411

Fenton, M.B. (1983) *Just Bats*. University of Toronto Press, Toronto.

Filippucci, M.G., Simson, S., Nevo, E. and Capanna, E. (1988) 'The chromosomes of the Israeli blacktailed dormouse, *Eliomys melanurus* Wagner, 1849 (Rodentia, Gliridae)', *Boll. Zool.*, 55:31–33.

Filippucci, M.G., Simson, S. and Nevo, E. (1989) 'Evolutionary biology of the genus *Apodemus* Kaup, 1829 in Israel. Allozymic and biometric analyses with description of a new species: *Apodemus hermonensis* (Rodentia, Muridae)', *Boll. Zool.*, 56:361–376.

Flower, S.S. (1932) 'Notes on the Recent Mammals of Egypt, with a list of species recorded from that Kingdom', *Proc. Zool. Soc., London*, pp. 369–450.

Frankenberg, E. and Pevzner, D. (1988) 'Canids and calves predation in the Golan', Nature Reserves Authority of Israel, Jerusalem. Unpublished report (in Hebrew).

Furley, C.W. (1986) 'Reproductive parameters of African gazelles: gestation, first fertile mating, first partuition and twinning', *Afr. J. Ecol.*, 24:121–128.

Gaisler, J. and Kowalski, K. (1986) 'Results of the netting of bats in Algeria (Mammalia, Chiroptera)', *Vestnik Ceskoslovenske Spolecnost Zoologicke*, 50:161–173.

Gaisler, J., Madkour, G. and Pelican, J. (1972) 'On the bats (Chiroptera) of Egypt', *Acta Sci. Nat. Brno*, 6:1–40.

Galil, J. (1967) 'On the dispersal of the bulbs of *Oxalis cernua* Thunb. by mole-rats (*Spalax ehrenbergi* Nehring)', *J. Ecol.*, 55:787–792.

Garfunkel, Z.(1988) 'The pre-Quaternary geology of Israel', in: Yom-Tov, Y. and Tchernov, E. (eds), *The Zoogeography of Israel*. W. Junk Publ., Dordrecht, pp. 7–34.

Garrod, D.A.E. and Bate, D.M.A. (1937) *The Stone Age of Mt. Carmel*. Vol. 2, Clarendon Press, Oxford. 237 pp.

Gavish, L. (1995) 'Observations on the behaviour and ecology of free-living populations of the subspecies *Sciurus anomalus syriacus* (golden squirrel) on Mount Hermon, Israel', *Israel J. Zool.*, 41:85–86.

Geffen, E. (1990) 'The behavioural ecology of the Blanford's fox, *Vulpes cana*, in Israel', Ph.D. thesis, Oxford University, England.

Geffen, E. (1994) '*Vulpes cana*', *Mammalian Species*, 462:1–4.

Geffen, E., Degen, A.A., Kam, M., Hefner, R. and Hagy, K.A. (1992c) 'Daily energy expenditure and water flux of free-living Blanford's foxes (*Vulpes cana*): a small desert carnivore', *J. Anim. Ecol.*, 61:611–617.

Geffen, E., Hefner, R., Macdonald, D.W. and Ucko, M. (1992a) 'Diet and foraging behaviour of Blanford's fox in Israel', *J. Mammal.*, 73:395–402.

Geffen, E., Hefner, R., Macdonald, D.W. and Ucko, M. (1992b) 'Morphological adaptations and seasonal weight changes in Blanford's fox, *Vulpes cana*', *J. Arid Envir.*, 23:287–292.

Geffen, E. and Macdonald, D. (1992) 'Small size and monogamy: spatial organization of the Blanford's fox, *Vulpes cana*', *Anim. Behav.*, 44:1123–1130.

Geffen, H. (1995) 'Ecological and physiological aspects of the mountain gazelle (*Gazella gazella gazella*) population in Ramat HaNadiv, Israel', M.Sc. thesis, Department of Zoology, Tel Aviv University (in Hebrew).

Getraida, S. and Perevolotsky, A. (1990) 'The ecology and biology of the mountain gazelle

(*Gazella g. gazella*) in Ramat HaNadiv', Society for the Protection of Nature in Israel and Yad HaNadiv. 44 pp. (in Hebrew).

Geva, E. (1980) 'Dispersion and movement of mountain gazelle (*Gazella gazella* Pallas 1766) in Naftali Mountains'. Research Thesis, The Technion, Haifa (in Hebrew, with English summary).

Ghobrial, L.I. (1974) 'Water relation and requirement of the dorcas gazelle in the Sudan', *Mammalia*, 38:88–107.

Ghobrial, L.I. (1976a) 'Observations on the intake of sea water by the dorcas gazelle', *Mammalia*, 40:489–494.

Ghobrial, L.I. (1976b) 'Daily cycle of activity of the dorcas gazelle in the Sudan', *J. Interdisciplinary Cycle Research*, 7:47–50.

Ghobrial, L.I. and Cloudlsey-Thompson, J.F. (1966) 'Effect of deprivation of water on the dorcas gazelle', *Nature*, 212:306.

Glickstein, H. (1982) 'Resource partitioning among four species of rodents inhabiting a rocky slope in Mitzpe Ramon', M.Sc. thesis, Environmental Biology Program, The Hebrew University of Jerusalem (in Hebrew, with English summary).

Golani, I. and Keller, A. (1974) 'A longitudinal field study of the behaviour of a pair of golden jackals', in: Fox, M.W. (ed.), *The Wild Canids*. von Nostrand Reinhold.

Golani, I. and Mendelssohn, H. (1971) 'Sequences of precopulatory behaviour of the jackal (*Canis aureus*)', *Behaviour*, 38:192–196.

Granot, Y. (1978) 'Comparative biology of two sympatric species of woodmice, genus *Apodemus*, on Mount Meron', M.Sc. thesis, Department of Zoology, Tel Aviv University, Israel (in Hebrew, with English summary).

Grau, G.A. and Walther, F.R. (1976) 'Mountain gazelle agonistic behaviour', *Anim. Behav.*, 24:626–636.

Greenberg-Cohen, D., Alkon, P.U. and Yom-Tov, Y. (1994) 'A linear dominance hierarchy in female Nubian ibex', *Ethology*, 98:210–220.

Groves, C.P. (1974) *Horses, Asses and Zebras in the Wild*. Ralph Curtis Books, Hollywood, Florida.

Günther, A. and Gebauer, H.J. (1982) 'The chromosomes of the otter, *Lutra lutra* (Carnivora: Mustelidae: Lutrinae)', *Mammalian Chromosome Newsletter*, 23:97–101.

Habibi, K. (1992) 'Arabian gazelles', National Commission for Wildlife Conservation and Development. Publication No. 4, Riyadh, 131 pp.

Haim, A. (1991) 'Behavior patterns of cold-resistant golden spiny mouse *Acomys russatus*', *Physiology and Behavior*, 50:641–643.

Haim, A. and Borut A. (1974) 'Differences in heat production among populations of the golden spiny mouse *Acomys russatus*', *Israel J. Med. Sci.*, 10:286.

Haim, A. and Borut A. (1975) 'Size and activity of a cold resistant population of the golden spiny mouse (*Acomys russatus*: Muridae)', *Mammalia*, 39:605–612

Haim, A. and Borut A. (1986) 'Reduced heat production in the bushy-tailed gerbil *Sekeetamys calurus* (Rodentia) as an adaptation to arid environments', *Mammalia*, 50:27–33.

Haim, A. and Rubal, A. (1992) 'The coexistance of two *Apodemus* species in the Mediterranean woodlands of Israel', in: Costas, A.T. (ed.) MEDECOS VI. University of Athens, Greece, pp. 127–132.

Haim, A. and Rubal, A. (1995) 'Thermoregulation and rhythmicity in *Eliomys melanurus* from the Negev highlands, Israel', *Hystrix,* 6:209–216.

Haim, A., Rubal, A. and Harari, J. (1993) 'Comparative thermoregulatory adaptations of field mice of the genus *Apodemus* to habitat challenges', *J. Comp. Physiol.,* B. 163:602–607.

Haim, A. and Tchernov, E. (1974) 'The distribution of myomorph rodents in the Sinai Peninsula', *Mammalia,* 38:201–223.

Hakham, E. (1981) 'Foraging pressure of the Nubian ibex (*Capra ibex nubiana*) on the vegetation of the 'En Gedi Nature Reserve', M.Sc. thesis, Hebrew University, Jerusalem (in Hebrew with English abstract).

Hakham, E. (1983) 'Crèche of young Nubian ibex in 'En Gedi', *Re'em,* 2:26–29 (in Hebrew).

Haltenorth, T. and Diller, H. (1980) *A Field Guide to the Mammals of Africa, including Madagascar.* Collins, London.

Happold, D.C.D. (1968) 'Observations on *Gerbillus pyramidum* (Gerbillinae: Rodentia) at Khartoum, Sudan', *Mammalia,* 32:44–53.

Harmata, W. (1969) 'The thermopreferendum of some species of bats (Chiroptera)', *Acta Theriol.,* 5:49–62.

Harrison, D.L. (1964) *The Mammals of Arabia: Insectivora, Chiroptera, Primates.* Vol. 1. E. Benn, London.

Harrison, D.L. (1968) *The Mammals of Arabia: Carnivora, Hyracoidea, Artiodactyla.* Vol. 2. E. Benn, London.

Harrison, D.L. (1972) *The Mammals of Arabia: Lagomorpha, Rodentia.* Vol. 3. E. Benn, London.

Harrison, D.L. (1975) 'Scientific results of the Oman Flora and Fauna survey, 1975. Description of a new subspecies of Botta's serotine (*Eptesicus bottae* Peters 1869, Chiroptera: Vespertilionidae) from Oman', *Mammalia,* 39:415–418.

Harrison, D.L. (1977) 'Mammals obtained by the expedition with a checklist of the mammals of the Sultanate of Oman', in: The scientific results of the Oman flora and fauna survey 1975. *Journal Oman Stud. spec. Rep.,* pp. 13–26.

Harrison, D.L. and Bates, P.J.J. (1991) *The Mammals of Arabia.* Harrison Zoological Museum, Sevenoaks, Kent, England, 364 pp.

Harrison, D.L. and Lewis, R.E. (1961) 'The large mouse-eared bats of the Middle East with a description of new subspecies', *J. Mammal.,* 42:372–380.

Harrison, D.L. and Makin, D. (1988) 'Significant new records of vespertilionid bats (Chiroptera: Vespertilionidae) from Israel', *Mammalia,* 52:593–596.

Hefetz, A., Ben-Yaakov, R. and Yom-Tov, Y. (1984) 'Sex specificity in the anal gland secretion of the Egyptian mongoose, *Herpestes ichneumon*', *J. Zool. Lond.,* 203:205–209.

Heim de Balsac, H. (1965) 'Quelques enseignments d'ordre faunistique tirés de l'étude du régime alimentaire du *Tyto alba* dans l'ouest de l'Afrique', *Alauda,* 33:309–322.

Hellwing, S. (1970) 'Reproduction in the white-toothed shrew *Crocidura russula* Thomas in captivity', *Israel J. Zool.,* 19:177–178.

References

Hellwing, S. (1971) 'Maintenance and reproduction in the white-toothed shrew *Crocidura russula monacha* Thomas in captivity', *Z. Säugetierkunde,* 36:103–113.

Hellwing, S. (1973a) 'Husbandry and breeding of white-toothed shrews in the Research Zoo of Tel Aviv University', *International Zoo Yearbook,* 13:127–134.

Hellwing, S. (1973b) 'The post natal development of the white-toothed shrew *Crocidura russula monacha* in captivity', *Z. Säugetierkunde,* 38:257–270.

Helversen, von O. (1989) 'New records of bats (Chiroptera) from Turkey', *Zoology of the Middle East,* 3:5–18.

Herselman, J.C. (1980) 'The distribution and status of bats in the Cape Province', Internal Report, Cape Department of Nature and Environmental Conservation.

Herter, K. (1965) *Hedgehogs.* Phoenix House Publ., London. 69 pp.

Hill, J.E. and Harrison, D.L. (1987) 'The baculum in the Vespertilionidae (Chiroptera) with a systematic review, a synopsis of *Pipistrellus* and *Eptesicus,* and the descriptions of a new genus and subgenus', *Bull. Br. Mus. nat. Hist. Zool.,* 52:225–305.

Hill, J.E. and Smith, J.D. (1985) *Bats: A Natural History.* British Museum (Natural History), London.

Hooijer, D. (1961) 'The fossil vertebrates of Kasr 'Akil, a Paleolithic rock-shelter in Lebanon', *Zoologische Verhandel (Leiden),* 49:1–67.

Horowitz, O. (1989) 'Habitat separation of two sympatric species of wood mice (*Apodemus,* Muridae) in Israel: A morphological approach', M.Sc. thesis, The Hebrew University, Jerusalem (in Hebrew).

Hsu, T.C. and Arrighi, F.E. (1966) 'The karyotype of 13 carnivores', *Mammalian Chromosome Newsletter,* 21:155–160.

Hungerford, D.A. and Snyder, R.L. (1969) 'Chromosomes of the rock hyrax, *Procavia capensis* (Pallas, 1767)', *Experientia,* 25:870.

Ilan, M. (1984) 'Biology and ecology of *Psammomys obesus* in Israel', M.Sc. thesis, Zoology Department, Tel Aviv University, Israel (in Hebrew, with English summary).

Ilan, M. and Yom-Tov, Y. (1990) 'The ecology of the fat sand rat, *Psammomys obesus,* in the Negev Desert, Israel', *J. Mammal.,* 71:66–69.

Ilani, G. (1979) 'A zoogeographical and ecological survey of the carnivores of Israel', Report no. 3. Nature Reserves Authority of Israel. Jerusalem. Research Report and surveys (in Hebrew).

Ilani, G. (1986) 'The ecology of the leopards of the Judean Desert', Progress Report No. 4 for 1984–1985. Submitted to the Nature Reserves Authority of Israel. Jerusalem (in Hebrew).

Ivanitskaya, E., Shenbrot, G. and Nevo, E. (1996a) '*Crocidura ramona* sp.n. (Insectivora, Soricidae): a new shrew species from the central Negev Desert, Israel', *Z. Säugetierkunde,* 61:93–103.

Ivanitskaya, E., Gorlov, I., Gorlova, O. and Nevo, E. (1996b) 'Chromosome markers for *Mus macedonicus* (Rodentia, Muridae) from Israel', *Hereditas* 124:145–150.

Jaffe, S. (1988) 'Climate of Israel', in: Yom-Tov, Y. and Tchernov, E. (eds), *The Zoogeography of Israel.* W. Junk Publ., Dordrecht, pp. 79–94.

Jotterand, M. (1971) 'La formule chromosomique de quatre espèces de Felidae', *Mammalia,* 78:1248–1251.

Kamali, M. (1975) 'Sex-chromosome polymorphism in *Nesokia indica* from Iran', *Mammalian Chromosome Newsletter*, 16:165–167.

Kingdon, J. (1974) *East African Mammals*. Vol. 2. Academic Press, London.

Kooler, Z., Ben-Arie, Y. and Naveh, U. (1995) 'Counting gazelles in the dunes of the southern coastal plain', Nature Reserves Authority of Israel, Information Bulletin no. 51. Appendix. pp. 710.

Krasnov, B.R., Shenbrot, G.I., Khokhlova, I.S., Degen, A.A. and Rogovin, K.A. (1996) 'On the biology of Sundvall's jird (*Meriones crassus* Sundvall, 1842) (Rodentia: Gerbillidae) in the Negev Highlands, Israel', *Mammalia*, 60:375–391.

Krasnov, B.R., Shenbrot, G.I., Medvedev, S.G., Vatschenok, V.S. and Khokhlova, I.S. (1997) 'Host-habitat relations as an important determinant of spatial distribution of flea assemblages (Siphonaptera) on rodents in the Negev desert', *Parasitology*, 114:159–173.

Krebis-Peterhans, J. and Kolska Horwitz, L. (1992) 'A bone assemblage from a striped hyaena (*Hyaena hyaena*) den in the Negev Desert, Israel', *Israel J. Zool.*, 37:225–245.

Kulzer, E. (1958) 'Untersuchungen ueber die Biologie von Flughunden der Gattung *Rousettus* Gray', *Z. Morph. Ökol.Tiere*, 47:374–402.

Lavappa, K.S. (1977) 'Chromosome banding patterns and idiograms of the Armenian hamster *Cricetulus migratorius*', *Cytologia*, 42:65–72.

Lay, D.M. (1967) 'A Study of the Mammals of Iran, resulting from the Street Expedition of 1962–1963', *Fieldiana, Zool.*, 54:1–282.

Le Berre, M. (1990) *Faune du Sahara*. Vol. 2. *Mammifères*. Raymond Chabaud-Lechevalier, Paris.

Leatherwood, S. and Reeves, R.R. (1983) *The Sierra Club Handbook of Whales and Dolphins*. Sierra Club Books, San Francisco.

Levi, N. and Bernadsky, G. (1991) 'Crèche behavior of Nubian ibex *Capra ibex nubiana* in the Negev Desert highlands, Israel', *Israel J. Zool.*, 37:125–138.

Lewis, R.E. and Harrison, D.L. (1962) 'Notes on bats from the Republic of Lebanon', *Proc. Zool. Soc. London*, 138:473–486.

Lewis, R.E., Lewis, D.L., and Atallah, S.I. (1967) 'A review of Lebanese mammals. Lagomorpha and Rodentia', *J. Zool. Lond.*, 153:45–70.

Lewis, R.E., Lewis, D.L., and Atallah, S.I. (1968) 'A review of Lebanese mammals. Carnivora, Pinnipedia, Hyracoidea and Artiodactyla', *J. Zool. Lond.*, 154:517–531.

Lindsay, I.M. and Macdonald, D.W. (1986) 'Behaviour and ecology of the Ruppell's fox, *Vulpes rueppellii*, in Oman', *Mammalia*, 50:461–474.

Macdonald, D.W. (1978) 'Observations on the behaviour and ecology of the striped hyaena, *Hyaena hyaena*, in Israel', *Israel J. Zool.*, 27:189–198.

Macdonald, D.W. (1979) 'The flexible social system of the golden jackal, *Canis aureus*', *Behavioural Ecology and Sociobiology*, 5:17–38.

Madkour, G. (1977) 'Further observations on bats (Chiroptera) of Egypt', *Agr. Res. Rev.*, 55:173–184.

Makin, D. (1977) 'Distribution and biology of Microchiroptera in Israel', M.Sc. thesis, Department of Zoology, The Hebrew University, Jerusalem (in Hebrew).

References

Makin, D. (1987) 'The insectivorous bats (Microchiroptera) of Israel: distribution and biology', *Re'em (Oryx)*, 6:12–76 (in Hebrew, with English summary).

Makin, D. (1990) 'Aspects of the biology of the fruit bat *Rousettus aegyptiacus* in Israel', Ph.D. thesis, Tel Aviv University, Israel (in Hebrew, with English summary).

Mann, S. (1986) 'Rodent distribution and comparative biology of spiny mice (*Acomys*) in the oasis of 'En Gedi', M.Sc. thesis, Environmental Biology Program, The Hebrew University, Jerusalem (in Hebrew, with English summary).

Matthey, R. (1954) 'Chromosomes et systematique des Canides', *Mammalia*, 18:225–230.

Meinertzhagen, R. (1954) *Birds of Arabia*. Oliver and Boyd, London, 624 pp.

Meltzer, A. (1967) 'The rock hyrax (*Procavia capensis syriaca* Schreber, 1784)', M.Sc. thesis, Department of Zoology, Tel Aviv University, Israel (in Hebrew, with English summary).

Mendelssohn, H. (1965) 'Breeding the Syrian hyrax', *International Zoo Yearbook*, 5:116.

Mendelssohn, H. (1972) 'Ecological effects of chemical control of rodents and jackals in Israel', in: Farvar, T.M. and Milton, J.P. (eds), *The Careless Technology: Ecology and International Development*. Natural History Press, New York, pp. 527–544.

Mendelssohn, H. (1974) 'The development of the populations of gazelles in Israel and their behavioral adaptations', in: International Symposium on the behavior of ungulates and its relation to management, IUCN, Morges, Switzerland, pp. 122–143.

Mendelssohn, H. (1982) 'Wolves in Israel', in: Harrington, F.H. and Paquet, P.C. (eds), *Wolves of the World*. Noyes Publications, New Jersey, pp. 173–195.

Mendelssohn, H., Golani, I. and Marder, U. (1971) 'Agricultural development and the distribution of venomous snakes and snake bites in Israel', in: de Vries, A. and Kochva, E. (eds), *Toxins of Animal and Plant Origin*. Gordon and Breach, London, pp. 3–15.

Mendelssohn, H., Groves, C.P. and Shalmon, B. (1997) 'A new subspecies of *Gazella gazella* from the southern Negev', *Israel J. Zool.*, 43:209–215.

Mendelssohn, H., and Yom-Tov, Y. (1987) *Mammals*, in: Alon, A. (ed.), *Plants and Animals of the Land of Israel. An Illustrated Encyclopedia*. Ministry of Defence Publishing House and Society for the Protection of Nature, Tel Aviv, Israel, Vol. 7, 295 pp. + Appendix 111 pp. (in Hebrew).

Mendelssohn, H., Yom-Tov, Y. and Groves, C.P. (1995) '*Gazella gazella*', *Mammalian Species*, 490:1–7.

Mendelssohn, H., Yom-Tov, Y., Ilani, G. and Meninger, D. (1987) 'On the occurrence of Blanford's fox, *Vulpes cana* Blanford 1877, in Israel and Sinai', *Mammalia*, 51:459–462.

Meylan, A. (1968) 'Formules Chromosomiques de quelques petits mammifères Nord-Americains', *Revue suisse Zool.*, 75:691–696.

Misonne, X. (1959) 'Analyse zoogéographique des mammifères de l'Iran', *Mém. Inst. Roy. Sci. Nat. Belgique*, 2nd ser., 59:11–57.

Müller, D.M., Kohlmann, S.G., and Alkon, P.U. (1995) 'A Nubian ibex nursery: crèche or natural trap?' *Israel J. Zool.*, 41:163–174.

Murray, G.W. (1912) 'The hamada country', *Cairo Sci. J.*, 6:264–273.

Muskin, Y. (1993) 'Conservation of insectivorous bats by means of nature reserve managment', M.Sc. thesis, Department of Evolution, Systematics and Ecology, The Hebrew University, Jerusalem (in Hebrew, with English summary).

Nadler, C.F. and Hoffmann, R.S. (1970) 'Chromosomes of some Asian and South American squirrels (Rodentia, Sciuridae)', *Experientia,* 26:1383–1386.

Naftali, Y. and Wolf, Y. (1974) 'Mammals and agriculture', *Hasadeh,* 55:1–47 (in Hebrew).

Neuman, M., and Nobel, T.A. (1978) 'Oesophageal worm (*Gongylonema pulchrum* Molin 1857) in a mountain gazelle (*Gazella gazella*)', *Acta Zoologica et Pathologica Antwerpiena,* 70:149–151.

Nevo, E. (1969) 'Mole Rat, *Spalax ehrenbergi*: Mating behaviour and its evolutionary significance', *Science,* 163:484–486.

Nevo, E. (1989) 'Natural selection of body size differentiation in Spiny mice, *Acomys*', *Z. Säugetierkunde,* 54:81–99.

Nevo, E. and Amir, E. (1961) 'Biological observations on the forest dormouse (*Dryomys nitedula* Pallas), in Israel (Rodentia, Muscardinidae)', *Bull. Res. Counc. Israel,* 9B(4).

Nevo, E., Beiles, A., Heth, G. and Simson, S. (1986) 'Adaptive differentiation of body size in speciating mole rats', *Oecologia,* 69:327–333.

Nevo, E., Heth, G. and Beiles, A. (1982) 'Population structure and evolution in subterranean mole rats', *Evolution,* 36:1283–1289.

Niazi, A.D. (1976) 'On the Mediterranean horseshoe bat from Iraq', *Bull. Nat. Hist. Res. Cent. Univ. Baghdad,* 7:167–176.

Nizan, R. (1997) 'Microhabitat preference of two species of woodmice (*Apodemus*) on Mount Carmel', M.Sc. thesis, Department of Zoology, Tel Aviv University (in Hebrew, with English summary).

Nobel, T.A. (1972) '*Fasciliasis gigantica* in a nature reserve in Israel', *Refuah Veterinarit,* 29:33–34 (in Hebrew).

Nobel, T.A., Klopfer, U. and Neuman, M. (1969) 'Toxoplasmosis in wild animals in Israel', *J. Comp. Pathology,* 79:127–129.

Ognev, S.I. (1940) *Mammals of the USSR and Adjacent Countries.* Vol 4. *Rodents.* Moscow. 626 pp. (in Russian).

Or, Y. (1974) 'The underground burrow of *Psammomys obesus*', *Teva Va'Aretz,* 16:280–284 (in Hebrew).

Osborn, D.J. and Helmy, I. (1980) 'The contemporary land mammals of Egypt (including Sinai)', *Fieldiana, Zool.,* (N.S.) 5. Field Mus. nat. Hist., Chicago.

Pathak, S. and Wurster-Hill, D. (1977) 'The distribution of constitutive heterochromatin in carnivores', *Cytogenet. Cell Genet.,* 18:245–254.

Peshev, D.T. and Al-Hossein, K. (1989) 'Karyology and biochemical characteristic of the polecat (*Vormela peregusna syriaca* Pocock) (Carnivora, Mustelidae) from Syria', *Acta zool. Bulgarica,* 38:54–57.

Peters, G. and Rodel, R. (1994) 'Blanford's fox in Africa', *Bonn. zool. Beitr.,* 45:99–111.

Peterson, R.L. and Nagorsen, D.W. (1975) 'Chromosomes of fifteen species of bats (Chiroptera) from Kenya and Rhodesia', *R. Ont. Mus. Life Sci. Occasional Papers,* 27:1–14.

Pitcher, D.A. (1976) 'The long eared hedgehog of Arabia', *J. Saudi Arab. Nat. Hist. Soc.,* 16:22–23.

Pye, J.D. (1972) 'Bimodal distribution of constant frequencies in some Hipposiderid bats', *J. Zool. Lond.,* 166:336.

References

Qumsiyeh, M.B. (1985) 'The bats of Egypt', Special Publications. The Museum of Texas Tech University, No. 23.

Qumsiyeh, M.B. (1991) 'Karyotype of the East European hedgehog, *Erinaceus concolor*, from Jordan', *Z. Säugetierkunde*, 56:375–377.

Qumsiyeh, M.B. (1996) *Mammals of the Holy Land*. Texas Tech University Press, Lubbock, Texas.

Qumsiyeh, M.B. and Baker, R.J. (1985) 'G- and C- banded karyotypes of the Rhinopomatidae (Microchiroptera)', *J. Mammal.*, 66:541–544.

Qumsiyeh, M.B.and Bickham, J.W. (1993) 'Chromosomes and relationships of long-eared bats of the genera *Plecotus* and *Otonycteris*', *J. Mammal.*, 74:376–382.

Qumsiyeh, M.B., Owen, R.D. and Chesser, R.K. (1988) 'Differential rates of genic and chromosomal evolution in bats of the family Rhinolophidae', *Genome*, 30:326–335.

Qumsiyeh, M.B., Schlitter, D.A. and Disi, A.M. (1986) 'New records and karyotypes of small mammals from Jordan', *Z. Säugetierkunde*, 51:139–146.

Rado, R., Wollberg, Z. and Terkel, J. (1992) 'Dispersal of young mole rats (*Spalax ehrenbergi*) from the natal burrow', *J. Mammal.*, 73:885–890.

Reichman, A. and Shalmon, B. (1992) 'Leopard survey in the Negev', Unpublished report. The Society for the Protection of Nature in Israel, Mammal Information Center, and the Nature Reserves Authority of Israel, pp. 1–10 (in Hebrew).

Renzoni, A. (1967) 'Chromosome studies in two species of rodents', *Mammalian Chromosome Newsletter*, 8:111–112.

Renzoni, A. (1970) 'The karyotypes of two wild carnivores', *Mammalian Chromosome Newsletter*, 11:26.

Ritte, U. (1964) 'Population dynamics of a community of small rodents in the Judean Hills', M.Sc. thesis, The Hebrew University, Jerusalem (in Hebrew).

Rosenfeld, A. (1996) 'The effect of hunting on populations of wild boar *Sus scrofa lybicus* Gray, 1868)', M.Sc. thesis The Hebrew University, Jerusalem (in Hebrew with English summary).

Rosevear, D.R. (1965) *The Bats of West Africa*. British Museum (Natural History), London.

Rotary N. (1983) 'Ecosociology of the white-toothed shrew *Crocidura russula monacha* Thomas 1906', Ph.D. thesis, Tel Aviv University, Israel (in Hebrew).

Saltz, D and Alkon, P.U. (1989) 'On the spatial behaviour of Indian crested porcupines (*Hystrix indica*)', *J. Zool. Lond.*, 217:255–266.

Saltz, D. and Rubenstein, D.I. (1995) 'Population dynamics of reintroduced Asiatic wild ass (*Equus hemionus*) herd', *Ecological Applications*, 5:327–335.

Sambraus, H.H. (1973) *Das Sexualverhalten der domestizierten einheimischen Wiederkäver*. Fortschrift der Verhaltensforschung. H. 12. Verlag Paul Parey, Berlin, 100 pp.

Schauenberg, P. (1974) 'Données nouvelles sur le Chat des sables *Felis margarita* Loche, 1858', *Revue suisse Zool.*, 81:949–969.

Schlein, Y., Warburg, A., Schnur, L.F. and Gunders, A.E. (1982) 'Leishmaniasis in Jordan Valley', *Israel J. Med. Sci.*, 18:312.

Schmidt, U. and Joergmann, G. (1986) 'The influence of acustical interferences on ecolocation in bats', *Mammalia*, 50:379–384.

419

Schober, W. and Grimmberger, E. (1987) *A Guide to the Bats of Britain and Europe*. Hamlyn, England.

Schoenfeld, M. and Yom-Tov, Y. (1985) 'The biology of two species of hedgehogs, *Erinaceus europaeus concolor* and *Hemiechinus auritus aegyptius*, in Israel', *Mammalia*, 49:339–355.

Schlawe, L. (1981) 'Materials, localities, bibliographical and iconographical sources as a basis for a species arrangement and a future revision in the genus *Genetta* G. Cuvier, 1816', *Zoologische Abh. st. Mus. Tierk., Dresden* 37:85–182.

Searight, A. (1987) 'The Golden Spiny Mouse, *Acomys russatus lewisi* in the wild and in captivity: Part 2', *Ratel,* 14:40–46.

Sever, Z. and Mendelssohn, H. (1988) 'Copulation as a possible mechanism to maintain monogamy in porcupines, *Hystrix indica*', *Anim. Behav.*, 36:1541–1542.

Sever, Z. and Mendelssohn, H. (1991) 'Spatial movement patterns of porcupines (*Hystrix indica*)', *Mammalia*, 55:187–205.

Shaham, Y. (1977) 'Physiology of reproduction of the female *Psammomys obesus* under laboratory conditions', Ph.D. thesis, Zoology Department, Tel Aviv University, Israel (in Hebrew).

Shalmon, B. (1988) 'The Arava gazelle', *Israel Land and Nature,* 13:15–18.

Shalmon, B. (1988) 'Microchiroptera of Israel: update of checklist and distribution', in: Shalmon, B., Simon, D. and Barak, Y. (eds), *Israel Bats*. Society for the Protection of Nature in Israel, Tel Aviv, pp. 65–69 (in Hebrew).

Shalmon, B. (1993) *A Field Guide to the Land Mammals of Israel: Their Tracks and Signs*. Keter Publ. House, Jerusalem (in Hebrew).

Shargal, E. (1997) 'Population biology and ecophysiology of coexisting *Acomys cahirinus* and *A. russatus*', M.Sc. thesis, Department of Zoology, Tel Aviv University (in Hebrew, with English summary).

Shargal, E., Kronfeld, N. and Dayan, T. (1998) 'A note on the population biology of the bushy-tailed jird (*Sekeetamys calurus*)', *Israel J. Zool.*, 44:61–63.

Shenbrot, G., Krasnov, B. and Khokhlova, I. (1994) 'On the biology of *Gerbillus henleyi* (Rodentia: Gerbillidae) in the Negev Highlands, Israel', *Mammalia*, 58:581–589.

Shenbrot, G.I., Krasnov, B.R. and Khokhlova, I.S. (1997) 'Biology of Wagner's gerbil *Gerbillus dasyurus* (Wagner 1842) (Rodentia: Gerbillidae) in the Negev highlands, Israel', *Mammalia*, 61:467–486.

Shimshony, A., Orgad, U., Baharav, D., Prudovsky, S., Yakobson, B., Bar-Moshe, B. and Dagan, D. (1986) 'Malignant foot-and-mouth disease in mountain gazelle', *Veterinary Record*, 19:175–176.

Shkolnik, A. (1971) 'Diurnal activity in a small desert rodent', *International J. Biomet.,* 15:115–120.

Shkolnik, A. (1988) 'Physiological adaptations to the environment: the Israeli experience', in: Yom-Tov, Y. and Tchernov, E. (eds) *The Zoogeography of Israel*. W. Junk Publ., Dordrecht, pp. 487–496.

Shkolnik, A. and Borut, A. (1966) 'Investigations in the water balance of the spiny mice (genus *Acomys*) of Israel', *Israel J. Zool.,* 15:31.

Shkolnik, A., Maltz, E. and Choshniak, I. (1979) 'The role of the ruminant's digestive tract as

References

a water reservoir', in: Ruckebusch, Y. and Thivend, P. (eds), *Digestive Physiology and Metabolism in Ruminants*. Proc. 5th Symposium on Ruminant Physiology. MTP Press, pp. 731–742.

Skinner, J. D. (1979) 'Feeding behaviour in the caracal *Felis caracal*', *J. Zool. Lond.*, 189:523–557.

Skinner, J.D. and Ilani, G. (1980) 'The striped hyaena *Hyaena hyaena* of the Judean and Negev deserts and a comparison with the brown hyaena *H. brunnea*', *Israel J. Zool.*, 28:229–232.

Slaughter, L. (1971) 'Gestation period of the dorcas gazelle', *J. Mammal.*, 52:480–481.

Smithers, R.H.N. (1983) *The Mammals of the Southern African Subregion*. University of Pretoria, Pretoria.

Stavy, M., Yom-Tov, Y. and Dotan, A. (1982) 'Notes on the taxonomy of the genus *Lepus* in Israel'. Abstracts of the Third International Theriological Congress, Helsinki.

Stebbings, R.E. (1977) 'Order Chiroptera, Bats', in: Corbet, G.B. and Southern, H.N. (eds), *The Handbook of British Mammals*. Blackwell Scientific Publication, Oxford, pp. 68–128.

Stubbe, M and Chotolchu, N. (1968) 'Zur Säugetierfauna der Mongolei', *Mittellungen aus dem Zoologischen Museum in Berlin*, 44:5–121.

Talbot, L.M. (1960) *A Look at Threatened Species*. Fauna Preservation Society.

Tchernov, E. (1988) 'The paleobiogeographical history of the southern Levant', in: Yom-Tov, Y. and Tchernov, E. (eds), *The Zoogeography of Israel*. W. Junk, Dordrecht, pp. 159–250.

Tchernov, E., Dayan, T. and Yom-Tov, Y. (1986/7) 'The biogeography of *Gazella gazella* and *Gazella dorcas* in the Holocene of Israel', *Israel J. Zool.*, 34:51–59.

Theodor, O. and Costa, M. (1967) *Ectoparasites. A Survey of the Parasites of Wild Mammals and Birds in Israel*. Part I. The Israel Academy of Sciences and Humanities, Jerusalem, 117 pp.

Thomas, O.F. (1900) 'On the mammals obtained in south-western Arabia by Messers Percival and Dobson', *Proc. Zool. Soc. London*, (1900):95–104.

Tristram, H.B. (1884) *Fauna and Flora of Palestine*. The Palestine Exploration Fund, London.

Tristram, H.B. (1886) *The Land of Israel; A Journal of Travels in Palestine, undertaken with special reference to its physical character*. Society for the Promotion of Christian Knowledge Pub,, London.

Tristram, H.B. (1888) *The Survey of Western Palestine. The Fauna and Flora of Palestine*. The Palestine Exploration Fund, London.

Uerpmann, H-P. (1986) 'Remarks on the prehistoric distribution of gazelles in the Middle East and Northeast Africa', *Chinkara*, 2:2–11.

Van Aarde, R.J., Skinner, J.D., Knight, M.H. and Skinner, D.C. (1988) 'Range use by a striped hyaena (*Hyaena hyaena*) in the Negev Desert', *J. Zool. Lond.*, 216:575–577.

Vesey-Fitzgerald, D. F. (1952) 'Wildlife in Arabia', *Oryx*, 1:232–235.

Vogel, P. and Sofianidon, T.S. (1996) 'The shrews of the genus *Crocidura* on Lesbos, an eastern Mediterranean island', *Bonn. zool. Beitr.*, 46:339–347.

Wahrman, J. and Goitein, R. (1972) 'Hybridization in nature between two chromosome forms of spiny mice', *Chromosomes Today*, 2:228–237.

Wahrman, J., Goitein, R. and Nevo, E. (1969) 'Mole rat *Spalax*: Evolutionary significance of chromosome variation', *Science,* 164:82–84.

Wahrman, J. and Gurevitz, P. (1973) 'Extreme chromosome variability in colonizing rodents', in: Wahrman, J. and Lewis, K.R. (eds), *Chromosomes Today,* 4:339–424.

Wahrman, J., Richler, C., Goitien, R., Horowitz, A. and Mendelssohn, H. (1973) 'Multiple sex chromosome evolution, hybridization and differential X chromosome inactivation in gazelles', *Chromosomes Today,* 42:434–435.

Wahrman, J., Richler, C. and Ritte, U. (1988) 'Chromosomal considerations in the evoluion of the Gerbillinae of Israel and Sinai', in: Yom-Tov, Y. and Tchernov, E. (eds), *The Zoogeography of Israel.* W. Junk, Dordrecht, pp. 439–485.

Waisel, Y. (1984) *Vegetation of Israel,* in: Alon, A. (ed.), *Plants and Animals of the Land of Israel. An Illustrated Encyclopedia.* Ministry of Defence Publishing House and Society for the Protection of Nature, Tel Aviv, Israel, Vol. 8, 262 pp. (in Hebrew).

Ward, D. and Saltz, D. (1994) 'Foraging at different spatial scales: Dorcas gazelles foraging for lilies in the Negev Desert', *Ecology,* 75:48–58.

Weisbein, Y. (1989) 'The biology and ecology of the caracal (*Felis caracal*) in the Arava Valley of Israel', M.Sc. thesis, Department of Zoology, Tel Aviv University (in Hebrew).

Whitaker, J.O., Shalmon, B. and Kunz, T.H. (1994) 'Food and feeding habits of insectivorous bats from Israel', *Z. Säugetierkunde,* p. 59:74–81.

Wilson, D.E. (1973) 'Bat faunas: a trophic comparison', *Syst. Zool.,* 22:14–29.

Wilson, D.E. and Reeder, D.M. (eds) (1993) *Mammal Species of the World.* 2nd Edition. Smithsonian Institution Press. Washington and London.

Woods, C.A, Contreras, L., Willner-Chapman, G. and Whidden, H.P. (1992) '*Myocastor coypu*', *Mammalian Species,* 398:1–8.

Yaakobi, D. and Shkolnik, A. (1974) 'Structure and concentrating capacity in kidneys of three species of hedgehogs', *Am. J. Physiol.,* 226:948–952.

Yahav, S. and Haim, A. (1980) 'Activity and thermoregulation of *Apodemus mystacinus* (Mammalia: Rodentia) of Mount Carmel', *Israel J. Zool.,* 29:200.

Yaseen, A. E., Hassan, H. A. and Kawashti, L. S. (1994) 'Comparative studies of the karyotypes of two Egyptian species of bats, *Taphozous perforatus* and *Taphozous nudiventris* (Chiroptera, Mammalia)', *Experientia,* 40:1111–1114.

Yerbury, J.W. and Thomas, O. (1895) 'On the mammals of Aden', *Proc. Zool. Soc. London,* pp. 542–555.

Yom-Tov, Y. (1967) 'On the taxonomic status of the hares (Genus *Lepus*) in Israel', *Mammalia,* 31:246–259.

Yom-Tov, Y. (1988) 'The zoogeography of the birds and mammals of Israel', in: Yom-Tov, Y. and Tchernov, E. (eds.). *The Zoogeography of Israel.* W. Junk Publ., Dordrecht, pp. 389–410.

Yom-Tov, Y. (1991) 'Character displacement in the psammophile Gerbillidae of Israel', *Oikos,* 60:173–179.

Yom-Tov, Y. (1993) 'Character displacement among the insectivorous bats of the Dead Sea area', *J. Zool. Lond.,* 230:347–356.

References

Yom-Tov, Y., Ashkenazi, S. and Viner, O. (1995) 'Cattle predation by the Golden Jackal *Canis aureus* on cattle in the Golan Heights, Israel', *Biological Conservation,* 73:19–27.

Yom-Tov, Y., and Ilani, G. (1987) 'The numerical status of *Gazella dorcas* and *Gazella gazella* in the southern Negev Desert, Israel', *Biological Conservation,* 40:245–253.

Yom-Tov, Y., Makin, D. and Shalmon, B. (1992a) 'The insectivorous bats (Microchiroptera) of the Dead Sea area, Israel', *Israel J. Zool.,* 38:125–137.

Yom-Tov, Y., Makin, D. and Shalmon, B. (1992b) 'The biology of *Pipistrellus bodenheimeri* (Microchiroptera) in the Dead Sea area of Israel', *Z. Säugetierkunde,* 57:65–69.

Yom-Tov, Y and Mendelssohn, H. (1988) 'Changes in the distribution and abundance of vertebrates in Israel during the 20th century', in: Yom-Tov, Y. and Tchernov, E. (eds), *The Zoogeography of Israel.* W. Junk Publ., Dordrecht, pp. 515–548.

Yom-Tov, Y., Mendelssohn, H. and Groves, C. P. (1995) '*Gazella dorcas*', *Mammalian Species,* 491:1–6.

Yom-Tov, Y. and Shalmon, B. (1989) 'First record of *Taphozous perforatus* in Israel', *Mammalia,* 53:661–662.

Yom-Tov, Y. and Werner, Y.L. (1996) 'Environmental correlates of the geographical distribution of terrestrial vertebrate species richness in Israel', *Israel J. Zool.,* 42:307–316.

Zafriri, A. (1972) 'Taxonomy and reproduction in the field of the shrew *Crocidura russula* in Israel', M.Sc. thesis, Department of Zoology, Tel Aviv University (in Hebrew).

Zahavi, A. and Wahrman, J. (1957) 'The cytotaxonomy, ecology and evolution of the gerbils and jirds of Israel, (Rodentia: Gerbillinae)', *Mammalia,* 21:341–380.

Zima, J. (1978) 'Chromosomal characteristics of Vespertilionidae from Czechoslovakia', *Acta Sci. Nat. Brno,* 12:1–38.

Zima, J. (1982) 'Chromosomal homology in the complements of bats of the family Vespertilionidae. II. G-band karyotypes of some *Myotis, Eptesicus* and *Pipistrellus* species', *Folia Zool. (Zool. Listy),* 31:31–36.

Zima, J. and Kral, B. (1984) 'Karyotype of European mammals III', *Acta Sci. Nat. Brno,* 18:1–51.

Ziv, Y., Abramsky, Z., Kotler, B.P. and Subach, A. (1993) 'The interference competition and temporal and habitat partitioning in two gerbil species', *Oikos,* 66:237–246.

Zohary, M. (1962) *Plant Life of Palestine (Israel and Jordan).* Ronald Press, New York.

Zuri, I. and Terkel, J. (1996) 'Locomotor patterns, territory, and tunnel utilization in the mole-rat *Spalax ehrenbergi*', *J. Zool. Lond.,* 240:123–140.

INDEX

Scientific and Common Names

PLATE I
Israeli habitats

A. Shoob

1. Naḥal Grizi, southern Negev: a wide wadi with various *Acacia* trees

A. Shoob

2. View of Har Ardon and the eastern part of Makhtesh Ramon, central Negev

A. Shoob

3. Naḥal Darga, Judean Desert. High cliffs near the Dead Sea

A. Shoob

4. ʿEn ʾAqrabbim in Naḥal Ẓin. A typical oasis

PLATE II
Israeli habitats

A. Shoob

1. Flint covered slope, central Negev near Sede Boqer

L. and O. Bahat

2. Sand dunes near Caesarea, Mediterranean Coastal Plain

A. Shoob

3. Mediterranean forest in Naḥal Beẓet, western Galilee

A. Shoob

4. Batha in eastern Samaria

PLATE III
Israeli habitats

A. Shoob

1. The Jordan River about 50 kms north of the Dead Sea

A. Shoob

2. Pine plantations near Lahav, northern Negev

L. and O. Bahat

3. Cultivated fields in Bet Ẓayda Valley, north-east of Lake Kinneret (Sea of Galilee)

A. Shoob

4. Cliffs at Naḥal ʿAmmud, eastern Galilee

PLATE IV
Extinct and threatened species
(1 and 2)

A. Shoob

1. *Ursus arctos syriacus* in captivity

A. Shoob

2. A mounted *Acinonyx jubatus* from the Schmitz Collection in Jerusalem. This specimen was killed in Jordan at the beginning of the 20th century

H. Mendelssohn

3. The results of cave fumigation aimed at exterminating fruit bats (*Rousettus aegyptiacus*) in Bitan Aharon, Mediterranean Coastal Plain: cadavers of 132 *Rhinolophus ferrumequinum* and 53 *Myotis emarginatus*

Z. Meshel

4. An ancient "desert kite" for capturing wild ungulates. Southern ʿArava Valley

PLATE V
Sirenia
Dugong dugon hemprichi

A. Shoob

1. Female which drowned after getting entangled in fishing nets in the Gulf of Elat

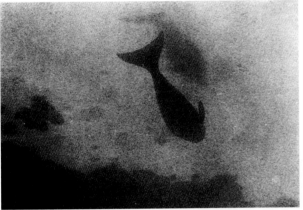

An unknown pilot of the Israeli Air force

2. Specimen photographed from the air at Nabq, Sinai, in 1971

A. Shoob

3. Head of male. Note the large incisors

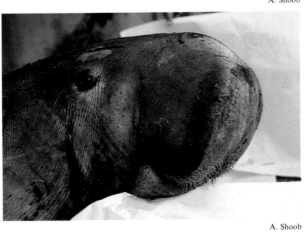

A. Shoob

4. Head of female

PLATE VI
Hedgehogs

A. Shoob

1. Skins of *Hemiechinus auritus aegyptius* (left), *Paraechinus aethiopicus pectoralis* (middle) and *Erinaceus concolor concolor* (right)

A. Shoob

2. A rolled-up *Erinaceus concolor concolor*

A. Shoob

3. *Hemiechinus auritus aegyptius*

A. Shoob

4. *Paraechinus aethiopicus pectoralis*

PLATE VII
Shrews

A. Shoob

1. *Crocidura suaveolens monacha*

D. Simon

2. *Crocidura leucodon judaica*

A. Shoob

3. *Suncus etruscus etruscus*

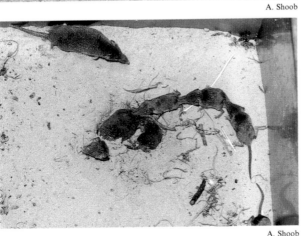

A. Shoob

4. A caravan of a female and young of
Crocidura suaveolens monacha

PLATE VIII
Bats

D. Bar-Shahal

1. *Rousettus aegyptiacus aegyptiacus*

A. Shoob

2. Face of *Rhinopoma microphyllum micro-phyllum*

A. Shoob

3. Face of *Rhinopoma hardwickei arabium*

4. *Rhinopoma hardwickei arabium*

D. Bar-Shahal

PLATE IX
Bats

A. Shoob

1. Face of *Rhinolophus ferrumequinum ferrumequinum*

2. Two captive *Rhinolophus ferrumequinum ferrumequinum*

A. Shoob

3. *Rhinolophus clivosus clivosus*

D. Afik

4. Face of *Asellia tridens tridens*. Note the three nasal lobes

H. Mendelssohn

PLATE X
Bats

1. *Asellia tridens tridens*　　H. Mendelssohn

2. *Pipistrellus rueppellii rueppellii*

A. Shoob

D. Bar-Shahal

3. *Pipistrellus kuhlii*

4. Face of *Otonycteris hemprichii jin*

H. Mendelssohn

Plate XI
Bats and otter

1. Face of *Plecotus austriacus christii*

H. Mendelssohn

H. Mendelssohn

2. *Tadarida teniotis rueppelli* feeding

H. Mendelssohn

3. *Lutra lutra seistanica*

M. Livne

4. A marking station of *Lutra lutra seistanica* near the Upper Jordan River

PLATE XII
Mustelids

A. Shoob

1. *Mellivora capensis wilsoni*

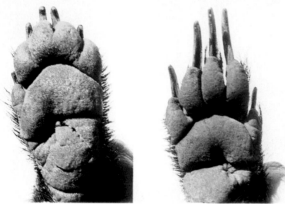

H. Mendelssohn

2. Front (with three long claws) and rear paws of *Mellivora capensis wilsoni*

H. Mendelssohn

3. Female and two large cubs of *Meles meles canescens*

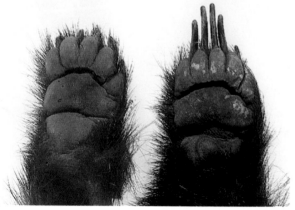

L. Maman

4. Front (with four long claws) and rear paws of *Meles meles canescens*

PLATE XIII
Mustelids and wolf

A. Shoob

1. *Vormela peregusna syriaca*

E. Bartov

2. Young *Martes foina syriaca*

Y. Eshbol

3. *Canis lupus pallipes* at a feeding station in the Negev

A. Shoob

4. Two captive *Canis lupus pallipes* in summer coat

PLATE XIV
Canids

A. Shoob

1. Male *Canis aureus syriacus*

A. Shoob

2. *Vulpes vulpes palaestina*. Note the black throat and abdomen

A. Shoob

3. *Vulpes vulpes arabica*. Note the white throat

A. Alon

4. Mosaic of *Vulpes vulpes* at an ancient synagogue in Gaza

PLATE XV
Canids and mongoose

I. Golani

1. *Vulpes rueppellii sabaea*

H. Mendelssohn

2. Frontal paw of *Vulpes rueppellii sabaea*.
Note the long hairs on the ventral side

A. Shoob

3. *Vulpes cana cana*

Y. Eshbol

4. *Herpestes ichneumon ichneumon*

PLATE XVI
Mongoose and hyaena

A. Shoob

1. *Herpestes ichneumon ichneumon* attacking a viper *Vipera palaestina*

A. Shoob

2. *Hyaena hyaena syriaca*

H. Mendelssohn

3. Foot prints of *Hyaena hyaena syriaca*. The front paw is larger than the rear one

A. Shoob

4. Juvenile *Hyaena hyaena syriaca* in its cave

PLATE XVII
Hyaena and wild cat

H. Mendelssohn

1. Bones, including remains of two human skulls, near a hyaena den in 'En Yahav, 'Arava Valley

H. Mendelssohn

2. Secretion of the anal gland of hyaena smeared on a concrete block in its cage

H. Mendelssohn

3. *Felis silvestris tristrami*. Male from the central Negev

Y. Eshbol

4. Female *Felis silvestris tristrami*

PLATE XVIII
Jungle cat, caracal and leopard

A. Shoob

1. Female *Felis chaus furax*

A. Shoob

2. Female *Felis margarita harrisoni*

3. *Caracal caracal schmitzi*

Y. Eshbol

Y. Yom-Tov

4. Leopard trap built and used by Bedouin in Wadi Watir, Sinai

PLATE XIX
Leopard and hyrax

E. Bartov

1. *Panthera pardus nimr* at ʿEn Gedi

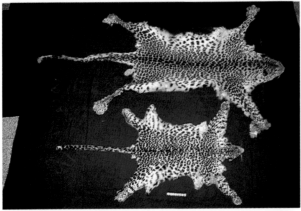

A. Shoob

2. Skins of female *Panthera pardus tulliana* (larger and darker) from the western Galilee and of female *Panthera pardus nimr* from the Judean Desert

Y. Eshbol

3. *Procavia capensis syriaca*. A female with two juveniles

G. Rubin

4. Territorial male *Procavia capensis syriaca* sings a territorial call

PLATE XX
Hyrax and wild ass

A. Loya

1. *Procavia capensis syriaca* feeding on *Acacia* at Naḥal ʿArugot, Judean Desert

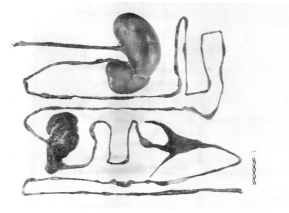

A. Shoob

2. Digestive system of *Procavia capensis syriaca*. Note the long intestines, the large stomach, the enlarged middle appendix and the double appendices near the rear part of the system

H. Mendelssohn

3. Light coloured *Procavia capensis syriaca* from Beʾerot ʿOded, Negev Desert

Y. Eshbol

4. *Equus hemionus* in the ʿArava Valley

PLATE XXI
Wild boar and gazelles

A. Shoob

1. Pellets of fresh grass chewed by *Sus scrofa lybicus*

D. Havlena

2. Moulting female and young of *Sus scrofa lybicus* on Mount Carmel

Y. Eshbol

3. Two males of *Gazella gazella acaciae* near Yotvata, southern 'Arava Valley. Note the dark patch on the nose and the long neck, ears, tail and legs

L. and O. Bahat

4. Male *Gazella gazella gazella* near Gamla, Golan Heights

PLATE XXII
Gazelles

H. Mendelssohn

1. Footprints of *Gazella gazella gazella* near Amaẓya, Judean Hills

H. Mendelssohn

2. Footprints of *Gazella dorcas littoralis* near Ḥaluẓa, western Negev. Narrower and more pointed than those of *Gazella gazella gazella*

A. Shoob

3. Marking station of *Gazella gazella gazella*, Ramot Yissakhar

A. Shoob

4. *Gazella gazella gazella*, female, young male in front and a fawn in Ramot Yissakhar

PLATE XXIII
Gazelles and ibex

1. Male *Gazella dorcas littoralis* grooming its fur

H. Mendelssohn

2. Three male *Gazella dorcas littoralis*   Y. Eshbol

3. Male *Capra ibex nubiana*   Y. Eshbol

4. Male *Capra ibex nubiana* courting a female. Note the erect penis and the protruded tongue

G. Rubin

PLATE XXIV
Nubian ibex and hare

A. Shoob

1. Hunter shoots an arrow at a male *Capra ibex nubiana*. An ancient rock painting from Wadi Zalaka, Sinai

U. Paz

2. Two male *Capra ibex nubiana*, fighting at ʿEn Gedi

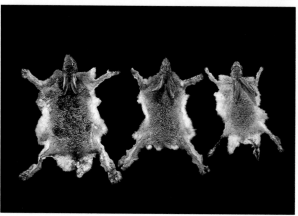

L. and O. Bahat

3. Two females and 16 kids of *Capra ibex nubiana*, in ʿEn Avedat, central Negev

4. Skins of *Lepus capensis* from the Golan (largest and darkest), central Israel (middle) and the ʿArava Valley

A. Shoob

PLATE XXV
Hare and rodents

A. Shoob

1. *Lepus capensis* from the Mediterranean region of Israel. Note the red tapetum

L. and O. Bahat

2. *Lepus capensis* from the Negev

A. Loya

3. *Sciurus anomalus syriacus.* Mount Hermon

A. Shoob

4. *Cricetulus migratorius cinerascens*

PLATE XXVI
Rodents

A. Shoob

1. *Microtus socialis guentheri*

H. Mendelssohn

2. Burrow openings of *Microtus socialis guentheri* in the Golan during a year of high population density (1971). Note that the vegetation near the burrow openings is grazed down to the ground

A. Shoob

3. *Microtus nivalis hermonis*

4. *Gerbillus pyramidum floweri*

A. Shoob

PLATE XXVII
Rodents

H. Mendelssohn

1. Footprints of *Gerbillus pyramidum floweri*

A. Shoob

2. *Gerbillus andersoni allenbyi*

A. Shoob

3. *Gerbillus gerbillus asyutensis*

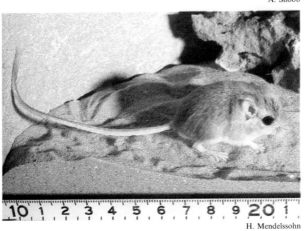

H. Mendelssohn

4. *Gerbillus henleyi mariae*

PLATE XXVIII
Rodents

H. Mendelssohn

1. *Gerbillus nanus arabium*

H. Mendelssohn

2. *Gerbillus dasyurus dasyurus* from the Golan

H. Mendelssohn

3. *Gerbillus dasyurus dasyurus* from Sinai

A. Shoob

4. *Sekeetamys calurus calurus*

PLATE XXIX
Rodents

A. Shoob

1. *Meriones tristrami tristrami*

A. Shoob

2. *Meriones sacramenti sacramenti*

A. Shoob

3. *Meriones crassus crassus*

H. Mendelssohn

4. Burrow openings of *Meriones crassus crassus*

PLATE XXX
Rodents

A. Shoob

1. *Psammomys obesus terrasanctae* feeding on leaves of *Atriplex halimus*

H. Mendelssohn

2. Burrow openings of *Psammomys obesus terrasanctae* at Ḥaẓeva, the 'Arava Valley

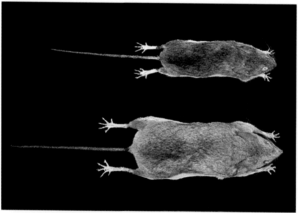

A. Shoob

3. Skins of *Apodemus mystacinus mystacinus* (larger) and *Apodemus sylvaticus arianus*

A. Shoob

4. *Apodemus sylvaticus arianus*

PLATE XXXI
Rodents

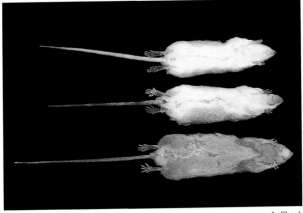

A. Shoob

1. Ventral views of skins of *Rattus rattus rattus*. Note the variety of colours

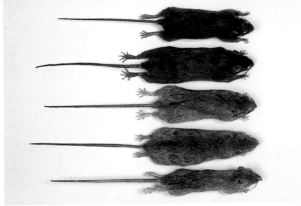

A. Shoob

2. Dorsal views of skins of *Rattus rattus rattus*. Note the variety of colours.

A. Shoob

3. Nest of *Egretta gularis* which served as a "feeding table" for *Rattus rattus rattus*, with remains of crabs (*Uca* sp.), near Nabq, Sinai

4. *Mus musculus praetextus*

A. Shoob

PLATE XXXII
Rodents

A. Shoob

1. *Acomys cahirinus dimidiatus*

A. Shoob

2. Skins of *Acomys cahirinus dimidiatus* from the Golan (dark coloured) and from Sinai

A. Shoob

3. *Acomys russatus russatus*

A. Landsman

4. *Nesokia indica bacheri* earth-brown and reddish-brown specimens, in captivity

PLATE XXXIII
Rodents

A. Shoob

1. *Spalax leucodon ehrenbergi*

R. Rado

2. Maternity hill of *Spalax leucodon ehrenbergi*

A. Shoob

3. *Eliomys melanurus melanurus*

H. Mendelssohn

4. Nest of *Dryomys nitedula phrygius* which was cut off a tree

PLATE XXXIV
Rodents

H. Mendelssohn

1. *Dryomys nitedula phrygius*

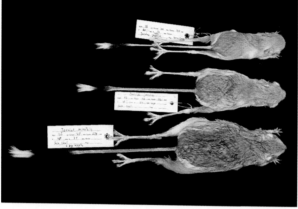

A. Shoob

2. Skins of jerboas: *Jaculus jaculus schlueteri* (lightest colour), *Jaculus jaculus vocator* (middle) and *Jaculus orientalis orientalis* (largest)

A. Shoob

3. *Jaculus orientalis orientalis* walking

4. Habitat of *Jaculus orientalis orientalis* near 'Arad

H. Mendelssohn

PLATE XXXV
Rodents

H. Mendelssohn and A. Shoob

1–3. Jumping sequence of *Jaculus jaculus*

H. Mendelssohn and A. Shoob

H. Mendelssohn and A. Shoob

H. Mendelssohn and A. Shoob

4. *Jaculus jaculus* turning during a jump

PLATE XXXVI
Rodents

D. Bar Shahal

1. A 34 day old, captive, *Hystrix indica indica*. Note that the thick, back spines have not yet grown

L. Maman

2. Front (on the right, with 4 digits) and rear (on the left, 5 digits) paws of *Hystrix indica indica*

3. Footprints of *Hystrix indica indica*

4. Hollow tail spines of *Hystrix indica indica*

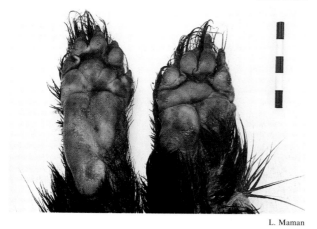

Y. Yom-Tov

A. Shoob

PLATE XXXVII
Rodents

Z. Sever

1. Copulation of captive *Hystrix indica indica*. Note the erect penis. The spines of the female are painted red for individual identification

A. Loya

2. *Myocastor coypu* in the Hula Valley

A. Shoob

3. *Myocastor coypu* eating a carrot

כתבי האקדמיה הלאומית הישראלית למדעים
החטיבה למדעי-הטבע

———

החי של ארץ-ישראל

יונקים בישראל
(MAMMALIA OF ISRAEL)

מאת

היינריך מנדלסון ויורם יום-טוב

ירושלים תשנ"ט